Modern Ireland is beset by economic, social, and environmental problems. This is in stark contrast with its promoted image as a land of rolling green pastures and a timeless way of life based on a self-sustaining rural economy.

While many of Ireland's problems are indigenous (the 'troubles' of the North), others are part of general European and world trends. This contemporary study looks at these problems from geographic and environmental perspectives, placing them within their regional, national and international context.

The volume covers topics from population characteristics, crime, and rural problems, to retailing, tourism, regional development and environmental resource issues, including energy and water resource management. This comprehensive survey will prove invaluable to students, decision-makers, and all those interested in the current situation in Ireland and its future.

IRELAND

CONTEMPORARY PERSPECTIVES ON A LAND AND ITS PEOPLE

EDITED BY
R.W.G. CARTER AND A.J. PARKER

London and New York

First published 1989
by Routledge
11 New Fetter Lane, London EC4P 4EE
New in paperback 1990

Simultaneously published in the USA and Canada
by Routledge
a division of Routledge, Chapman and Hall, Inc.
29 West 35th Street, New York, NY 10001

© 1989 R.W.G. Carter and A.J. Parker

Printed and bound in Great Britain by
Mackays of Chatham PLC.Chatham, Kent

British Library Cataloguing in Publication Data

Ireland: contemporary perspectives on a land and its people
   1. Ireland
   I. Carter, Bill II. Parker, A.J. (Anthony
John), *1944*–
   941.50824
   ISBN 0-415-05294-7 (Pbk)

Library of Congress Cataloging in Publication Data
has been applied for

For Anna, Sophie and Jane

For Ben and Helen

CONTENTS

CONTRIBUTORS

PREFACE

ACKNOWLEDGEMENTS

1   INTRODUCTION   Bill Carter
and Tony Parker   1

2   EUROPEAN ECONOMIC POLICIES
AND IRELAND   Frank J. Convery   9

3   PARTITION, POLITICS AND
SOCIAL CONFLICT   Dennis Pringle   23

4   IRISH POPULATION
PROBLEMS   John Coward   55

5   CRIME IN GEOGRAPHICAL
PERSPECTIVE   David Rottman   87

6   THE HISTORICAL LEGACY IN
MODERN IRELAND   Stephen A. Royle   113

7   THE PROBLEMS OF RURAL
IRELAND   Mary Cawley   145

8   AGRICULTURAL DEVELOPMENT
Desmond A. Gillmor   171

9   THE NEW INDUSTRIALISATION
OF IRELAND   Barry Brunt   201

10   THE CHANGING NATURE OF
IRISH RETAILING   Tony Parker   237

11   TRANSPORTATION
James Killen and Austin Smyth   271

12   PATTERNS IN IRISH TOURISM
John Pollard   301

13   IRISH ENERGY : PROBLEMS
AND PROSPECTS   Palmer Newbould   331

14   WATER RESOURCE
MANAGEMENT   David Wilcock   359

15   RESOURCES AND MANAGEMENT
OF IRISH COASTAL WATERS
AND ADJACENT COASTS   Bill Carter   393

16   AIR POLLUTION PROBLEMS
IN IRELAND   John Sweeney   421

17   REGIONAL DEVELOPMENT
STRATEGIES   James A. Walsh   441

INDEX   473

CONTRIBUTORS

Dr. Barry Brunt is Lecturer in Geography at University College, Cork.

Dr. Bill Carter is Head of the Department of Environmental Studies at the University of Ulster, Coleraine.

Dr. Mary Cawley is Lecturer in Geography at University College, Galway.

Professor Frank Convery is Director of the Resource and Environmental Policy Centre and Head of the Department of Geography at University College Dublin.

The late Dr. John Coward was Lecturer in Geography at the University of Ulster, Coleraine.

Professor Desmond A. Gillmor is in the Department of Geography, Trinity College, Dublin.

Dr. James Killen is Senior Lecturer in Geography at Trinity College, Dublin.

Professor Palmer Newbould is Emeritus Professor of Environmental Science from the University of Ulster, Coleraine and formerly Acting Vice-Chancellor of The New University of Ulster.

Dr. Tony Parker is Director of the Centre for Retail Studies and a Statutory Lecturer in the Department of Geography at University College, Dublin.

Dr. John Pollard is Senior Lecturer in Geography at the University of Ulster, Coleraine.

Dr. Dennis Pringle is Lecturer in Geography at St. Patrick's College, Maynooth.

Dr. Dave Rottman is a Senior Researcher with the Economic and Social Research Institute, Dublin and is currently a

Researcher with the National Center for State Courts, Williamsburg, Va.

Dr. Steve Royle is Lecturer in Geography at The Queen's University, Belfast.

Professor Austin Smyth is Professor of Transport Studies at the University of Ulster, Coleraine.

Dr. John Sweeney is Lecturer in Geography at St. Patrick's College, Maynooth.

Mr. Jim Walsh is Lecturer in Geography at St. Patrick's College, Maynooth and was formerly Head of the Department of Geography, Carysfort College, Blackrock, Co. Dublin.

Professor David Wilcock is in the Department of Environmental Studies at the University of Ulster, Coleraine.

PREFACE

In the last few decades Ireland has changed substantially, as a result of demographic, social, economic and political pressures, both internally and externally. The Irish landscape - in the broadest sense - is quite different to even twenty years ago, and change has brought considerable conflicts and problems. A rapidly changing economy, new demographic structures, increasing urbanisation, changing technology, the deterioration in land, air and water quality, have all contributed to the new Ireland that is facing into the twenty-first century. There is a heightening of awareness in environmental, social and economic matters and Irish people are becoming more concerned about the implications of change, at a time when financial constraints leave little room for manouevre.

Geographers and environmental scientists are uniquely placed not only to study but to take an influential role in planning the future of Ireland. The recent report of the United Nations on World Development and Environment (1987) emphasises the radical shift in our lifestyles that must come in the twenty-first century if we are to sustain our planet, and its distinctive nations.

Generally, texts on the geography of Ireland have taken a somewhat traditional approach. However there is a clear need for a text which considers the changes and problems that have beset modern Ireland. This book adopts this approach, examining the changing, current social, political, economic and environmental state of Ireland. The approach is multi-disciplinary set within the framework of considering the similarities and differences of both Northern Ireland and the

Republic of Ireland. In as much as the pressures and changes that are affecting Ireland are not unique, the text relates the Irish situation to the general European and international situation. As such, the book can be regarded not only as a modern geography of Ireland, but also as a case study of a peripheral European community within the framework of a wider world.

The Introduction takes up these themes in greater detail, while Frank Convery (Chapter 2) sets the Irish situation in the context of membership of the European Community. Thereafter the book is divided into three broad sections considering the social and political geography of Ireland; the economic resources of the island; and environmental resource perspectives, respectively. Socio-political aspects of Ireland are then considered, with the scene being set by Dennis Pringle (Chapter 3), population characteristics and problems being considered by John Coward (Chapter 4) and David Rottman discussing crime (Chapter 5). The historical dimension in Ireland is considered from a modern resource perspective by Stephen Royle (Chapter 6), while Mary Cawley's chapter on the problems of rural Ireland concludes the initial section of the book (Chapter 7).

The economic resource perspectives section of the book also contains five chapters which consider developments in agriculture (Desmond Gillmor, Chapter 8); the new industrialisation of Ireland (Barry Brunt, Chapter 9); and the changing nature of retailing (Tony Parker, Chapter 10). James Killen and Austin Smyth address the problems of transportation in Chapter 11, while tourism forms the subject of Chapter 12 by John Pollard.

The final section of the book is concerned with environmental resource issues and regional development. Palmer Newbould examines energy planning and policy (Chapter 13), while David Wilcock considers water resource management (Chapter 14) and Bill Carter discusses the resources and management of Irish coastal waters (Chapter 15). Air pollution forms the subject of Chapter 16 (John Sweeney). The concluding chapter presents an overview of regional planning and development policies throughout Ireland (James Walsh, Chapter 17).

The book should prove essential reading for both the 600 or so students who graduate in Geography or Environmental Science each year in Ireland, and the 10 to 20 times that number who take some formal examination in the areas addressed by the text. These students are the the future decision makers of Ireland and must be equipped with the knowledge and skills to tackle the impending problems described and discussed in this book. We would like to think that our future graduates benefit in some small way from our endeavours in writing and compiling this text.

Bill Carter / Tony Parker

Coleraine and Dublin

ACKNOWLEDGEMENTS

We are grateful to the following authors and publishers for permission to reproduce the following figures:

Figure 1, Chapter 7 - Dr. David Eastwood and the Geographical Society of Ireland.

Figure 2, Chapter 10 - The Centre for Retail Studies, Department of Geography, University College Dublin.

Figure 2, Chapter 17 - Dr. Tony Hoare and Academic Press Ltd, London.

The editors would like to express their thanks to our Authors for bearing with us through the long gestation period that the book has experienced. We would also like to thank Sheila Flanagan who acted as production assistant for her sterling work and the staff of the Computer Centre, University College Dublin for their assistance with the word processing package. Our thanks are due also to technical staff at both University College Dublin and the University of Ulster, Coleraine, and postgraduates in University College Dublin who assisted at various stages of the production. Finally our thanks must go to the forbearance of our families who, now that the book is completed, might see us around the house again!

1 INTRODUCTION

Bill Carter and Tony Parker

Images of Ireland abound. The Ireland of the tourist brochure is an idyllic landscape populated by friendly people with a thousand welcomes. The Ireland of the media is all too often one of divided communities, political intransigence, terrorism and crime. Industrial Ireland is a European base, with generous tax concessions, low overheads, a large pool of unemployed labour and factories that can be abandoned without conscience when times are hard. Agricultural Ireland is green fields and dairy cows, a bottomless pit for European Community funds. While such images are not wholly unrepresentative, they do reflect many complex and changing facets that comprise Modern Ireland.

The essential aim of this book is to explore the dynamic spatial patterns that form the physical and social environments of Ireland. Of particular interest are Irish resources, ranging from people and their skills, to lands and minerals; the ways in which these resources are being allocated or used, and the conflicts that arise during these processes. Over the last 40 years, Ireland has undergone major changes in social and economic conditions. By and large these have been achieved at the expense of environmental quality. However, we are entering an era when the need to replace the old concepts of continual economic growth with ideas of 'sustainable economics', in which pragmatic economics, social justice and environmental well-being are cornerstones. In many ways such a strategy finds the geographer with his/her eclectic

1

skills, embracing physical, biological, social, economic and cultural systems, exceptionally well-placed to determine national and international futures. Whether geographers will grasp this nettle, or even have the confidence to try is another matter!

As most people are aware, Ireland comprises two states, the Republic of Ireland and Northern Ireland (Figure 1). The

Figure 1. Ireland

former is an autonomous republican democracy, while the latter is part of the United Kingdom, a monarchial democracy, governed from Westminster. While details of events leading up to partition in 1922 are well recorded elsewhere, the consequences of this action have had a fundamental impact on Irish geography ever since. The divergence of political systems, the dichotomy of approach to common geographical problems (like industrialisation, rural deprivation and land reform) has added an additional veneer to already complex circumstances. In many ways the presence of the UK's only land border - 410 km long - is both inflammatory and intimidatory, yet arguably provides the economic stimulus along an otherwise deprived and marginal zone. It is a paradox that many of the people who condemn and ignore the border, are those who benefit most, in financial terms, from its presence. Whether one is dealing with acts of political terrorism or doing the family shopping, the border is a prominent feature in the human geography of Ireland.

The political border between the states is a geographical nonsense, based on nothing more than Anglo-Norman county lines, often cutting roads, communities, farms and even buildings. There are many other, clearer spatial contrasts. Perhaps the most acute is the division between the relatively affluent east and relatively poor west. By an accident of geography, the east of Ireland, both in the Republic and Northern Ireland is endowed preferentially with a less extreme climate, better soils and high biological diversity and productivity. Also the east is nearer the dominant transport routes to Britain and Europe, and has developed the largest urban centres - Dublin and Belfast - with the more sophisticated economic, cultural and social infrastructures. As a result there are strong east to west falling gradients in terms of disposable income, educational opportunity, employment, economic prosperity and resource marginality. Such gradients are often self-reinforcing as people move towards the urban east to avail themselves of perceived advantages. Attempts to stem these migrations, especially in the Republic of Ireland, for reasons ranging from linguistic preservation to the mitigation of urban over-crowding, have met with variable success.

As will become clear in the various chapters of this book, Ireland is an island under stress. These stresses emanate not only from religious, cultural and political divides, but also from a tendency to 'overreach' in environmental terms in order to meet economic goals. All too often political expediency has determined policy, leaving confused and even chaotic situations. For example, there is no coherent energy policy (although Ireland is not alone in this deficiency), rather there has been a series of tactics, in which various options have been promoted and then discarded over the last 30 years. Environmental stresses in Ireland have combined with social ills to reduce the quality, if not the quantity, of life. Thus Dublin suffers the last chronic urban pollution in Europe, again resulting from inadequate long-term policy in terms of housing, fuel supply and service provision. Deprived and stressed communities are prone to react. This reaction may be confined to specific cohorts, for example unemployed adolescents, but nevertheless is usually associated with general social and environmental malaises. Thus urban Dublin and Belfast, and even some rural areas, have witnessed staggering increases in crime, self-abuse (drug taking, alcoholism), community denigration (littering, vandalism, intimidation) and political radicalism.

As well as spatial division and gradation, one must add social discrimination, founded not only on value-based social stratification, religious beliefs or social class, but also on newer schisms of educational and vocational mobility, psychological attainment and welfare provision. Added together these spatial and social dimensions constitute an extremely complex human ecology of Ireland.

Perhaps the most significant date in Irish history since 1922 is 1973, when Ireland, both the Republic and Northern Ireland, was accepted as a member of the European Economic Community (now shortened to European Community (EC)). Reasons for joining the Community were twofold. First there was a hedonism to be part of a common economy in which trade and personal mobility barriers were negligible, and second, there was a need to retain important traditional markets (mainly that between the Republic and the UK). So far both parts of Ireland have fared well from EC membership, to some extent living vicariously on the successes of the core member states (West Germany, France and Britain). Differential application of EC tariffs and intervention prices across the border have tended to encourage smuggling - for example it is thought about 90,000 head of Northern Irish cattle crossed the border illegally in 1986 to qualify for the higher beef tariffs in the Republic. The very fact that EC affairs in Northern Ireland have to be channelled through London is also a handicap and may have disadvantaged some economic sectors in the Province.

The EC has moved, through the Common Agricultural Policy (CAP) and the Common Fisheries Policy (CFP), to protect Irish Farming and Fisheries, while membership has opened new markets, not only for agricultural goods, loosening the economic ties with Great Britain. Much of Western Ireland has benefitted from the Regional Development fund, which has financed many infrastructural improvements, from new harbours and roads to the replacement of fishing boats and the promotion of tourism. All of Ireland is considered 'deprived' in EC terms, along with southern Italy, Greece and parts of Spain, Portugal and France. Specific EC enhancement projects, like STAR, are aimed at improvement of facilities and services, and at creating opportunities for development, especially in the creation and encouragement of small and medium sized enterprises.

Yet the style of job creation and provision is markedly different in the two parts of Ireland despite broadly similar objectives. Northern Ireland has, until recently, concentrated on large, often prestige, investments, often coercing British 'mainland' firms to open new plants. Much of the new 1960s industry has now closed, only one man-made fibre plant, Du Pont in Londonderry, remains out of six, and there have been some spectacular collapses including heavy investment in two American projects in cars (De Lorean) and aircraft (Lear Fan). The investment per job is high (over £25,000) and only 25% of those created provide long-term secure employment. The industrial geography of Northern Ireland is largely one of chaos and trauma, with often nebulous criteria dictating decisions on location and funding. Meanwhile, the Republic of Ireland has adapted a more flexible, and to the annoyance of its northern counterparts, a less bureaucratic and more generous strategy, enabling them to attract more jobs over a wider range of skills and locations. There have been failures here as well, but by-and-large they have been fewer and less well publicised.

In addition to direct benefits, EC membership has opened labour, education and leisure to wider markets, and fostered development of new transport links in air and shipping. The Republic of Ireland, in particular, has achieved a new assertiveness within Europe, matched, to some extent, by an evolving identity and a new sense of place, all qualities worthy of study by geographers.

Simultaneously, Ireland has become pivotal to non-EC countries wishing to enter or expand in Europe. Most significantly, Ireland is often closer, in all senses (time, distance and culture), to eastern North America than to many parts of the European Community, thus making the country attractive for investment.

The detriments of belonging to a pan-European community are perhaps more subtle than the benefits. The economic map of Europe has been redrawn, many would welcome redrafting of the political one, with Ireland retaining only a nominal representation in a strengthened Euopean Parliament. Ireland is peripheral to Europe's core, and in the case of Northern Ireland, denied direct access. Such peripherality has led to a 'deskilling' of Ireland, both by the gravitation of the better trained towards the European 'core' (assisted by the abolition of work permits within the EC), and the reduced demand for skilled labour in what is becoming a 'branch plant' economy. It could be argued that Europe has a vested interest in maintaining this *status quo,* preferring to balance the outflow of talent with inflow of supporting funds to develop the infrastructure, but not the economic base.

The late-twentieth century finds Ireland a much altered country. Demographic and social turbulence is not new, but significant new trends have emerged. The burgeoning population, one in three is under twenty, coupled to a reawakening of nationalistic and religious beliefs, has led to a retrenchment of distinctive values, and a reborn opposition to politically scripted social change.

Towns, villages, families and individuals have undergone upheavals in lifestyle within the last two generations. Towns and villages have changed functions and the waxing and waning of regional urban centres has led to changes throughout Ireland. The building of new roads, faster trains, and even the establishment of an internal air service, has reinforced the dominance of a few major centres, at the expense of small towns. Some survive by metamorphosis, switching from small district centres to commuter centres, including Antrim, Newtownards and Lisburn around Belfast, and Bray, Lexlip and Clondalkin around Dublin. Others towns have developed as tourist centres including Wexford, Sligo, Killarney and Enniskillen. Many urban centres continue to survive with a mixture of small industry, service economies and visitors.

Rural Ireland presents the greatest paradox. It is promoted as a major attraction for all the qualities - relaxing, old-fashioned, quiet - that make it hard to live in, especially for the young. The decline of the traditional nuclear family, most noticeable since 1950, has presaged all manner of social problems, from a breakdown in kinship to a neglect of the elderly. Rural Ireland is dominated by farmers; some are very wealthy - rural Northern Ireland boasts more BMWs than any other comparable area of the UK - but many are still poor, particularly those farming marginal land, where profitable years are less than one in three. Today, the 'forty shades of green' tend to reflect the the level of fertiliser application, rather than natural variations in the agricultural landscape. A journey from the outskirts of Dublin to north Co. Donegal, or Belfast to west Tyrone, crosses one of the most abrupt human gradients in Europe. At the extremes, rural life is sustained only by state financial support and in some cases 'money from America'.

The late twentieth century transformation of Irish society and landscape owes as much to the surge of economic activity in the 1960s and early 1970s as it does to the slump in the late 1970s and 1980s. Attempts to stave-off economic disaster only result in the penalty of environmental demurrage at a later date. The tactic of encouraging economic growth and exports will, for example, often be at the expense of laxity in pollution abatement. Too much Irish waste is just dumped into the sea, rivers, onto land or into the air. Such tension may manifest itself in the setting of environmental standards which reflect the judgement of society as to what is an 'acceptable' level of environmental degradation commensurate with particular economic or social benefits. Although such

7

standards (often imposed through EC directives) have their origins in dosage response relationships, they are usually set at a politically expedient level based on the consensus of the electorate. (A complication in this statement has arisen over the release by Britain of nuclear waste into the Irish Sea, provoking an international disagreement between the two countries.)

It is interesting that there is no effective 'Green' political party in Ireland, especially in the Republic of Ireland where the Proportional Representation electoral system might allow participation in government. While there is undoubtedly a 'green conscience' among many elected representatives, it is not particularly evident at party political level. The decision in 1987 to dismantle the State planning and conservation service, An Foras Forbartha, is symptomatic of this total lack of commitment, as well as an act to be deplored. It may be that the style of politics, often of the 'pork-barrel' variety in the Republic, and secular in Northern Ireland, mitigate against an effective environmental grouping. Notwithstanding this, it is clear from interest and action that the Irish population is becoming environmentally sensitive and this trend may give rise to an effective lobby in due course.

Ireland, both north and south, is a land in a state of rapid change, against a backdrop of long-term conservatism and tradition. The framework of the EC with a common market emerging in just four years time is increasingly dominant in Irish affairs, and the problems faced by the island are similar to those in other peripheral lands of Europe and elsewhere. This book is written by professional geographers and environmental scientists who share an affection for Ireland and who wish to focus attention upon the many-faceted aspects of Ireland today and in so doing, if possible, help navigate a passage towards the potentially calmer waters of the twenty-first century.

2 EUROPEAN ECONOMIC POLICIES AND IRELAND

Frank J. Convery

In this chapter, it is assumed that 'European Economic Policy' comprises an amalgam of economy-related financial and institutional policies applied by the European Community (EC) to Ireland, the United Kingdom and Denmark since their acceptance for membership in 1973. The purpose of this chapter is to trace the economic implications for Ireland of these policies over the 1973-87 period. Most attention is devoted to the agricultural sector but other resources are considered. The regional policy of the EC is also a minor but positive force in the Irish economy, and is dealt with elsewhere in this volume (see Chapter 17).

It is impossible to discover unambiguous cause and effect relationships between initiatives and their effects. The potential for mischief in this regard is beautifully captured by Stephen Leacock: 'when I state that my lectures were followed almost immediately by the Union of South Africa, the banana riots in Trinidad and the Turco-Italian War, I think the reader can form some opinion of their importance'.

The difficulties of analysis in the case of Europe are compounded by the secular shifts which are characteristic of the period 1973-1987 which owe very little to European policy *per se*. These include the dramatic energy price increases which occurred in 1974 and 1979, the consequent global economic dislocations and recessions, the emergence of right-of-centre economic policies in several nations, the emergence of

9

self sufficiency in food in India and China, and the blossoming of the information revolution. In Ireland itself, the 'troubles' in Northern Ireland and the budget deficit of the government in the Republic of Ireland created 'local' difficulties unrelated to European policy.

While the effects of EC economic policy cannot therefore be credibly quantified in a classic 'with-without' type analysis, the *direction of change* implied by such policies can certainly be identified. For the following Irish resources - agriculture, forestry, marine fisheries, natural and built environment - the nature of the EC policies bearing upon them can be summarised and their time span examined. Discussion will be confined to the European dimension since full chapters of this book are devoted to individual resources.

AGRICULTURE

Policies for agriculture overshadow all other programmes of the EC in terms of budget share and political resources devoted to their resolution. Over 70% of the Community Budget is directed to the support of the Common Agricultural

Table 1: **European community transfers to agriculture, Republic of Ireland.**

Year	FEOGA Guarantee Section (Price support)	FEOGA Guarantee Section Statutory Support)	Total
	Millions IR £		
1982	342.7	59.6	402.3
1983	436.7	63.2	499.9
1984	649.7	45.1	694.8
1985	834.0	52.0	886.0
1986	834.0	58.0	892.0

Source: Pre-budget Tables, Deptartment of Finance, Dublin.

Policy (CAP). This finance is provided mainly in order to support prices but also to assist modernisation, re-structuring etc. Table 1 summarises the involvement in the Republic.

The impact of EC membership on agricultural prices can be discerned from the real price trends for milk shown in Figure 1. The implications for this form of output mix and real family farm income - aggregate and *per capita* - have been described by Boyle and Kearney (1983) and Sheehy (1984) for the Republic of Ireland, and by Stainer (1985) for Northern Ireland.

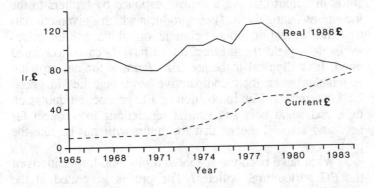

Figure 1 Milk prices in the Republic of Ireland, 1965-83

The pattern of development in the Republic has been as follows: Rapid growth in real income and investment up to 1978 followed by a sharp fall, with only a partial recovery thereafter; expansion concentrated on grass-based enterprise, with milk output growing particularly fast. In Northern Ireland, because of UK policy *vis-à-vis* prices, the income benefits to farmers have been much less pronounced. Sheehy (1984) observes that 'Gross Agricultural Product at constant prices in 1978 was 32% higher in the South than in 1970, but was 13% lower in the North'.

The effects on enterprise mix has been as follows (Sheehy *et al.*, 1981): 'Tillage continued its rapid decline in the North, whereas in the South cereal acreage actively

11

increased, and total tillage declined only slightly. The farm-yard enterprises declined in both areas, but the rate of decline, particularly in pigs was again greatest in the North. Cows and cattle increased North and South with the highest rate in increase in the South'.

The expansion in cattle numbers has been spread fairly evenly throughout the country, but pig and poultry numbers have been concentrated in a few areas. In the Republic, pig numbers have grown rapidly in the border region of Donegal - Cavan - Leitrim - Monaghan, while poultry production is now concentrated in Monaghan and Limerick. These relative shifts in output mix are a logical responce by farmers to the incentives facing them. These products which - given our nat-ural endowment of soil and climate, our skills, and the prices available - yield the greatest return have been concentrated upon. It is illogical to berate Irish farmers for not growing vegetables when their comparative advantage lies in grass-land enterprises. The imposition of EC production quotas on milk and sugar beet and similar restrictions in prospect for beef and cereals, means that the future will not be like the past.

What have been the spatial and physical implications of the EC agricultural policies? The profits generated in the 1974-1978 period provided a windfall surplus. Those farmers who used the surplus to expand their capacity rapidly - espe-cially dairy capacity - and who now have a large milk or sug-ar beet quota and very low bank borrowings, have been the primary beneficiaries. During and after this period, there was a dramatic and very visible private investment in housing which - combined with public investment North and South - has resulted in a dramatic improvment in housing quality in rural areas. However, much of this construction is 'one-off' and dispersed development which is difficult and expensive to service, and which is too often of a design that is intrusive and unsympathetic to the landscape.

The intensification of milk and cattle production, com-bined with the vagaries of the Irish weather which make hay production so undependable, have resulted in a dramatic expansion in silage-making. The liquid by-product of silage has an exceptionally high BOD (Biochemical Oyxgen Demand) requirement and if it leaks into water courses can be lethal to fish and other aquatic organisms; in the summer of 1987, there were numerous media reports of fish kills. It is probably incorrect to attribute EC policy *per se* to this increased pollution. Indeed it might be argued, based on our experience with pigs and poultry, that a lower level of subvention would result in larger more concentrated production units, as producers were forced to capture economies of scale to stay in business. Because of the high costs of feed in the country, pigs suffer a competitive disadvantage relative to grass fed cattle. Sixty seven% of pigs in the Republic are in herds of 1,000 or more, while only 24% of cattle are found in the largest herd category of >100.

The concentration of pig production on the wet mineral soils of the border region has proved unfortunate for the environment as the capacity of these soils to absorb and assimilate waste is very limited, so that polluted groundwater enters rivers and lakes which are among the best freshwater and fishing sites in Europe. As the lakes have become polluted as a result of animal waste run-off, the fisheries have been diminished. However, such impacts owe little directly to EC policy.

In addition to the price support, the Community has also provided financial support on a matching fund basis for measures to improve farm structures. As is clear from Table 1, this form of EC support only amounted in 1986 to 7% of the subvention for price support in the Republic. However, because they are directed to specific projects, and involve overt cash transfers to farmers, the impact of structure programmes is easier to identify on the ground. State and EC transfers to agriculture in the Republic of Ireland (1982-1986) are shown in Table 2, while transfers per acre per employee and per agricultural holding for 1985 are presented in Table 3. The transfers of IR£1,222 million comprise 65% of gross net output in agriculture in 1985 (Central Bank, 1985).

In terms of physical impact, it is the EC support of infrastructural investment - notably of land drainage - which has been most significant. Under the provisions of a special EC programme for the accleration of drainage works in the West of Ireland, to operate from 1979 to 1986, 50% of the cost of drainage schemes for the Corrib-Mask-Robe, Boyle and Bonet catchments has been met by the EC, but with a ceiling of IR£15.3 million.

Table 2: State and European Community support in relation to agriculture

millions IR£ current

	1982	1983	1984	1985	1986 (Prov)
State					
Direct Aids reducing production and overhead costs and production incentives(1)	101	129	133	136	151
Schemes operated under EC regulations and directives(2)	89	68	68	53	68
Long-Term Development Aids(3)	27	23	25	22	21
Sub-total Direct	217	220	226	211	240
Indirect Education, research advisory and inspection	59	63	67	67	66
Disease eradication	15	21	15	16	12
Marketing aids	1	1	1	1	1
General administration and overhead costs	36	36	37	41	43
Sub-total Indirect	111	121	120	125	122
Total State	329	342	346	336	362
European Community					
FEOGA Guarantee Section	343	437	650	834	834
FEOGA Guidance Section	60	63	45	52	58
Total	403	500	695	886	892
Grand Total	732	842	1,041	1,222	1,254

Source: Pre-Budget Tables, 1986, Dept. of Finance, (pp.7-8).

Table 3: Provisional average transfers

Transfers from:	State			European Community			Total
	Direct	Indir-ect	Total	Guar-antee	Guid-ance	Total	
Total amount (IR£M)	521	309	830	2,060	128	2,189	3,019
Employment (Two estimates)							
Labour Force Survey(1984)[a] 181,600 persons employed.							
Per person	2,171	1,700	4,571	11,349	707	2,506	16,627
Community Survey(1980)[b] 304,300 persons employed.							
Per person	17,121	1,015	2,727	6,773	422	7,195	9,923
Area 4.69 million ha							
per acre	44	27	72	177	10	187	259
Holdings Number 223,500							
Per holding	944	559	1,503	3,732	233	3,964	5,468

[a]Includes forestry and fishing
[b]Includes the full-time equivalent of all part-time work.
Sources: Table 1: Labour Force Survey 1984, CSO, Dublin 1985 (p.23); Community Survey on the Structure of Agricultural Holdings 1979/80, Vol. 11, Eurostat, Luxembourg, 1985, p.201.

The Blackwater cross-border scheme attracts a special grant, while up to 30% of other schemes is recouped from the European Regional Development Fund (Stationery Office, 1986). The Western Drainage Scheme provided funds for field drainage. All 150,000 ha provided for in this EC Directive were taken up under this programme. In making a judgement as to how much of such programmes would have been embarked upon in the absence of EC membership, it is helpful to examine their costs per ha. Agricultural land prices per ha have fallen in the range of IR£2,890 to IR£3,565 from 1981 to 1985 (Kelly, 1986). It is unlikely that the State would have been willing to finance entirely out of its own resources drainage schemes which were not much cheaper per acre - and in the case of the River Blackwater (Ulster) were dramatically more expensive - than the market value of the agricultural land.

The economic impact of European agricultural policies to Ireland can be summarised as follows. They dominate all other Community influences, EC price and structures support averaged IR£78 per acre and IR£3,732 per holding in 1985. The effect of the price and intervention support has been to intensify the competitive advantage of grass agriculture, notably via milk production. Large 'wind-fall' gains accrued to farms between 1974 and 1978 but margins since then have been squeezed.

Programmes of arterial drainage and field clearance were maintained and expanded as a result of EC subvention. These created the potential for increased income on the benefitting farms. The arterial drainage programmes also imposed costs in terms of disruption and/or destruction of fisheries and wildlife habitats.

FORESTRY

In both Northern Ireland and the Republic of Ireland, forestry has been predominantly a State activity for most of this century.

The European Community does not have an active forest policy *per se*. However, membership of the EC has influenced forestry in two ways. First the rapid rise in the price of farm output and net income in the mid 1970s resulted in a dramatic rise in the real price of farmland (Figure 2). This in turn made it difficult to justify the acquisition and planting of land for forestry. Partly as a result, State planting programmes in both parts of Ireland were reduced. Second, and counterbalancing this negative impact is the grant support provided by the EC in the Republic of Ireland for afforestation under the Programmes for Western Development (1981-1991).

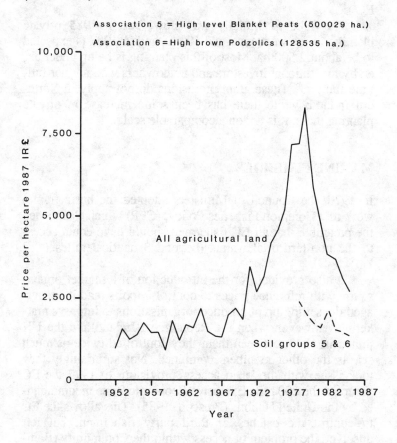

Figure 2 Price of farmland in the Republic of Ireland, 1952-87

The objectives of the forestry scheme are to stimulate and encourage private sector planting of land which is marginal for agriculture but suitable for forestry. Grant assistance is available for expenditure on ground preparation, planting, forest roads and the protection and maintenance of plantations. The level of grants is generous, covering 85% of approved costs in the case of farms and 70% in the case of non-farmers, subject to a maximum of IR£800 per hectare (Grant assistance is financed equally by the State and by the EC.)

The 'take-up' initially was slow, but in 1985 private planting in the west exceeded 600 ha and in 1987 is expected to be about 2,500 ha. Most of this planting is being undertaken by institutional investors and landowners who are not full-time farmers. These grant provisions do not apply in Northern Ireland; while there has been some increase in private planting there, it is not on a comparable scale.

MARINE FISHERIES

In 1970, the Council of Ministers adopted the basic framework of a Common Fisheries Policy, (CFP) which established the preamble that all EC fishermen should have equal access to the non-territorial waters (beyond 3 nautical miles (c. 7 km)).

It also provided for the introduction of a market organisation with reference prices to control imports, and it encouraged the setting-up of producer organisations to improve marketing. However, when the Republic of Ireland and the UK joined in 1973, between them they controlled twice as much sea as the other members combined. Not surprisingly they took issue with the 'open access' provision. In 1977 the EC accepted the 200 mile Exclusive Economic Zone recommended by the United Nations (Prescott, 1985). This allows the EC to control the catches of third party fishermen, but left unsolved the problem of access within the Community (Bamford and Robinson, 1983). Agreement was finally reached on a Total Allowed Catch for the EC as a whole, quotas for various species have been allocated by country, and transfers from the Community have helped finance protection vessels.

On balance, there has probably been some modest short-term economic advantage to Irish fishermen from membership of the Community, but this has been achieved at the cost of limiting the potential - from a very low base - of expansion in the future. The EC funds have been used to re-equip fishing fleets both in the Republic of Ireland and in Northern Ireland.

NATURAL AND BUILT ENVIRONMENT

The various EC environmental initiatives which have been implemented have had relatively little economic impact in Ireland, North and South. Various EC directives on water quality have not resulted in demonstrable significant costs accrued, or benefits conferred. EC membership has probably resulted in a more overt attention to standards and procedures than would have obtained otherwise.

The European Community's 1980 directive on air quality limit and guide values for sulphur dioxide and suspended particulates specifies standards to be met on a daily basis. Periodically, air quality in Dublin fails to meet the limit value standards for suspended particulates. As a result air quality legislation has been enacted, but it remains to be seen whether it will be implemented. Belfast is pursuing a smoke control policy which owes its origins and implementations to UK policy. However, there is a Commission proposal to (in effect) reduce sulphur emissions from power stations. If instigated, this will guarantee substantial costs and benefits.

Community policies concerning the built environment are, as yet, at the experimental stage. Belfast has benefitted from EC funding as a consequence of its designation - along with Milan - as an example of an integrated urban renewal programme. The extent to which European funding was in fact 'additional', or was simply used to replace funds which would anyway have been spent by the British Exchequer, is a matter of debate. There is no doubt that the European dimension has given the renewal process a higher profile, and placed it in a wider context than would have been the case in the absence of Community support.

INDUSTRY

In their review of Irish industry, Blackwell and O'Malley (1984) conclude that membership of the European Community has stimulated foreign investment in Irish industry, and contributed to a relatively strong overall performance. However, they also point out that the indigenous industrial base has failed to expand, and that the potential implicit in access to large markets remains unrealised. The nature of much of foreign industrial investment is such that it is very vulnerable to shifts in consumer preferences or relative costs.

CONCLUSIONS

Membership of the European Community has affected the economic life of Ireland mainly through the workings of the Common Agricultural Policy. For a 5-6 year period this provided substantial increases in real income, with the greatest income being captured by those who could expand output to 'take advantage of the higher prices before rises in input costs eliminated much of the gain'. The recent imposition of quotas has ensured that the advantage will continue to lie with this group, so that in effect they now have 'property rights' to a high (relative to the world level) price for a substantial volume of output. Forestry suffered from the high land prices which followed the rapid income growth in agriculture, but benefitted from generous grants in the Republic of Ireland to the extent that private forestry is expanding for the first time in 100 years. The effects of Community policies in the economy of coastal communities has been probably mildly positive, but at the cost of inhibiting long-term development.

Community policies concerning the environment have had very little effect on the Irish economy. Membership has obviously made Ireland attractive as a location for some industry needing access to the Community, but has apparently been of little benefit to most indigenous industries.

REFERENCES

Bamford, C.J. and Robinson, H. (1983) *Geography of the EEC*, MacDonald and Evans, Plymouth.

Blackwell, J., and O'Malley, E., (1984) The impact of EEC membership on Irish industry, in Drudy, P.J. and McAleese, P. (eds) *Ireland and the Economic Community*, University Press, Cambridge, 107-144.

Boyle, G.E. and Kearney, B. (1983) Intensification in agricultural trends and prospects, in Blackwell, F.J., and Convery, F. (eds) *Promise and Performance:Irish Environmental Policies Analysed*, Resource and Environmental Policy Centre, University College, Dublin, 77-102.

Central Bank (1985) *Annual Report 1985*, Dublin.

Kelly, P. (1986) Land prices, *Farm and Food Research*, 53-54.

Prescott, J.V.R. (1985) *The Maritime Political Boundaries of the World*, Methuen, London.

Sheehy, S.J. (1984) The Common Agricultural Policy and Ireland, in Drudy, P.J. and McAleese, D. (eds) *Ireland and the European Community*, University Press, Cambridge, 79-105.

Sheehy, S.J., O'Brien, J.T. and McClelland, S.D. (1981) *Agriculture in Northern Ireland and the Republic of Ireland*, paper 3, Co-operation North, Dublin and Belfast.

Stainer, T.F. (1985) *An Analysis of Economic Trends in Northern Ireland Agriculture since 1970*, Department of Agriculture, Belfast.

Stationery Office (1986) *Comprehensive Public Expenditure Programmes 1986*, Government Publications, Dublin.

3 PARTITION, POLITICS AND SOCIAL CONFLICT

Dennis Pringle

Ireland, although a fairly small island, is politically partitioned between two sovereign states: the Republic of Ireland, which has sovereignty over 26 'southern' counties; and the United Kingdom which has sovereignty over the other 6 counties (which together constitute Northern Ireland). Since its formal introduction, under the terms of the Government of Ireland Act (1920), partition has been a source of considerable dissension and conflict, ranging from the civil war in the south (1922-3) to the present-day 'troubles' in Northern Ireland. Partition, however, is not the only source of conflict in Irish society. Both north and south, society is divided into conflicting interest groups by a variety of social, ethical and distributional issues, including the control of space and other limited resources. These issues have no obvious unanimously acceptable solutions - otherwise they would cease to be a source of conflict - but the likelihood of a rational democratic solution (loosely defined as one which is acceptable to as many people as possible, without impinging upon the socially accepted rights of the minority) is in many instances considerably reduced by the existence of partition, and by the conflicts which it has generated. This chapter examines some aspects of the impact of partition upon politics and social conflicts in both Northern Ireland and the Republic.

23

SOCIAL CONFLICT, POLITICS AND SPACE

The utilization of human and natural resources, which is essential to meet the material needs of society, necessitates a high degree of social and spatial organization. However, within any given context there are numerous alternative forms which this organization could take, each of which confers different benefits and costs upon different sections of society. The social and spatial organisation of society is consequently the subject of considerable social and political conflict between various interest groups and the form which it assumes at any particular time can be regarded as a reflection of the balance of power arising out of previous conflicts.

For example, under capitalism space is territorially divided for ownership purposes. This territorial division confers advantages upon landowners, but it also gives rise to a variety of potential conflicts between landowners and other groups (e.g. between landowners and dispossessed landowners over ownership rights, between landowners and non-landowners over inequalities in the distribution of wealth, and between landowners and environmentalists or conservationists over how the land should be used). Each group attempts to further its own particular interests using whatever power is at its disposal. In a democratic society, these conflicts are partly conducted through the medium of politics, thereby placing politicians in a position of considerable power.

However, the power of politicians extends beyond the power to frame legislation or to make executive decisions if voted into government. They also play an active role as shapers of public opinion. Whether it be through high media exposure, or (if in government) by their control over the ideological apparatus of the state, politicians help to define which issues are regarded by society as the most important. Political power may therefore be used to block social change in certain directions by ensuring that potentially threatening issues are given low priority and hence never come up for discussion - a mechanism which Schattschneider (1960) referred to as 'non-decision making'. Controversial issues, whose resolution might threaten the interests of the ruling parties or their supporters, can in this way be 'organised off the agenda' (Taylor, 1985), provided that alternative issues can be promoted as requiring higher priority.

Conflicts relating to the spatial aspects of the human environment are in some instances major issues in their own right, but in other instances they may be relatively minor issues which are deliberately manipulated by powerful interest groups to organise more important issues off the agenda. Partition has emerged as probably the most controversial spatial issue in Ireland. This chapter examines some of the ways in which the partition of Ireland between two sovereign states has influenced the nature of politics on both sides of the border.

The remainder of the chapter is divided into four sections. The first section reviews some of the historical reasons for partition. The following two sections examine some of the consequences of partition for politics and social conflicts within Northern Ireland and the Republic respectively. The final section then speculates about the likelihood of changes in the nature of Irish politics in the future.

THE BACKGROUND TO PARTITION

The background to partition is treated in more depth elsewhere (e.g. Pringle, 1985), but it is necessary to review some of the more salient points for present purposes.

Ireland, in the eighteenth century, was a British colony with its own parliament centred in Dublin, but governed by an executive controlled by Westminster. Following the Act of Union (1800), Ireland became an integral part of the United Kingdom. Under the terms of the Act, Ireland was to be represented in the British House of Commons by 100 Members of Parliament (Beckett, 1966). However, during the course of the nineteenth century, paralleling the growth of nationalism elsewhere in Europe, the majority of Irish public opinion, fuelled by resentment caused by appalling material conditions, especially in rural Ireland (epitomised by rack-renting, evictions, famine, and large-scale migration), supported demands for a repeal of the Act of Union and, later, for the establishment of home rule.

These demands met with opposition within Ireland from unionists who were particularly numerous in east Ulster. The

25

concentration of unionists in Ulster was largely, though indirectly, due to the area's proximity to Scotland, which is only 13 miles away at its closest point. Movement between Ulster and Scotland was common throughout history, but the floodgates were opened following the defeat of the great Gaelic clans in central and western Ulster in 1603. Large numbers of land-hungry Scots tenants were enticed to Ulster throughout the seventeenth century by the relatively attractive conditions of tenancy offered by the new English and Scottish landlords who had been granted confiscated land after the flight of the Gaelic leaders in 1607. Given that most of the Scots had been converted to Calvinism by the early seventeenth century, whereas the native Irish remained Catholic, the seventeenth century influx of Scots is reflected today by a high concentration of Protestants, especially Nonconformists, in Ulster.

In terms of folkculture, the Scots settlers were not as different from the pre-existing population as many unionists would probably like to believe (Buchanan, 1982); but neither was the pre-existing population as homogeneously Gaelic as many nationalists would like to believe (Stewart, 1977). However, the two groups were sufficiently different, especially with regards to religion, for most Protestants to feel alienated by the Gaelic Catholic ethos espoused by the Irish national movement in the nineteenth century. Ulster Protestants consequently found it much easier to identify, in terms of national identity, with people living elsewhere in the United Kingdom than with the Catholic majority in the rest of Ireland.

Support for the union was also preconditioned by the perceived need for the highly developed Ulster industries, especially linen, shipbuilding, and engineering, to retain unhampered access to their British markets and external sources of essential raw materials. Irish nationalism, in contrast, was rooted in the industrially underdeveloped agricultural economy of rural Ireland. Although the large grazier farmers in southern and eastern Ireland had little to lose from a continuation of free trade with Britain, more radical sections within the nationalist movement proposed establishing protective tariff barriers to facilitate the growth of indigenous industries. This made sense for the under-industrialised parts of Ireland, but it would have resulted in the destruction of the then thriving Ulster industries. Unionism, in short, was based not only upon a Protestant desire to avoid being religiously and culturally dominated in a Gaelic Catholic Ireland, but also to retain free trade with the United Kingdom because it offered positive material advantages.

The nationalist demands for home rule remained frustrated for a variety of reasons until after the end of World War I, by which time they had escalated into a concerted campaign for full national independence. Attempts by the British authorities to re-establish control by force, resulting in the Anglo-Irish War (1919-1921), failed miserably, with the result that the British were forced to concede independence for 26 southern counties in the Anglo-Irish Treaty in 1921.

The Treaty, accepted by the majority faction within the new Irish parliament (Dáil Éireann), also ratified the partition of Ireland along the lines defined by the Government of Ireland Act which had been passed in the previous year. This Act had created local parliaments for two areas referred to as Northern Ireland and Southern Ireland. Southern Ireland, under the terms of the Treaty, became independent as the Irish Free State (later renamed the Republic of Ireland in 1948), whereas Northern Ireland remained a part of the United Kingdom. Unlike other parts of the United Kingdom, Northern Ireland retained its own regional parliament, housed after 1932 at Stormont.

It is often argued that Ireland was partitioned so that Britain could retain control over as much of the island as possible. However, an alternative explanation is that partition was forced upon the British government by the intense, and armed, opposition to Irish nationalism by unionists in Ulster. Although the unionists represented a minority within Ireland as a whole, they were spatially concentrated in an area where they formed a substantial local majority. Any attempt to include the unionists in an independent united Ireland would have resulted in a very bitter and destructive civil war between unionists and nationalists, which would probably have resulted in some form of partition anyway. British interests in Ireland could therefore best be served by a compromise aimed at avoiding a civil war by partially satisfying the conflicting demands of both sides through partition.

NORTHERN POLITICS

The border between Northern Ireland and the Republic is usually viewed by Irish nationalists as an undesirable and artificial *barrier* between nationalists living in the north from those living in the south. However, the border can also be viewed as a *container* in the sense that it defines the extent of the two areas over which the respective governments have direct authority. As a container it also defines the character of the electorates from whom the politicians must seek support. Whilst the divisive effects of partition have had a very obvious impact upon Irish politics, the compositions of the respective electorates have also exercised a very profound impact upon politics on both sides of the border.

Although partition left Northern Ireland with a large unionist majority, it also had a substantial nationalist minority. Unionist candidates, for example, won 66.9% of the votes cast in the 1921 general election in Northern Ireland, whilst Sinn Féin and Nationalist candidates combined won 32.3%. Given the immediacy of the national question, less than one% of the votes in 1921 went to Labour and Independent candidates. The existence of a large nationalist minority, which never really accepted the legitimacy of the Northern Ireland state, has helped ensure that the national question - i.e. whether Northern Ireland should be united with the Republic of Ireland or remain part of the United Kingdom - has remained the dominant issue in Northern Ireland politics. Had the unionist majority been larger, and therefore more secure, or had the nationalist minority dropped its aspirations for a united Ireland, the national question might have become a dead issue, and politics might have developed along totally different lines.

The relations between unionists and nationalists were further complicated by the spatial distributions of the two communities within Northern Ireland (Figure 1). Protestants, and hence unionists, form a substantial majority in most of what might be termed the greater Belfast region; i.e. areas within about 30 miles radius of Belfast. The major exception is west Belfast which is mainly Catholic and strongly nationalist. Outside the greater Belfast region, population densities are lower, and the population is more evenly divided between unionists and nationalists. Most of the areas at the very edge of Northern Ireland, such as south Armagh, south Fermanagh, and Derry west of the river Foyle, have substantial nationalist majorities. Some form of partition may have been necessary to circumvent the need to force unionists into a united Ireland, but the present location of the border would be very difficult to justify in terms of the national identities and aspirations of people living in border areas. Northern Ireland, in short, is an over-bounded state.

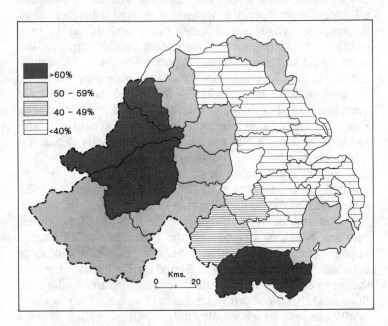

Figure 1. Percentage of Catholics in each District Council Area, 1981 (based on estimates by Compton and Power, 1986)

This was realised when Ireland was initially partitioned under the Government of Ireland Act, and a Boundary Commission was consequently proposed under article 12 of the Anglo-Irish Treaty to 'determine in accordance with the wishes of the inhabitants, so far as may be compatible with economic and geographic conditions, the boundaries between

Northern Ireland and the rest of Ireland ' (Fanning, 1983, 24). The Boundary Commission did not in fact meet until 1924, due to the unsettled political climate, and when it did meet it proposed only minor adjustments of the border. Although these would have resulted in the transfer of 183,290 acres and 31,319 people to the Free State, they would also have resulted in the transfer of 49,242 acres and 7,594 people to Northern Ireland (Harkness, 1983).

The proposals, however, were never implemented. Before the Commission's report could be officially published, a London newspaper, the *Morning Post,* published a map of the proposed changes. Although only three of the 17 proposed land transfers were shown more or less correctly, the main thrust of the newspaper report was correct (Andrews, 1968). Nationalist public opinion was outraged by the leaks, not only because the transfers to the Free State would have been much smaller than had been anticipated, but because the Free State would also lose some of its prime agricultural land in Donegal. The Dublin government consequently demanded a three way conference between representatives of the Belfast, Dublin and London governments (Fanning, 1983). This resulted in the signing of a Tripartite Agreement in 1925 which ratified the existing location of the border.

The long delay in formalising the location of the border between Northern Ireland and the Free State left a lasting impact upon politics within Northern Ireland. Nationalists had been optimistic that the Boundary Commission would transfer very large areas within Northern Ireland to the Free State, and that Northern Ireland would consequently become totally unviable as a separate entity. Such an outlook betrays a nationalist misunderstanding of the material basis of unionism. Unionists required Northern Ireland to remain part of the United Kingdom; they did not intend it to be a separate self-contained entity. Any consideration of the economic viability of Northern Ireland would therefore have had to consider the viability of Northern Ireland within the context of the United Kingdom. The loss of even large areas in the border areas might have enraged unionists, but it would probably not have proved fatal to the viability of Northern Ireland.

Irrespective of the arguments as to what the implications might have been, the uncertainty generated by the boundary commission encouraged a belief amongst nationalists that partition was only an interim measure. Many northern nationalists consequently regarded the Dublin government as the 'real' government for the whole island, and refused, with the support of the Dublin government and the Catholic Church, to recognise the Unionist government in Belfast. Teachers in Catholic schools in the north, for example, refused to recognise the Belfast Ministry of Education and for a period had their salaries paid by Dublin.

The lack of co-operation by sections of the Catholic minority reinforced the unionist belief that all Catholics in Northern Ireland should be regarded as potential fifth columnists intent on overthrowing the state. The unionists therefore reacted by systematically excluding Catholics, even those with distinguished service records in the British army or civil service, from positions within the Northern Ireland state apparatus where they could do damage. Discrimination against Catholics by the state in turn reinforced the nationalist belief that social justice was impossible within Northern Ireland and that social progress of any type was contingent upon an end to partition. Thus, by the time that the uncertainty over the location of the border had been resolved by the Tripartite Agreement in 1925, politics within Northern Ireland had become strongly polarised around the central issue of the national question. All other issues were forced to assume a secondary position.

The impact of the nationalist boycott was most clearly seen in the case of local government. Although nationalists formed a minority within Northern Ireland as a whole, the fact that they were a local majority in many areas meant that a large number of local authorities (including Fermanagh and Tyrone county councils) were under nationalist control. The legitimacy of the Northern Ireland government was consequently seriously challenged when these local authorities began to pass resolutions refusing to recognise the Belfast government and pledging their allegiance to the Dublin government. It therefore became imperative for the Unionists to re-establish Unionist control over the errant local authorities.

This was achieved by a number of 'reforms' in local government. Proportional representation was abolished in 1922 for local government election, necessitating a comprehensive redistricting of electoral areas. The new boundaries resulted in many accusations of gerrymandering, especially in the border counties. The franchise was also altered to exclude non-property owners, and later extended to give companies up to 6 additional business votes. By 1924 only 77 per cent of the Northern Ireland electorate were entited to vote in local government elections (Douglas, 1982). Given that unionists were more strongly represented in the more affluent sections of society, there were many complaints that the changes in the franchise discriminated against nationalists. Curiously enough, there were comparatively few complaints that it discriminated even more directly against the urban working classes, irrespective of their religion or national allegience - a reflection of the extent to which class politics had become almost totally obliterated by the existence of the national question.

The net effect of these changes in local government was that Unionists gained control of many of the local authorities where there was a nationalist majority. Perhaps the most striking case was Londonderry Corporation which, following a further re-organisation in 1936, was consistently retained under Unionist control despite the presence of a substantial nationalist majority. In the 1967 elections, for example, the Unionists won 12 of the 20 seats on the Corporation with only 38 per cent of the vote, thanks to a combination of gerrymandering, malapportionment and company votes (see Busteed (1972) for details of the election results).

Control over the local authorities was more important to the Unionists than might be immediately apparent. The local authorities were important as sources of employment and had responsibility for the provision of many important services (including housing). Control over a local authority therefore conferred powers to decide who should get jobs and houses. In a depressed economy, where jobs and houses were in short supply, this effectively meant that local authorities had the power to decide who should be forced to emigrate. Unionist politicians were consequently able to offset the traditionally higher Catholic birthrate, which might have eventually threatened the Protestant majority in Northern Ireland as a whole, by 'encouraging' a higher rate of out-migration amongst Catholics at the local level. It also, it might be noted, gave the

Protestant bourgeoisie, who dominated the Unionist party, considerable scope to manage potential conflicts with the Protestant proletariat through patronage in the public sector as well as in the private sector.

Unionist politics remained fairly united and paternalist in the south and west of Northern Ireland where unionists were constantly on the defensive because of the fairly even balance between unionists and nationalists. However, the large size of the unionist majority created more scope for internal dissension in the east. This was highlighted in the 1920s when a deterioration in the economy resulted in the growth of opposition to the government along class lines. The Unionist party had won 40 out of 52 seats in the 1921 general election for the Northern Ireland parliament contested using proportional representation, but in 1925 they won only 32 seats. Seven of the eight seats lost were in Belfast (where the Unionists only retained 8 out of a total of 16 seats). None of the seats were lost to nationalists; rather, the challenge came from Labour and Independent Unionist candidates who challenged the official Unionist party on class issues.

The response of the Unionist government was to abolish proportional representation for the next general election in 1929. New single seat constituency boundaries had to be drawn to replace the former multiple seat constituencies. Although nationalists complained of gerrymandering, there was (with arguable exceptions in Fermanagh and Armagh) little ground for complaint (Osborne, 1979). The unionist majority at the Northern Ireland level was too large to require any artificial inflation by electoral malpractices. The main objective of the 1929 reforms was to squeeze out the smaller parties and to stifle the challenge from within the Protestant ranks (Pringle, 1980). Under the new plurality electoral system, the number of issues was effectively reduced to one - the national question. Despite a further 4.3% decline in the vote, the Unionist party, contesting the elections under the slogan 'Safety First - Vote Unionist', gained 5 seats. One by-product of the new electoral system was that more then half the seats were rarely contested over the next 40 years because as long as the national question retained its central position the result was a foregone conclusion.

By keeping the national question to the fore, the Unionist government, ably assisted by the Dublin government's territorial claims to the whole of Northern Ireland, was able to keep other issues off the agenda. However, when the post-war economic boom began to fizzle out in the 1950s, the Unionists again found their position being challenged along class

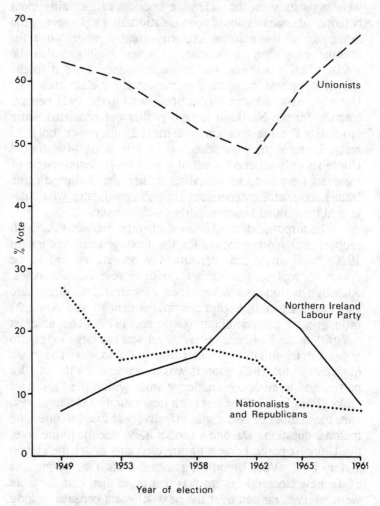

Figure 2. The percentage of votes for each of the major parties in Northern Ireland general elections, 1949-69

lines. For example, although its support was not translated into seats because of the plurality electoral system, the Northern Ireland Labour Party steadily increased its share of the vote to 26% in 1962, at the expense of both nationalists and unionists (Figure 2).

The declining fortunes of the old style of Unionist government gave rise to a new form of unionism, symbolised by the succession of Captain Terence O'Neill to leadership in 1963. The government intensified its campaign, begun in the 1940s, to attract overseas industrial investment to Northern Ireland to replace the jobs being lost because of a decline in the traditional industries, such as agriculture, shipbuilding, engineering and textiles. In addition to providing advance factories and grants, the government also embarked upon a programme of modernisation to make Northern Ireland more attractive to investors. Part of this programme focussed upon improving the spatial infrastructure. Plans were drawn up to limit the growth of Belfast, build the new town of Craigavon, initiate a motorway system, and establish a second university. The other part of the programme was to modernise Northern Ireland society by trying to make Northern Ireland more acceptable to the Catholic minority through a series of moderate liberal reforms.

Given the polarised nature of Northern Ireland society, the policies of the government were treated with suspicion by both the Catholic minority, who regarded them as little more than window dressing, and by the more conservative elements within the Protestant majority who felt that they opened a Pandora's box. Nationalists complained that the new industries were being directed by the government to the Protestant core area in the east; the government argued that the industries themselves showed a preference for the east for a variety of reasons, including the availablity of a skilled workforce and better infrastructure (Hoare, 1981). Nationalists complained that the second university should have been located in Derry, Northern Ireland's second city where there already was a university college, rather than at Coleraine which lies within the unionist heartland. The government argued that Coleraine had been selected by the independent Lockwood Commission, although Osborne (1982) suggests that the Derry proposal was opposed by elements within the local Unionist party in Derry who were worried that an influx of students might upset the finely tuned gerrymander in the city. Nationalists also complained that resources were to be ploughed into building the new city of Craigavon, controversially named after Northern Ireland's first Unionist Prime Minister, 30 miles west of Belfast, rather than being used to revitalise Derry. Unionists argued that Craigavon was better located to attract Belfast's surplus population. Irrespective of the rights and wrongs of these arguments, their net effect was to undermine the minority's confidence in meaningful reforms by the government.

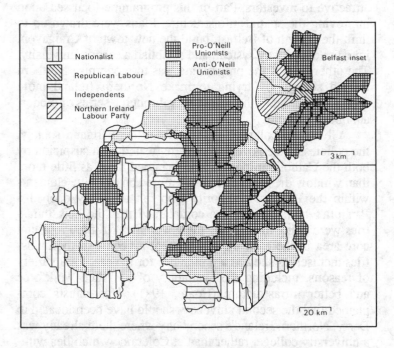

Figure 3. Distribution of seats won by pro- and anti-O'Neill Unionists, 1969

Unionist opposition to O'Neill came from two main sources. Many working class Protestants alarmed at the very rapid collapse of the traditional industries - 11,500 jobs were lost in shipbuilding alone between 1960 and 1964 (Mac-Laughlin and Agnew, 1986) - and finding themselves placed in open competition with unemployed Catholics for the new jobs, yearned for a return to a traditional paternalistic form of unionism in which loyalty to the state brought material rewards. Protestant working class areas subsequently proved to be a very fertile breeding ground for Ian Paisley's new Democratic Unionist Party in the 1970s. The other main source of opposition to O'Neill, and initially much more serious, came from the more conservative elements within his own party. This opposition was strongest in the south and west of the province, where the unionist majorities were marginal and where liberal reforms, especially with regard to gerrymandering, were therefore likely to cause most damage. Support for O'Neill, as indicated by the results of the 1969 general elections, was strongest in the east where sectarian conflict was least pronounced because of the higher level of prosperity and the large size of the Protestant majorities (Figure 3).

Catholic disenchantment grew as the rate at which reforms were introduced by the government was hindered by opposition from within the unionist ranks. The Northern Ireland Civil Rights Association (NICRA) was consequently founded in 1967 and began to hold demonstrations demanding the implementation of fundamental reforms. Although initially supported by some liberal Protestants, the vanguard within NICRA was provided by a new generation of articulate and, as a result of the introduction of free education by the 1947 Education Act, educated working class Catholics. Conservative unionists, wishing to avoid the real issue, resorted to traditional tactics by playing up the national question and NICRA was accused of being a front for the paramilitary Irish Republican Army (IRA).

NICRA was obliged by law to provide notice of public demonstrations. Many of these demonstrations, however, resulted in violence because of counter-demonstrations by Paisley and his supporters. The authorities consequently reacted by banning both sets of demonstrations. This was interpreted by NICRA as an attempt to stifle all legitimate protest, thereby reinforcing the nationalist belief that social justice could not be achieved without an end to partition. Banned from holding legal demonstrations, NICRA had little alternative but to hold illegal demonstrations, which brought it into direct conflict with the authorities. This, in turn, merely served to reinforce the unionist suspicion that the real objective of NICRA was to overthrow the Northern Ireland state.

Thus, whilst there may have been some behind the scenes organization by republicans within the civil rights movement, what was basically a demand for civil rights within Northern Ireland became diverted into an increasingly violent conflict over the national question, reflected by the renaissance of the IRA. Space does not permit a detailed blow by blow account of the deepening crisis, but it probably suffices for present purposes to make a few general points.

First, as conflict over the national question intensified in the late 1960s, other issues within Northern Ireland politics became increasingly peripheralised. The Northern Ireland Labour Party, which had been the second largest party in the early 1960s, was squeezed out of existence by the 1970s as politics became polarised along sectarian lines. The traditional Unionist and Nationalist parties did not survive intact either. The Official Unionist Party is now challenged by the Democratic Unionist Party for the unionist vote, whilst the old Nationalist Party has been replaced by the Social Democratic and Labour Party (SDLP) and Sinn Féin. However, in each instance the differences between the parties is mainly one of tactics on how to tackle the national question. Even the Alliance party - the only party able to transcend the sectarian divide with any real credibility - is mainly noted for the fact that it is non-sectarian, rather than for any policies which it holds on social issues.

The second point is that the strategy of the conservative unionists backfired completely. Most of the reforms demanded by NICRA, including an overhaul of the local government system, were implemented by the early 1970s. The local government system was reorganised and streamlined into a new system based upon 26 district councils: gerrymandering and other electoral abuses were eliminated; and STV proportional representation was reintroduced to ensure a fairer representation for minorities. Control over public sector housing was taken away from the local authorities and vested in the centralised Northern Ireland Housing Executive, where it was rendered less controversial by the introduction of a standardised points scheme for housing allocations. Control over other social services was likewise centralised to minimise the likelihood of malpractices at local level, and, although accused of being ineffectual, various watchdog bodies, such as the Fair Employment Agency, have been set up to investigate cases of alleged discrimination.

However, a more serious setback for the unionist majority is that, as a result of the 'troubles' - as the present period of violence is euphemistically referred to - they have lost virtually all control over their own destiny. Control over security was greatly reduced by the introduction of the British army in 1969, executive control was removed by the introduction of 'direct rule' from Westminster in 1972, and more recently the remaining, fairly feeble, legislative powers were eliminated by the disbandment of the Northern Ireland Assembly. Northern Ireland is now ruled directly from Westminster, via the Northern Ireland Office, in more or less the same way as any other part of the United Kingdom - with the important exception that the citizens of Northern Ireland are denied the right to vote for any of the parties capable of forming the next British government, because the major parties each refuse to organise and contest elections in Northern Ireland.

The political impotence of unionism was highlighted in 1985 when the British government, without consulting the unionist majority, signed an agreement with the Dublin government, which (to add insult to injury) consulted fully with representatives of the Northern Ireland minority - the SDLP. The Anglo-Irish Agreement apparently gives the the Dublin government a say in Northern Ireland's internal affairs, which, if correct, means that the electorate in the Republic of Ireland has, at least in theory, a greater say in the administration of Northern Ireland than the Unionist majority in Northern Ireland. The absurdity of this situation has not only given rise to a concerted unionist campaign to wreck the Anglo-Irish Agreement, but also to a demand for equal citizenship with the rest of the United Kingdom, including the right to vote for or against the government.

The third, if somewhat obvious, general point to be noted is that, although the violence may have contributed to some positive changes, such as the pedestrianisation of town centres for security reasons (Brown, 1983), the net impact of the violence upon Northern Ireland society has been overwhelmingly negative. Apart from costing the lives of over 2,500 people, and resulting in the injury of ten times as many others, the troubles have undoubtedly aggravated Northern Ireland's already serious economic problems. The rate of growth in the Northern Ireland Gross Domestic Product in the ten years before 1969 was 40 per cent higher than the rest of the United Kingdom; in the 1970s it was 60% lower (New Ireland Forum, 1984). Translated into jobs, it is estimated that the troubles cost Northern Ireland 39,000 jobs in the decade 1970-80 (Rowthorn, 1981). Arguments over the location of new factories become somewhat academic when there are very few new factories to locate.

The social costs of the violence are probably even more damaging. Polarisation within Northern Ireland society was already reflected by a high degree of religious segregation and territorial behaviour before the troubles began (e.g. see Boal, 1969; Poole, 1982, 1983; Poole and Boal, 1973), but minorities on both sides of the sectarian divide have since been forced to seek refuge with their co-religionists, either due to direct intimidation or simply a fear of attack. Catholics have suffered more in this regard than Protestants, mainly due to the fact that before the troubles began more Catholics lived in Protestant dominated areas than vice versa (Poole and Boal, 1973). Spatial segregation, whilst providing greater security in the short term, greatly reduces the likelihood of a compromise through mutual understanding in the long term. It is therefore especially sad, though understandable, to see spatial segregation being formalised by the building of large walls along the so-called 'peace lines' (i.e. conflict interfaces) and by the construction of new self-contained inward-facing housing estates designed to provide defensible space (Dawson, 1984).

SOUTHERN POLITICS

Southern Irish politics have been influenced by partition almost as much as Northern Irish politics, although in a less direct and less obvious way. Partition has shaped the course of southern politics in two main ways.

First, by excluding the industrial and largely Protestant north-east, partition left the southern state with a relatively homogeneous and distinctive population. Following partition, 92.6% of the population was Catholic; 67.7% lived in rural areas or in towns with 1,500 people or less (14.4% of the remainder lived in Dublin); and only 80,000 skilled workers, out of a total of 300,000 in Ireland as a whole, lived within the Free State according to the 1926 Census of Population. A very high percentage of the rural population were property owners, as a result of a series of land reforms in the late nineteenth century culminating with the Wyndham Act in 1903. These had assisted 370,000 tenants to buy out their farms (Lee, 1973). The Free State could therefore be described, with a fair degree of accuracy, as Catholic, under-industrialised, rural and property owning. These features exercised a profound impact upon the dominant ideology within the new state.

Second, by leaving the national question unresolved, partition has provided southern Irish politics with a major issue. Indeed, possibly because of the relative homogeneity of the population, the national question quickly emerged as the single most divisive issue in southern politics. Although accepted by the majority faction in the Dáil, IRA opinion was deeply divided on the terms of the Anglo-Irish Treaty, resulting in a civil war between the pro-Treaty Free State government and the anti-Treaty republican opposition (1922-3). The two major political parties in the Republic today, Fianna Fáil and Fine Gael, can trace their origins to the anti-Treaty and pro-Treaty factions respectively. Although partition was not the major issue in the civil war - Commonwealth status, and the oath of allegiance to the British crown were much more contentious issues at the time - partition subsequently provided the prime focus of differentiation between Fianna Fáil and Fine Gael, whose policies are basically devoid of any major ideological differences on other issues. Fianna Fáil's policy on partition has always tended to be more idealist and dog-matic; Fine Gael's has tended to be more compromising and pragmatic. By stressing their differences on the national question, giving rise to what is sometimes referred to as 'civil war politics', the major parties have in effect organised other issues off the agenda.

Given the very low percentage of non-Catholics in the population, it was inevitable that civil legislation would become imbued by a Catholic ethos. However, the influence of the Catholic Church upon issues which many non-Catholics would regard as secular was not simply a case of numerical superiority: it was strengthened by an education system in which primary and secondary level education was left under the control of the Churches (of all denominations). The overwhelming majority of Irish schoolchildren consequently received an education which stressed a Catholic perspective on life, with the result that Catholic values consequently remained a very important component within the dominant ideology of the southern population. Catholic Church doctrine became enshrined within civil legislation on issues such as censorship, divorce, contraception, and abortion, not only because the Catholic bishops formed a politically powerful 'lobby group', which politicians could only afford to ignore at their peril, but because the legislation on these issues accurately reflected the sincerely held belief of the majority that these issues were too fundamental to permit compromise.

The Catholic majority in the south, with a few exceptions in the immediate post-partition period, did not consciously discriminate against its non-Catholic minority. However, by uncompromisingly creating the southern state in its own image, the Catholic majority tended to alienate non-Catholics from the mainstream of southern society. The southern state consequently became an uncomfortable place in which to live for non-Catholics. The number of people affiliated to the three main Protestant denominations declined from 203,313 (7.5%) in 1926 to 115,411 (3.4%) in 1981, whilst the number of Catholics increased from 2,751,269 (92.6%) to 3,204,476 (93.1%) in the same period. Although differences in birth-rates and assimilation of the minority through inter-marriage (especially when the Ne Temere decree, which required the children in a mixed marriage to be brought up as Catholics, was rigorously enforced) need to be taken into consideration, much of the decline can be attributed to outmigration.

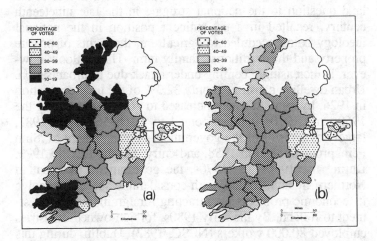

Figure 4. Distribution of the 'secular' vote in (a)The Abortion Referendum, 1983, and (b) The Divorce Referendum, 1986

Attitudes within the Republic have changed considerably over the last two decades. There has been a very marked trend towards secularisation since the late 1960s, as reflected by a relaxation of the laws governing censorship and contraception, and by a campaign to remove the constitutional ban on divorce. However, the impact of secular thinking has so far been limited and spatially uneven, as indicated by the votes cast in recent referenda on the introduction of a constitutional ban on abortion, and on the abolition of the existing ban on divorce. In each instance the 'secular' lobby was soundly defeated by the 'traditionalists'. The voting patterns in the two referenda were very similar (r=0.96) and suggest that secularism is diffusing from east to west, and down the urban hierarchy from Dublin (Figure 4). Nevertheless, the

43

dominant ideology in the Republic still remains that of the traditionalists in the rural west ('the real Ireland'), rather than that emanating from Dublin ('Britain's last colonial outpost').

The dominant ideology also reflects the concerns of a rural outlook on life. The initially large majority of people from a rural background, coupled with the central role of the land question in the national struggle in the late nineteenth century, resulted in a pre-eminent position in the national ideology being accorded to agriculture, the rights of private property and the 'traditional' family farm. This outlook, however, is increasingly coming under attack due to urbanisation. Urban dwellers constituted only 32.3% of the total population in 1926; by 1961 this had increased to 46.1%; and during the 1960s and 1970s it mushroomed to reach 56.6% by 1981, largely as a result of the government's abandonment of 'Sinn Féin' protectionism in 1958, and entry into the EEC in 1973. During the 1960s and 1970s, the government, like that in Northern Ireland, sought, with considerable success, to solve its economic problems by attracting foreign industrial investment to Ireland. By the early 1980s, foreign-owned industries employed 80,000 workers (NESC, 1982). Dublin, during this period, was possibly the fastest growing major city in Europe, but, due to a regional policy of dispersal, the medium sized towns grew at an even faster rate, resulting in a diffusion of urban values from east to west (Pringle, 1986).

Rapid urbanization has been reflected in recent years by an intensification of urban-rural conflict within southern society. There is a growing feeling amongst urban workers, for example, that farmers are receiving more than their fair share of the benefits of EC membership and of state assistance in general, whilst urban workers are being forced to pick up more than their fair share of the bill through higher taxation. PAYE workers (i.e. salaried and wage workers) on average paid £2,490 tax in 1985 as compared to an average of only £672 by farmers. However, non-payment of tax must also be taken into account. All PAYE workers pay tax, and have taxes deducted at source to minimise the possibility of non-payment, whereas farmers and the self-employed are billed for tax in arrears if their estimated incomes exceed a certain level. In 1984 only 127,000 farmers out of a total of 194,000 were deemed eligible for tax, and of these only 17,500 actually paid the tax which they owed (*Irish Times*, 6 November, 1986). Farmers at the same time also owed about £24M in uncollected levies and health charges, which are deducted automatically from PAYE workers. Despite protests by urban workers in the late 1970s and early 1980s at the apparent inequalities in the tax system, tax payments by the PAYE sector actually increased between 1981 and 1985 at almost twice the rate as that for farmers.

The rapid growth of urbanisation has also generated a variety of planning problems which the southern state has been poorly equipped to deal with because of the prevailing rural ethos. Urban planning, for example, in effect did not exist before 1963 when the government passed the Local Government (Planning and Development) Act which required local authorities to produce a plan for their area within three years, and to update it every five years subsequently. Although local authorities had been empowered to engage in planning by the 1934 and 1939 Town and Regional Planning Acts, only one local authority (Dublin C.B.) actually managed to produce and adopt a plan during the next 30 years. However, even under the 1963 Act, and its subsequent modifications, the powers of planners to control urban development are very severely curtailed by the Irish constitution which, reflecting a traditional rural outlook, defends the rights of private property. Under the 1963 Act, property owners who are refused planning permission, except under specific circumstances, are entitled to seek compensation from the local authority for the loss of profits which they would have made if their lands had been developed. Given the shortage of local authority funding, planners have been forced in many instances to give permission for developments which are not in the public interest.

The constitutional defence of private property has also compounded problems arising from property speculation in urban areas. A government commission, set up under the chairmanship of Justice Kenny to investigate problems arising from the high cost of building land, recommended in 1973 that local authorities should be entitled to buy land for development at its present value plus 25% (as opposed to the exhorbitant prices demanded by speculative landowners). These recommendations, however, have never been implemented, supposedly because the proposals would be expensive to implement and because they are repugnant to the constitution. Whether the recommendations of the Kenny Report actually are repugnant to the constitution or not is open to question, but a perhaps more telling point in this instance is that neither of the major political parties, when in government, made any serious effort to find out. Even if the Supreme Courts decided that the Kenny proposals are repugnant to the constitution, the constitution, as highlighted by no fewer than six constitutional referenda since 1973, can always be changed. It is difficult to avoid the suspicion that neither Fianna Fáil nor Fine Gael have any real interest in curbing property speculation, and are therefore quite content to see the issue organised off the agenda.

Space does not permit consideration to be given to all the conflicts in southern Irish society. Suffice it to say that, in a society plagued with social and economic problems such as high prices, a high rate of inflation, high unemployment, a housing shortage, drug abuse and a high crime rate, it is inevitable that the burdens imposed by these problems will be unevenly borne by different sections of society. The disadvantaged sections, whether they be the working classes, urban dwellers, women, school leavers or the very old, would naturally like to see positive action to remedy their problems, but in most instances all they seem to get from politicians are pious platitudes because their particular grievances, which can only be solved at the expense of some other section of society, are organised off the agenda. Southern society may be undergoing fundamental changes due to industrial urbanization, but the political system still seems to be geared, whether by design or by accident, towards a maintenance of the *status quo*.

The inertia in southern politics can be partly attributed to the electoral system. The single transferable vote (STV) method of proportional representation (which was originally introduced by the British to protect the interests of the small Protestant minority) in theory increases the likelihood of minority interests winning parliamentary seats, and should therefore increase the number of important issues in elections. However, electoral reforms, introduced over the years by both the major parties when in power, reduced the mean number of seats in each constituency and thereby reduced the likelihood of representation for minority interests (Paddison, 1976).

A more important consideration, however, is the large seats to votes ratio (set by the constitution at not less than one seat for every 30,000 people, but in practice one seat for roughly each 14,000 members of the electorate). This, coupled with the multiple seat nature of STV, which places candidates in competition with other candidates from the same party, has resulted in a very high degree of localism in voting patterns (Parker, 1982, 1986). A candidate's perceived ability to fix the problems of individual constituents at the local level is therefore usually a more decisive factor in elections than party policies on national issues. Once elected, politicians must expend an incredible amount of time acting as unofficial ombudsmen on behalf of their constituents if they are to retain electoral support. Small parties, with little resources and even less power, consequently find it extremely difficult, irrespective of how popular their policies might be, to establish sufficient credibility as potential problem solvers to break the hold of the major parties and gain representation. The smaller parties therefore rarely get an opportunity to place new items on the agenda, whilst the elected politicians from the larger parties are usually too busy doing constituency work to devote much time to broader issues even if they wanted to.

The brokerage nature of southern politics tends to produce politics without issues. Political parties which actively side with one section of society on a particular issue at national level are likely to alienate potential support from competing sections of society nationwide, whereas the rules of the game dictate that the candidates must attempt to sell themselves at local level as being capable of solving everyones' problems. It is therefore better not to have any well defined policies on anything that might adversely affect a section of southern society. This enables local candidates from both the major parties to promise, for example, to do something about inequitable taxation in urban constituencies, whilst other candidates from the same parties can simultaneously assure farmers in rural constituencies that their position will not be undermined. Given that they cannot possibly keep both sets of promises, the party in power usually tries to avoid implementing major reforms of any kind unless really forced to by public or outside pressure.

Partition consequently provides a very convenient issue which, if skilfully manipulated, can be used to push other potentially controversial, and therefore damaging, issues off the agenda. It also has the added advantage that government failures to achieve any progress on the national question can be blamed on factors outside the jurisdiction of the state, such as the intransigence of the unionists or the British government. The insult provided by these failures to the national integrity in turn lends credibility to the supposed need to resolve the national question, even if it means downgrading political activity on more 'mundane' issues within the jurisdiction, such as rampant unemployment or mass emigration.

DISCUSSION

The existence of partition has provided politics in both parts of Ireland with an issue of such apparent importance that all other issues are either squeezed off the agenda or else distorted to fit into a nationalistic framework. The impact of partition, however, is different in each state. Society in Northern Ireland is polarised into largely self-contained and spatially segregated unionist and nationalist blocs in which the defence or destruction of the state is of such immediate concern that it completely overshadows all other issues. Society in the Republic, on the other hand, is characterised by a fairly high degree of homogeneity and a very pervasive dominant ideology. Challenges to the dominant ideology (and thereby to the dominant classes) are deflected by tactical squabbles over how to bring about an end to partition.

Both societies are characterised by major distributional inequalities, material deprivation and social problems. Given that an end to partition seems highly unlikely in the forseeable future, due to the fact that there has been no discernible change in the single most important barrier to a united Ireland - namely the determined resistance of almost one million unionists - it is pertinent to consider whether political debate on other important issues will continue to be squeezed off the agenda.

The indications are that a change, if it comes, is more likely to occur in the south. Public opinion in the south would appear to be lukewarm on the importance of the national question in a period of economic crisis. An opinion poll conducted before the last general election found that only 3% of those interviewed spontaneously mentioned Northern Ireland as a major issue (*Irish Times*, 28 November, 1987). Issues perceived as being more important included unemployment (85 per cent), taxation (45%), prices (26%), government finances (20%), and crime and vandalism (11%). However, even in Northern Ireland, a review of opinion polls conducted between 1974 and 1982 found that the percentage of Catholics who favoured a united Ireland as their preferred solution to the Northern Ireland problem was never higher than 41% (*Irish Times*, 22 August, 1984). Indeed, the highest percentage of Catholics who even regarded a united Ireland as an 'acceptable' solution was only 66 per cent, whereas as many as 45% said they would regard total integration into the United Kingdom as acceptable. Opinion polls must always be treated with caution, but they would seem to suggest that nationalist Ireland is not nearly as nationalistic as politicians, both nationalist and unionist, would lead one to believe.

The advent of a new party, the Progressive Democrats, committed to breaking the stranglehold of civil war politics in the south could prove significant. The PDs have already overtaken the Labour party as the third largest party in the Republic. If they continue to grow, the PDs, although a conservative party, could possibly benefit the socialist parties in the long term by re-orientating southern politics along class lines. However, against this, the recent decision by Sinn Féin to take their seats if elected to the southern parliament is likely to damage the socialist parties, not only by re-emphasising the national question at the expense of other issues within southern politics, but also by directly competing with the socialist parties for the votes of the more alienated sections of Irish society.

It is more difficult to envisage how changes might arise in the north. One possibility might be if the major British political parties decided to organise and contest elections within Northern Ireland, as suggested by the Campaign for

Equal Citizenship. Given favourable conditions, 'normal' class politics might then replace sectarian divisions on the national question. However, given the perceived importance of the national question within Northern Ireland, it is quite likely that traditional voting patterns would be maintained as long as the plurality system of voting was used for Westminster elections because of a fear among voters of splitting the vote and letting the 'other lot' in by mistake. The situation might be different under the STV method of proportional representation (as used in Northern Ireland local government elections), but it is difficult to see how proportional representation could be introduced for Westminster elections in Northern Ireland without acceding to the long-standing British Liberal Party demand to have proportional representation introduced throughout the whole of the United Kingdom.

Another possibility might be to try to take the national question out of 'normal' Northern Ireland politics by holding a periodic plebiscite on partition. This, however, is unlikely to get much support from the existing political parties within Northern Ireland because it would be tantamount to asking them to vote themselves out of existence given that, like the major parties in the south, they do not have clearly articulated views on very much other than the national question.

Overall the hopes for change do not look too bright. Gerry Adams, the President of Sinn Féin, was recently reported as saying: 'I don't think that socialism is on the agenda at all at this stage on the agenda now is an end to partition. You won't even get near socialism until you have national independence; it's a prerequisite' (*Irish Times*, 10 December, 1986). Gerry Adams is unfortunately probably correct about socialism, and other issues, not being on the agenda because of the national question. However, national independence - i.e. an end to partition - is not on the agenda at present either, and probably never will be as long as the IRA (generally regarded as Sinn Féin's military wing) continues to wage a war against the unionist population. The great tragedy is that socialism and other issues have *never* been on the agenda: they have always been pushed off the agenda by those who regard the national question as having higher priority; or else manipulated off by elements wishing to retain the *status quo* both north and south. Perhaps it is time for everyone in

Ireland, especially those in the more deprived sections of society, to re-evaluate the true importance of the national question.

REFERENCES

Andrews, J.H. (1960) The *Morning Post* Line, *Irish Geography*, 4(2), 99-106.

Andrews, J.H. (1968) The papers of the Irish Boundary Commission, *Irish Geography*, 5(5), 477-81.

Beckett, J.C. (1966) *The Making of Modern Ireland, 1603-1923*, Faber and Faber, London.

Boal, F.W. (1969) Territoriality on the Shankill-Falls Divide, *Irish Geography*, 6(1), 30-50.

Brown, S. (1983) Central Belfast's security segment - an urban phenomenon, *Area*, 17(1), 1-9.

Buchanan, R. (1982) The Planter and the Gael: cultural dimensions of the Northern Ireland problem, in Boal, F.W. and Douglas, J.N. (eds) *Integration and Division: Geographical Perspectives on the Northern Ireland Problem*, Academic Press, London.

Busteed, M.A. (1972) *Northern Ireland: Geographical Aspects of a Crisis*, Research Paper 3, Department of Geography, University Of Oxford.

Compton, P.A. and Powell, J.P. (1986) Estimates of the religious composition of Northern Ireland Local Government Districts in 1981 and change in geographical pattern of religious composition between 1971 and 1981, *Economic and Social Review*, 17(2), 87-105.

Dawson, G.M. (1984) Defensive planning in Belfast, *Irish Geography*, 17, 27-41.

Douglas, J.N. (1982) Northern Ireland: spatial frameworks and community relations, in Boal, F.W. and Douglas, J.N. (eds) *Integration And Division: Geographical Perspectives on the Northern Ireland Problem*, Academic Press, London.

Fanning, R. (1983) *Independent Ireland*, Helicon, Dublin.

Harkness, D. (1983), *Northern Ireland since 1920*, Helicon, Dublin.

Hoare, A.G. (1981) Why they go where they go: the political imagery of industrial location *Transactions of the Institute Of British Geographers*, 6(2), 152-75.

Komito, L. (1983) Development plan rezonings: the political pressures, in Blackwell, J. and Convery, F.J. (eds) *Promise and Performance: Irish Environmental Policies Analysed*, Resource and Environmental Policy Centre, University College Dublin, Dublin.

Lee, J. (1973) *The Modernisation of Irish Society, 1848-1918*, Gill and Macmillan, Dublin.

MacLaughlin, J. and Agnew J.A., (1986) Hegemony and the regional question: the political geography of regional industrial policy in Northern Ireland, 1945-1972, *Annals of the Association of American Geographers*, 76(2), 247-61.

NESC (1982) *A Review of Industrial Policy*, National Economic and Social Council, Report No. 62, Dublin.

New Ireland Forum (1984) *The Cost of Violence arising from the Northern Ireland Crisis since 1969*, Government Stationary Office, Dublin.

Osborne, R.D. (1979) The Northern Ireland Parliamentary Electoral System: The 1929 Reapportionment, *Irish Geography*, 12, 42-56.

Osborne, R.D. (1982) The Lockwood Report and the location of a second university in Northern Ireland, in Boal, F.W. and

Douglas, J.N. (eds) *Integration and Division: Geographical Perspectives on the Northern Ireland Problem*, Academic Press, London.

Paddison, R. (1976) Spatial bias and redistricting in Proportional Representation Election Systems: a case study of the Republic of Ireland, *Tidjschrift voor Economische en Sociale Geografie*, 67(4), 230-41.

Parker, A.J. (1982) The friends and neighbours voting effect in the Galway West Constituency, *Political Geography Quarterly*, 1(3) 243-62.

Parker, A.J. (1986) Geography and the Irish electoral system *Irish Geography*, 19(1), 1-14.

Poole, M.A. (1982) Religious residential segregation in urban Northern Ireland, in Boal, F.W. and Douglas, J.N. (eds) *Integration and Division: Geographical Perspectives on the Northern Ireland Problem*, Academic Press, London.

Poole, M.A. (1983) The demography of violence, in Darby, J. (ed.) *Northern Ireland: The Background to the Conflict*, Appletree Press, Belfast and Syracuse University Press.

Poole, M.A. and Boal, F.W. (1973) Religious residential segregation in Belfast in mid-1969: a multi-level analysis, in Clarke, B.D. and Gleave, M.B. (eds) *Social Patterns in Cities*, Institute of British Geographers, London.

Pringle, D.G. (1980) Electoral systems and political manipulation: a case study of Northern Ireland in the 1920s, *Economic and Social Review*, 11(3), 187-205.

Pringle, D.G. (1985) *One Island, Two Nations? A Political Geographical Analysis of the National Conflict in Ireland*, Research Studies Press, Letchworth and John Wiley, New York.

Pringle, D.G. (1986) Urbanisation in modern Ireland, in Nolan, W. (ed.) *The Shaping of Ireland: The Geographical*

Perspective, Mercier Press, Cork.

Rowthorn, W. (1981) Northern Ireland: an economy in crisis *Cambridge Review of Economics*, 5.

Schattsneider, E.E. (1960) *The Semi-Sovereign People*, Dryden, Hinsdale, Illinois.

Stewart, A.T.Q. (1977) *The Narrow Ground: Aspects of Ulster, 1609-1969*, Faber and Faber, London.

Taylor, P.J. (1985), *Political Geography: World-Economy, Nation-State and Locality*, Longman, London.

4 IRISH POPULATION PROBLEMS

John Coward

Demographic trends and characteristics have many varied socio-economic, political and environmental repercussions and represent 'population problems' when they give rise to governmental or public concern. Within both Northern and Southern Ireland, for example, there has been particular concern expressed over the harmful social, economic and psychological effects of sustained emigration and the Report of the Commission on Emigration and Other Population Problems (1954) in the Republic was instrumental in influencing subsequent government economic and regional policy. A variety of contemporary demographic issues give rise to governmental and public concern in Northern and Southern Ireland and it is these that are discussed in this chapter. A broad geographical framework is adopted by emphasising contrasts at three spatial scales: by placing Irish population characteristics in a broader west European setting, by comparing problems between Northern and Southern Ireland and by examining the varying nature of problems within Ireland. The chapter first discusses population problems in general and then briefly assesses some major Irish demographic characteristics in a wider west European setting. It then examines a variety of population characteristics, such as population size, rates of growth and distribution, from a problem orientated perspective and concludes by providing a general review of current and possible future Irish population problems.

POPULATION PROBLEMS

Given the important two-way links that exist between demographic phenomena and wider social issues, there are, potentially, many population trends and relationships which can give rise to governmental and public concern. Berelson (1971), for example, examines four major demographic dimensions concerning population size, rates of growth, distribution and also population structure and composition. Each of these can have important economic, political, social and environmental implications and this classification therefore pinpoints sixteen important areas of interaction between demographic and other phenomena. Thus the scope for a study of population problems is, potentially, quite broad. 'Population problems', however, are not always easy to define and it is often difficult to classify a particular population's demographic characteristics according to whether or not they represent problems. This is partly because most population characteristics - whether population growth is rapid or slow, whether there is net immigration or emigration, etc. - have numerous implications and effects, some of which can be viewed as generally advantageous while others can be viewed as more harmful. Moreover, 'population' problems may be dependent on many non-demographic factors. For example, rapid population growth can prove economically beneficial in some circumstances and more harmful in other circumstances, depending on broader economic, political and resource factors. Thus there are important inter-linkages between demographic and non-demographic issues and many population problems have their roots in non-demographic issues.

Whether or not particular demographic characteristics represent population problems will vary in relation to the varying personal, political and ethical values of individuals and groups of individuals within society, whether they be government policy makers, non-governmental pressure groups, academic researchers or members of the general public. While a consensus approach is generally appropriate for some issues, there are other socio-political issues having important demographic implications, such as abortion legislation, which are dominated by conflict and controversy and where a conflict approach may be more appropriate.

For certain countries a guide to the nature of general population problems can be gauged through the government's stated population policy. However, neither the Republic of Ireland nor the United Kingdom have such overall policies and there is therefore no general statement or classification of the nature of the perceived problems from a governmental perspective. Moreover, legislation on specific issues which have a demographic component provides only a rather narrow and restricted picture of population problems because many demographic problems (such as an aging population) cannot be directly solved or reversed by governmental legislative action or policy. Indeed, in many cases, governments are hesitant to react to population issues either because it is felt that little can be done to influence particular trends and characteristics or because it is recognised that such characteristics have a variety of both benign and harmful effects.

Thus while the general definition of population problems encompasses those demographic issues which give rise to public and governmental concern, there is greater difficulty in classifying particular characteristics on this basis. However, this chapter will attempt to assess Irish population problems in their broader context, irrespective of whether such problems are directly associated with specific legislation.

RECENT DEMOGRAPHIC TRENDS IN IRELAND: A WEST EUROPEAN PERSPECTIVE

In some respects demographic characteristics in Northern and Southern Ireland differ from those exhibited by many countries of Western Europe and are thus likely to be associated with differing sets of problems. In terms of historical background, Ireland represents the exceptional example of sustained population decline such that the combined current population size of just over five million is much smaller than that of the pre-Famine population which exceeded eight million. Moreover, the actual size of population in Ireland is relatively small: the Republic of Ireland has the second smallest population of the European community next to Luxembourg, while the size of the population in Northern Ireland is the smallest of the standard regions in the United Kingdom. Coupled with this, overall population density in the Irish Republic is low by west European standards (Table 1) while population density in Northern Ireland is the lowest of the standard regions of the United Kingdom. Also, it can be seen from Table 1 that, by west European standards, levels of urbanisation are relatively low in Northern and Southern Ireland.

***Table 1*:   Comparative demographic indicators**

	A	B	C	D	E	F	G	H
Rep. of Ireland	3.5	19	1.0	50	56	31	11	72
Northern Ireland	1.5	8	0.7	106	68	26	12	61
GB	55.1	1	0.1	224	76	20	15	54
France	55.4	3	0.4	97	73	22	13	54
Netherlands	14.5	1	0.4	379	88	22	12	52
W.Germany	61.0	-3	-0.2	239	94	17	15	47
Spain	38.5	8	0.6	74	91	26	11	59
Sweden	8.3	-5	0.0	18	83	19	16	54

Columns: A Population size 1986 (Mill.)
B Projected increase 1985-2000 (%)
C Natural increase per 1,000 population p.a. 1982/83.
D Density per sq km
E Urban population (%)
F Age structure 1982/83 - under 15 (%)
G Age structure 1982/83 - over 64 (%)
H Age structure 1982/83 - dependency ratio (population under 15 plus over 64 per 100 persons aged 15-64 years)

Data Sources: 1985/86 World Population Data Sheets (Population Reference Bureau); Regional Trends (HMSO); Population Projections 1983-2023 (OPCS).
Definitions of 'Urban' vary from country to country.

The most distinctive contemporary feature of Irish demography is the relatively high rate of natural increase: thus Table 1 indicates that while most European countries exhibit natural increase close to or around zero, natural change in Ireland, although currently declining, is particularly high. Such characteristics reflect the relatively high birth rates which, in turn, reflect numerous socio-economic and cultural characteristics of the populations (Kennedy, 1973; Coward, 1980). It can also be seen from Table 1 that the relatively high levels of fertility have also brought about quite large proportions of the population aged less than 15 years and relatively high age dependency, whereas many of the countries of Western Europe have quite large proportions in the elderly groups, reflecting the more rapid aging of these populations. The high levels of natural increase indicate that there is considerable potential for population growth in both Northern and Southern Ireland, although emigration has generally led to much slower growth as well as population decline. However, as seen in Table 1, anticipated future growth rates are again high by west European standards and current population growth rates in the Republic are the highest in western Europe.

Thus while there are important demographic differences between Northern and Southern Ireland, the dominant feature is the distinctiveness of Irish demographic characteristics when viewed in a broader west European setting and it may be expected, therefore, that these will also give rise to a set of population problems that are quite distinctive by west European standards.

POPULATION SIZE AND RATES OF GROWTH

Over the twentieth century the populations of Northern and Southern Ireland have exhibited phases of overall decline, stability and growth, reflecting the varying intensity of natural increase and net migration. Each of these phases can be associated with population problems and this section examines the general features of population change and assesses their implications from a problem orientated perspective.

The differing trends in population size in Northern and Southern Ireland indicate that population changes have fluctuated less markedly in the North (and the six counties prior to partition), where numbers increased gradually up to 1971 but have since stabilised (Figure 1). In the Irish Republic, on the other hand, it can be seen from Figure 1 that population size declined up to 1961 (apart from a small increase 1946-51) but has since increased rapidly, particularly during the 1970s.

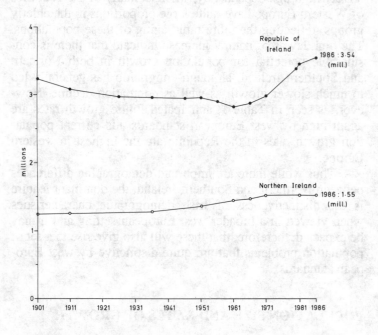

Figure 1. Population of Northern Ireland and the Republic of Ireland, 1901-86

It can be seen from Table 2 that the slow growth in the population of Northern Ireland has been brought about by a high rate of natural increase which has generally just exceeded losses through emigration. The stabilisation of numbers

Table 2: Population Change in Ireland, 1926-86

Period	Population at end of period (000s)	Population change	Natural increase	Net migration
Republic of Ireland				
1926-36	2,968	-0.1	5.5	-5.6
1936-46	2,955	-0.4	5.9	-6.3
1946-51	2,961	+0.4	8.6	-8.2
1951-61	2,818	-4.9	9.2	-14.1
1961-66	2,884	+4.6	10.3	-5.7
1966-71	2,978	+6.4	10.1	-3.7
1971-79	3,368	+15.4	11.1	+4.3
1979-81	3,443	+11.1	11.8	-0.7
1981-86	3,537	+5.4	9.7	-4.3
Northern Ireland				
1926-37	1,280	+1.7	5.8	-4.1
1936-51	1,371	+4.9	8.5	-3.6
1951-61	1,425	+3.9	10.5	-6.6
1961-66	1,485	+8.2	13.4	-5.2
1966-71	1,536	+5.9	10.1	-4.2
1971-81	1,534	-0.1	7.0	-7.1
1981-86	1,547	+2.6	7.2	-4.6

Figures on population change, natural increase
and net migration refer to average rates per 1,000
population per annum

since 1971 reflects the generally lower levels of natural change (brought about by reductions in the birth rate) and a considerable rise in net migration loss such that these components have more or less compensated one another (Table 2). In the Irish Republic prior to 1961 the quite high natural increase was more than off-set by migration losses, but the reduction of emigration in the 1960s and the unprecedented feature of migration gain during the 1970s, coupled with a sustained level of high natural increase, have resulted in the substantial growth since 1961. Rates of increase were particularly high during the 1970s, but a recent return to net migration loss and a reduction in natural increase (Table 2) make it unlikely that such rates will be achieved again.

There has been little discussion of population optima in Ireland, but actual rates of population change have received greater attention. Thus there has been general agreement concerning the harmful effects of population decline in the Irish Republic in terms of economic implications (a declining population limits the domestic market and could act as a deterrant to economic growth) and in the ways in which emigration (responsible for the declines) has been a selective process. Also, population decline represented a considerable symbolic loss, of a nation unable to keep its young and where dwindling numbers were equated with diminishing world power and status and, in some circles, the eventual possibility of the dying out of the Irish nation.

The effects of population growth, on the other hand, are more difficult to assess from a problem orientated perspective and it is hard to discern government perceptions concerning the merits of population growth and related problems, particularly in the Irish Republic. Thus in one sense the rapid population growth in the Republic during the 1970s was welcomed because it emphasised the change from the declines prior to the 1960s and symbolised the economic successes of the 1970s and a nation able to support its population. However, the population increases also generated considerable pressure on the demand for employment and service provision.

The major consideration involved in assessing population growth concerns its interrelationship with economic development. Potentially, population growth can have economic benefits in terms of increasing the size of the domestic market, increasing the demand for service provision and in providing an expanding labour force to utilise a nation's resources. On the other hand, rapid population growth can be harmful when it exceeds economic growth and thus leads to an overall *per capita* decline in the standard of living. However, economic development is influenced by a broader range of factors than purely demographic trends, including the natural resource base of a particular country, its political stability and the effects of broader external economic and political factors, such that it is difficult to isolate the overall demographic effects, or, for Ireland, assess whether population growth has been beneficial. Indeed Walsh (1968) argues that this issue is not particularly important because of the mediating effects of emigration. Thus migration has acted as a safety valve in the economic/demographic system and the generally high rates of migration loss are indicative of a demand for jobs which exceeds the rate at which they can be created as well as the relatively superior employment prospects in countries of destination such as Britain. While factors other than disparities in job opportunities influence migration trends, it is these that have proved most influential in explaining variations in emigration (Walsh, 1968; Slattery, 1978) and Walsh argues that the dominant causal direction of these influences is that population (migration) trends are influenced by economic conditions, rather than vice versa: 'The various arguments in the debate concerning the effects of population growth on growth in income per person are... irrelevant to the Irish situation... Irish population growth is a consequence of either domestic prosperity or British adversity, but in no way a cause of Irish prosperity in the short run'. It is, however, more difficult to ascertain if population growth has been similarly irrelevant in influencing the economic problems of the last decade. While the basis of these economic problems are non-demographic, the effects of rapid population growth during the recent period of recession and retrenchment have aggravated the economic situation by placing extra pressure on limited financial resources through the increased demand for employment and increasing investment in infrastructure and services (Courtney and McCashin, 1983; Raftery, 1983). Gillmor (1985), for example, states: 'the high birth rate has critical and daunting implications for future employment provision' while

63

Haughton (1983) argues: 'a nation is made up of people who, ultimately, represent its major resource. Like all resources, it has to be properly managed and this means investment in education and training, and in job creation'. Indeed, the potential for the rapid growth of the labour force is considerable in both Northern and Southern Ireland and has partly influenced the high levels of youth unemployment. The return to high levels of emigration in the 1980s in the Republic again reveals the mediating effects of migration, but it is possible that the dramatic increases in unemployment in Britain have altered the nature of these effects upon the Irish demographic/economic system.

Within Northern Ireland the less volatile rates of population change in recent decades have not been associated with the range of problems experienced in the Republic and the maintenance of a more or less stationery population offers certain advantages in terms of consolidation of service provision. Thus it has been stated: 'This period of slow population growth will present a unique opportunity for improving the physical and social environment of the Province. Financial and other resources can be directed to improving the living conditions of the existing population rather than towards meeting the needs of a rapidly increasing population' (Department of the Environment, 1977). However, the interplay between migration losses and natural increase ensure that the structural and compositional characteristics of the Northern Irish population have been changing, resulting in increasing age dependency, and, as in the Republic, the high levels of natural increase indicate the potential for labour force growth.

To summarise, there are important linkages between economic and demographic factors and economic conditions have an important influence on population characteristics, affecting, in particular, migration patterns which, in turn, influence population size and structure. However, depending on overall economic conditions, demographic characteristics can have a variety of effects on economic development. Both Northern and Southern Ireland illustrate that economic growth and population growth can be maintained in periods of economic expansion, as evident in the 1960s, but, in periods of recession, population growth, particularly rapid growth, can aggravate economic problems. The severe economic problems currently facing Northern and Southern Ireland are essentially a product of non-demographic factors, but these problems can been exacerbated by the potential for rapid population growth as well as by the effects of changing demographic structure.

EMIGRATION AND ITS EFFECTS

High and sustained rates of emigration represent a major demographic characteristic of the populations of Northern and Southern Ireland and have had a large number of socio-economic and psychological effects. Net loss of population has been the dominant pattern since the Plantations and in consequence large numbers of people of Irish descent live overseas, particularly in the United States of America and the United Kingdom. Generally, net emigration rates have been higher from the Republic than from Northern Ireland but there was a dramatic reversal in trends in the 1970s (Table 2), reflecting the economic expansion in the Republic in the early 1970s associated with net migration gain. Out migration from Northern Ireland reached high levels, particularly during the early 1970s, as a result of the onset of economic recession, further declines in manufacturing industry and the effects of the Troubles.

In some senses emigration can have beneficial effects. First, for example, because emigration is primarily a response to poor economic conditions, it has been instrumental in reducing the size of the labour force and in influencing the numbers unemployed. Second, many emigrants have achieved personal success and raised their standard of living and, in addition, sent remittances to their families in Ireland. Third, emigrants have helped spread knowledge about Ireland and Irish culture in general and '...given Ireland a significance abroad which is out of proportion to the size of the home population' (Commission on Emigration, 1954).

However, the overall economic, psychological and social consequences of emigration are generally adverse and many of the problems which were reviewed by the Commission on Emigration (1954) are still applicable today in both Northern and Southern Ireland. For example, the economic effects of population decline, brought about solely through emigration, have been discussed above. Moreover, migration has been a selective process which has not been compensated by large-scale counter-flows and this has created various problems. Thus the depletion of a nation's young population - most emigrants are in their late teens and twenties - removes future community and political leaders and the life-blood of a country. Many of those with particular skills and high levels of education emigrate and, in addition, it has been suggested (though not explicitly demonstrated) that emigration removes those individuals who are more innovative (Kennedy, 1973), thus leaving a residual population that is more conservative, less likely to question established authority and less receptive to social change in general. Finally, migration losses have several harmful social and psychological consequences on community life. Thus apart from the removal of the young, migration loss is sex-selective and rural areas in particular have been affected by the greater propensity of females to leave, producing distorted sex ratios which have important implications for the maintenance of community cohesion and marriage rates. Moreover the disruption of social ties caused by migration loss also affects the continuation of social life and communities which have experienced considerable migration loss and population decline can experience deep-rooted psychological stress caused by emigration (Brody, 1973). Thus there are a variety of ways in which emigration, essentially a response to relatively poor economic conditions, has caused numerous problems and given rise to much concern. Indeed, the return to quite high levels of emigration from the Republic in recent years has fostered comparisons with the 1950s, although, as Table 2 demonstrates, current levels of loss are much lower than thirty years ago.

POPULATION DISTRIBUTION AND REGIONAL CHANGE

The overall rates of population change obscure major geographical variations at finer spatial scales and this section examines trends in population change and overall population distribution from a problem orientated perspective. The patterns of population change and distribution in Ireland over this century generally reflect the processes of urbanisation and concentration. More recently, decentralisation has occurred within the metropolitan regions of Belfast and Dublin, although the trends toward counter-urbanisation and wider decentralisation are less advanced in Ireland compared with more developed Western countries.

Several features characterise regional population change over the last fifteen years. First, population decline has continued to occur in the inner urban areas of the major cities, reflecting the proceses of suburbanisation, decentralisation and, in Northern Ireland, the effects of civil strife. Such changes have been particularly marked for the larger urban areas but, more recently, have spread through the urban hierarchy to the cores of middle order towns in both Northern Ireland and the Republic. The population of Belfast Local Government District declined by 25 per cent between 1971 and 1981 (Compton, 1986) while Dublin County Borough's population declined by 7 per cent over the same period. The reductions in inner city population numbers were originally encouraged through decentralisation policies and have proved beneficial in terms of reducing urban congestion and providing the opportunity for reorganising land uses within the inner city. However, the scale and rapidity of these declines, combined with the selective nature of population loss in terms of age and social class, has led to problems in terms of an unbalanced social and demographic composition of the remaining population, a lowering of the tax and rating base of the inner city and an underutilisation of infrastructural provision in these areas.

Second, and partly as a response to the first trend, population growth around the major urban areas has been particularly rapid. This trend reflects the process of suburbanisation and, for the metropolitan regions of Belfast and Dublin, the decentralisation of employment opportunities into the wider commuting zone. In addition, the relatively high rates of natural increase of these populations, reflecting their young age structures, make a substantial contribution to population growth. In Northern Ireland recent rates of population increase have been particularly high in the commuting zone

67

beyond Belfast's fringe (especially the towns of North Down and South Antrim), contrasting with the large declines of the inner city and slower growth of the older inner suburbs (Figure 2). Moreover, the residential segregation of the population by religion is an important consideration in the planning of these areas and the development of Poleglass, just outside the Belfast urban area, represents an example of providing housing for Roman Catholics moving from inner West Belfast (Singleton, 1981).

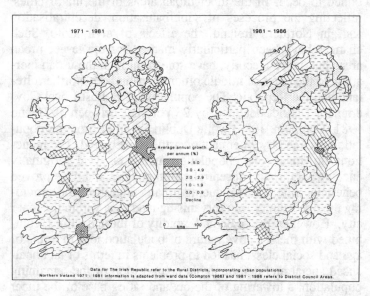

Figure 2. Population change in Northern Ireland and the Republic of Ireland, 1971-81 and 1981-86

In the Republic of Ireland rates of population growth have been particularly high in the suburban areas and towns to the west of Dublin and the wider commuting zone outside the city (Figure 2). Growth has been particularly marked in the 'satellite' towns of Blanchardstown, Clondalkin and Tallaght whose combined population size quadrupled to 100,000 during the 1970s (Hourihan, 1983; Horner, 1985). The rapid build-up of population in these areas has led to problems in terms of pressure on land and housing, the costs of infrastructure provision and the loss of agricultural land (Bannon, 1981; Horner, 1985; Drudy, 1986). Also, the large scale movement of city dwellers into outlying areas can create considerable social polarisation with the existing populations of these communities (Duffy, 1978; Cawley, 1979), while the spread of low density suburbs around Dublin has created problems in providing an adequate yet economically viable system of public transport (Hourihan, 1982), and the resulting greater dependence on private transport has led to congestion and pollution problems.

Recent population change in rural areas displays a varied pattern. Traditionally, rural areas, particularly those in remote and isolated locations and with poor economic prospects, have experienced sustained population losses and this has continued over the last fifteen years in some of the poorer western rural areas of both Northern and Southern Ireland (Figure 2). Thus some of the major problems associated with migration loss and population decline mentioned by the Commission on Emigration (1954) - concerning migration selectivity, disruption of social ties, low marriage rates and social isolation - still prevail in the poorer parts of Ireland. Moreover, the continuing population losses in such areas have important repercussions for the viability of certain services, influencing retailing, health, education and public transport provision (Cawley, 1986). Indeed, the self-perpetuating nature of decline in many poorer rural areas raises considerable concern over their long-term viability. Recent migration intention surveys in the Republic reveal a certain preference for areas of origin (Walsh, 1984), but high levels of out-migration are anticipated as a result of poor employment opportunities.

In contrast to the 'traditional' pattern of rural population change, however, over the 1970s many rural areas in the Irish Republic experienced population increases (Figure 2) and for some areas population increase was recorded for the first time since regular censuses have been conducted. This partly reflected the expansion of urban areas into their rural hinterland and the growth of commuting, but in other cases growth occurred in rural areas outside the influence of larger urban areas (Horner and Daultrey, 1980), reflecting, amongst other things, rural industrialisation policies. Such policies proved quite successful through the 1970s, but have been more difficult to sustain in the 1980s (Gillmor, 1986) and, as Figure 2 indicates, overall rural population gains have been less widespread in the 1980s. In Northern Ireland, however, most of the rural areas west of the Bann experienced further declines (Figure 2) and growth in this region was mainly confined to the areas in and around the larger towns. In consequence, the problems created by sustained population decline are particularly marked in these rural western areas.

The general processes of urbanisation, concentration and, more recently, decentralisation are reflected in overall population distribution. The major features here are the dominance of the eastern metropolitan regions, the relatively small proportions living in the western regions and the generally dispersed nature of rural settlement. In Northern Ireland the population of the greater Belfast region represented one half of the total population in 1981 whilst in the Republic the eastern region centred on Dublin constituted 38 per cent of the population. In general, population has become increasingly concentrated in these regions over the twentieth century - with considerable differences in growth between core and fringe - although the particularly large declines in Belfast over the 1970s and the spread of population beyond the urban fringe led to an absolute and relative decline of population of the Greater Belfast region over the 1970s (Compton, 1986). Moreover, population increase in the Dublin urban area slowed down in the early 1980s compared with the massive increases of the 1970s.

A further conseqence of urbanisation is that the rural populations have been declining in both absolute and relative terms and the generally dispersed nature of much of the rural settlement has exarcebated the problems of attempting to provide adequate service provision in such areas. Some of these problems connected with population change and distribution have been tackled through regional policy and planning legislation. Within Northern Ireland, planning policy in the 1960s (Matthew, 1964; Wilson, 1965) concentrated on restricting the expansion of Belfast and encouraging the growth of other key urban centres including the new town of Craigavon, while more recent policies have attempted to encourage economic growth throughout the province by concentrating development in the district towns (Department of the Environment, 1977). The latter policy may help arrest the decline in some of the western parts of the province, but, in order to succeed, greater numbers will live in the urban areas, towns and villages where it is easier to provide infrastructure, services and employment. It has also been recommended that a more flexible planning policy towards restrictions on rural house building might encourage larger numbers to remain in the countryside and country towns and villages (Cockcroft Report, 1978). Thus in Northern Ireland there has been a shift in policy away from the original growth and key centres (Singleton, 1981). Within the Republic of Ireland it was recommended that the development of Dublin be concentrated in selected areas to the west of the city (Wright, 1967), but there has been no effective stop-line policy around Dublin as for Belfast. Indeed, the current spread of the Dublin built up area, associated with low housing densities on much of the periphery, creates particular problems concerning loss of agricultural land and provision of such services as public transport (Bannon, 1981; Hourihan,1982). It has been estimated that the population of the Dublin urban area will continue to expand quite considerably (ERDO, 1985), and while these estimtes need to be revised in light of the much slower growth during the 1980s (due to a marked decline in fertility and increases in emigration), Bannon's (1985) prognosis is still relevant: 'Such a scale of development may well have profound implications for the quality of life of the citizens and for the use of resources in the East region, particularly in respect of land consumption'. Attempts to spread development throughout the Republic, either through concentrating development in regional centres and growth towns (Buchanan, 1968) or through the Industrial Development Authority's policy of rural industrialisation (Brennan, 1984) have achieved certain success and have been instrumental in encouraging development and checking population decline outside the Dublin region. However, it has been argued that rural industrialisation has lessened rural depopulation but not solved the problem.

Thus recent population change and overall distribution of population in Northern and Southern Ireland have given rise to a variety of problems concerned with the most efficient use of space, overcrowding and viability of areas with dispersed populations. In certain respects planning policy has ameliorated some of these problems, but planning policy, or sometimes the lack of it, has been criticised in many ways. Thus for both Northern and Southern Ireland, there have been pleas for a more coherent policy for the development of rural areas, for greater public participation in planning and criticism levelled at the adoption of planning policies based on the experiences of Britain and the United States (Caldwell and Greer, 1984; Duffy, 1986). In the Republic, particular criticism has concerned the lack of a national settlement strategy and the seemingly unordered growth of the Dublin built up area (Bannon, 1981; Drudy, 1986). Planning policy has helped modify some of the adverse consequences of trends in population change and distribution, but the basic problems associated with inner city losses, suburban spread and peripheral decline and redistribution are likely to remain.

POPULATION STRUCTURE AND COMPOSITION

Population structure and composition are influenced by, first, the fundamental demographic processes of fertility, mortality and migration which determine age and sex structure and, second, various broader socio-economic and cultural factors that influence, for example, marital status and household composition. This section pays particular attention to age structure and also briefly examines several features concerning population composition by sex and marital status.

Relatively high fertility in Northern and Southern Ireland has produced an age structure which is quite young by the standards of many West European countries and which also, as demonstrated earlier, displays relatively high age dependency. Age structure characteristics vary through time and the most dramatic changes have occurred recently in the Irish Republic where the turnaround in migration trends in the 1970s, coupled with continuing high fertility, has been associated with an age structure which is particularly young by West European standards (Courtney, 1986), with 48 per cent of the population aged less than 25 in 1981 (Table 3). While dependency ratios provide only a crude indicator of economic potential, high levels of dependency indicate that more financial resources are needed to support non-dependants. Also, the high birth rates of the recent past, coupled with the reduced levels of emigration from the Republic in the 1970s, indicate that, compared with countries such as Britain, the ratios of labour force entrants to leavers is quite large, particularly in the Republic (Table 3), and this currently raises considerable problems concerning the rising demand for employment.

Table 3: **Age structure indices for Ireland and Great Britain**

Index	Northern Ireland	Republic of Ireland	Great Britain
Dependency ratio (1983)	61	72	54
% Population <25 (1983)	44	48	36
Ratio of labour force entrants to leavers (1983)	1/1.9	1/2.3	1/1.4
% growth of population aged 15-19, 1961-81	20	40	25
% growth of population aged 65+, 1961-81	25	17	32
% growth of total population, 1961-81	7	22	4

Dependency ratio: persons aged under 15 plus over 64 per 100 persons aged 15-64.
Labour Force Ratio: Ratio of those aged 10-19 to those aged 55-64.

In a healthy economy large numbers of young people and a growing labour force are of importance in providing the means whereby resources can be exploited and production increased. However, in economies which are struggling with recession and the provision of adequate employment, a youthful population and increasing size of labour force are likely to exacerbate rather than ameliorate such problems. Similarly, the demand for higher education will be influenced by demographic factors such as the number of potential entrants - in addition to changing educational aspirations and participation rates - and the large increases in the population aged 15-19 in the Republic (Table 3) have important implications for the rising demand for third level education. Higher education provision will have important benefits for the future, but is currently placing considerable pressure on financial resources.

Longer term reductions in fertility and improvements in life expectancy influence the number and proportions of elderly within the population. Within the Republic the growth of the elderly has been less than that of the overall population (Table 3), but in Northern Ireland, where the birth rate has declined considerably over the last twenty years, the absolute and relative size of the elderly has increased substantially (Table 3) and while little can be done to alter the trend towards the aging of the population, the important implications concerning the changing demand for housing requirements and health provision must be recognised (Compton, 1986). Again these issues will be particularly problematical when overall financial resources are limited.

Fluctuations in age structure can also be associated with problems connecting the supply of resources with varying demand. For example, the decline in the birth rate in Northern Ireland from 1964 to 1977 had important implications for the declining need for primary school places and teachers, but the short-term increase in the birth rate from 1977 to 1982 placed considerable pressure on places and teachers. Thus changes in the structure of the population can create problems in terms of balancing fluctuating demand with the supply of certain key services.

Overall sex ratios in Ireland are more or less even and are broadly in line with the countries of mainland Europe, although the Republic prior to the mid 1980s was one of the few countries with slightly larger numbers of men than women. However, as seen in the discussion on emigration, spatial variations in the sex structure of populations at the local scale, particularly in those rural areas with very imbalanced ratios caused by high rates of female out-migration, can lead to problems associated with the maintenance of social life within a community.

In the past, Ireland's marriage patterns of relatively late ages at marriage and high levels of celibacy gave rise to considerable concern - 'one of our two great population problems' (Commission on Emigration, 1954) - and were associated with several broader social, economic and psychological problems (Commission on Emigration, 1954; Walsh, 1968). Such problems were particularly acute in rural areas and were generally more severe in the Republic compared with Northern Ireland. However, it can be noted that the Irish marriage pattern has generally been changing towards earlier and more universal marriage and now marriage patterns in Northern and Southern Ireland are more akin to West European trends (Walsh, 1968, 1985; Courtney, 1986).

A HEALTHY POPULATION?

The study of mortality patterns in Ireland has generally received less attention than that of migration or fertility, but there are some aspects of health and mortality that give rise to concern and these are briefly examined in this section. There have been major reductions in mortality rates over this century, particularly at younger ages, and current life expectancy in Northern and Southern Ireland is broadly similar to that of many other developed countries (Table 4). On the other hand, there are other features which indicate that, relative to other European countries, both Northern and Southern Ireland fare rather poorly on a variety of health indicators. Thus life expectancy in Ireland is generally below that of many West European countries, although the differences are relatively small, and this partly reflects the poorer socio-economic conditions in Ireland as well as variations in diet and lifestyle. Also, as Table 4 demonstrates, levels of infant mortality - often used as an indicator of general socio-economic conditions - in Northern and Southern Ireland are higher than many developed countries. A report on infant mortality in Northern Ireland (Baird, 1980) suggests that a variety of causal factors are of importance, including poor socio-economic conditions such as low wages and poor housing, quality of health care and demographic factors such as maternal age. Moreover, mortality from heart disease is relatively high in Ireland, particularly in Northern Ireland which exhibits some of the highest rates worldwide (Elwood *et al.*, 1986). There are obvious health implications in such a high incidence of heart disease, although it is difficult to determine the precise causal influences. Current research is attempting to unravel some of the various associated causal factors, associated with poor socioeconomic conditions and deprivation, quality of health care, environmental influences, genetic factors, diet or lifestyle factors such as smoking and exercise pattern. Similarly, mortality from congenital abnormalities is relatively high in Northern Ireland, along with male deaths from accidents, poisonings and violence. On a more optimistic note, the incidence of cancers in Ireland is relatively low compared with many European countries.

Table 4: Comparative mortality indices

Country	Date	Life expectancy at birth		Infant mortality rate
		Male	Female	
Republic of Ireland	1980-82	70.1	75.6	10.5
Northern Ireland	1983	69.2	75.6	11.9
England and Wales	1981-83	71.3	77.3	10.1
Scotland	1981-83	69.3	75.5	9.8
France	1981	70.4	78.5	9.0
The Netherlands	1982-83	72.7	79.5	8.4
W.Germany	1981-83	70.5	77.1	10.1
Spain	1983	70.4	76.2	9.6
Sweden	1983	73.6	79.6	7.0

Life expectancy for Spain refers to 1975

Apart from overall trends and characteristics, mortality differences by area are quite marked. Generally, overall mortality rates are highest in the main urban areas: for example it has been demonstrated that male mortality in Dublin is considerably higher than that of the overall male population (Ward *et al.*, 1978) and the urban-rural divide is clearly seen for many causes of death. Indeed, there is a sharply defined spatial pattern of mortality in the Republic, generally reflecting the lower mortality of the poorer rural north-west (Pringle, 1986), indicating that patterns of mortality are, surprisingly, inversely associated with level of living (Pringle, 1982). The high urban mortality rates may partly reflect the socio-economic composition of such areas - reflecting the inverse gradient between mortality and social status - as well as smoking patterns and environmental influences such as atmospheric pollution (Sweeney, 1982). In addition, Pringle (1983) has demonstrated that quite marked mortality differences exist within the Belfast urban area, where social class is again an important correlate of the spatial patterning. McCarthy and Reid's (1986) investigation of spatial variations in certain causes of age standardised mortality in Northern Ireland again illustrates the quite considerable degree of regional variation. Certain distributions, such as ischaemic heart disease, do not reveal clear-cut spatial variations - although Howe (1986) shows that most parts of Northern Ireland display much higher rates compared with Britain - while other causes of death are associated with a more clearly defined pattern. For example, mortality from other (non-ischaemic) heart disease is relatively low in and around the Belfast conurbation and proves to be signifantly related to unemployment and overcrowding, while the relative incidence of cancers, particularly lung cancer, is generally higher in the more urbanised parts of the province.

CONSENSUS OR CONFLICT?

As mentioned previously, individuals and groups perceive population characteristics and trends in relation to their varying needs and ideologies and it is thus to be expected that some population issues will be subject to controversy and conflict. In some cases it is possible to recognise a broad consensus of opinion as in, for example, the adverse effects of sustained emigration and population loss, but, in other cases, the existence and nature of the problem is open to much more controversy. Indeed, for the Irish Republic it has been suggested that '...we are moving from what was perceived as a consensus model of society... towards a conflict society' (Mulvihill and Van der Kamp, 1985). Such controversies are not new, but have come to the fore more recently and are particularly apparent where wider social change affects demographic behaviour. For example, many of the social changes which are reflected in patterns of marriage and fertility in Ireland are subject to controversy. These include the declines in average family size, greater reliance on artificial forms of family planning, an increasing incidence of divorce, separation and cohabitation, a recent marked increase in illegitimacy and an increasing recourse to procuring abortion in England. One perspective sees these as reflecting the increasing materialism and secularisation of Irish society and a gradual weakening of those traditional values which hold society together, whereas another perspective sees these as representing the development of more liberal societies with increasing freedom of choice for the individual. Recently, such issues have been raised in the political arena in the Irish Republic, as seen in the debates over family planning legislation and the national referendums on abortion and divorce. Within Northern Ireland there has been considerable debate over the provision of abortion facilities and whether legislation should be brought in line with that of the remainder of the United Kingdom. A recent survey of married women indicated that the majority (55 per cent) were not in favour of changing abortion legislation for Northern Ireland, although there were majorities for change amongst Protestants, the middle classes and those living in Belfast (Compton, Coward and Power 1986).

Moreover, some population trends have important political implications, the effects of which may be adverse or benign. In Northern Ireland, for example, the Roman Catholic population has had higher fertility and emigration rates than that of Protestants, and marginally higher rates of increase overall, and the different demographic behaviour of the two communities fuels speculation over their future relative sizes and the possibilities and implications of an eventual Catholic majority. For example, on the basis of rates of natural increase for 1977 and emigration for 1971-77, Compton (1982) estimates that a Roman Catholic majority could be expected by the middle of the next century. However, there have been considerable reductions in Roman Catholic fertility over the last decade (Compton, Coward and Wilson-Davis 1985) such that inter-denominational birth rates are converging and the possibility of an eventual Catholic majority will very much hinge on differential rates of emigration. Whether the higher growth potential of the Catholic population represents a demographic 'problem' again depends on personal and political perspectives. There has also been considerable debate concerning the extent to which the problem of the particularly high unemployment among Roman Catholics is a product of their higher rate of population growth (Compton, 1981; Osborne and Cormack, 1986).

CONCLUSIONS

Recent demographic trends and characteristics in Northern and Southern Ireland give rise to a variety of concerns. In some cases the precise nature of the population problem is difficult to define and varies in relation to the varying needs, attitudes and ideologies of differing social groups. However, in other cases the nature of the problem is less ambiguous and in this respect four main themes have been pinpointed. First, some problems are associated with the potential for quite rapid population growth in Northern and Southern Ireland and this, along with the associated changes in population structure, has given rise to concerns about the adequate provision and costs of services, infrastructure and employment. It has been argued that a potential increase in population may not always have adverse economic effects, indeed in some cases it could prove positively beneficial, but in an era of economic recession and limited financial resources population growth is likely to exarcebate existing economic problems. Thus the roots of this problem are economic rather than demographic. The second general problem concerns emigration which, while helping to counter population growth, is associated with numerous adverse consequences as a result of its selective nature and its consequences on communities which have experienced sustained population losses. The third set of problems concern the regional nature of recent population change and the associated feature of population distribution, whereby the differing characteristics of continued population decline in some areas and rapid growth in other areas create numerous problems concerning the provision and efficient utilisation of services and infrastructure and the varying quality of life in these areas. The fourth theme concerns some of the adverse health characteristics of the populations of Northern and Southern Ireland.

In some respects Irish population problems are distinctive from those experienced by other, more developed, West European countries and regions. Thus the more developed countries are currently coming to terms with population stability and its associated structural features of an aging population (as a result of very low natural increase and net migration close to zero) and also with the restriction of immigration from outside Europe. In contrast, Ireland faces problems connected with a considerable potential for growth - which, depending on migration, may be translated into rapid growth - and the diverse consequences of emigration. In this respect Ireland is more akin to the poorer areas of southern Europe than the more developed areas of northern and western Europe. However, in other respects common population issues emerge in that most countries experience problems connected with an uneven distribution of population, inequalities in health and controversy over the demographic expressions of social change.

As far as the future is concerned, much will depend on the economic progress of Northern and Southern Ireland and the financial resources available to provide for a potentially expanding population. On the demographic side, the recent declines in fertility can be expected to continue such that natural increase and the potential for growth will be much lower by the turn of the century. However, the quite large birth cohorts of the late 1970s and early 1980s will maintain the momentum for the demand for employment early into the next century. In the short term, therefore, relatively high natural increase will ensure that the potential for growth will be maintained. Given the poor economic outlook for Northern and Southern Ireland, emigration will probably continue at quite high levels and overall population growth will thus be below that of natural increase. It is estimated that Northern Ireland's population will increase slowly to 1.69 million by the year 2000 (OPCS, 1985) while in the Republic population growth is anticipated, albeit at a slower rate than the 1970s, such that the population will increase to 3.6 million in 1990 (Blackwell, 1985). Given the unlikely prospects of a major economic recovery in the short term, the problems associated with emigration and the pressure of population growth on financial resources will continue. It is also likely that population change and distribution will continue to provide a challenge to national, regional and local planning, along with concerns over health and the demographic ramifications of continued social change. Thus the nature and complexion of future Irish population problems, at least in the short term, can be expected to be broadly similar to those of today.

REFERENCES

Baird, T. (1980) You and your baby: *Report of the Advisory Committee on Infant Mortality and Handicap in Northern Ireland*, HMSO, Belfast.

Bannon, M. (1981) Urbanisation: problems of growth and decay in Dublin, *National Economic and Social Council, Report No. 55*, The National Economic and Social Council, Dublin.

Bannon, M. (1985) Irish Urbanisation: issues in the built environment, in *Ireland in the Year 2000: Urbanisation*, An Foras Forbartha, Dublin, 61-66.

Berelson, B. (1971) Population policy: personal notes, *Population Studies*, 25(2), 173-82.

Blackwell, J. (1985) The issue in perspective: demographic trends and forecasts, in *Ireland in the Year 2000: Urbanisation*, An Foras Forbartha, Dublin, 5-16.

Brennan, D. (1984) Economic planning for the rural areas in

the Republic of Ireland, in Jess, P., Greer, J., Buchanan, R. and Armstrong, J. (eds) *Planning and Development in Rural Areas*, Institute of Irish Studies, Queen's University Belfast, 31-40.

Brody, H. (1973) *Inishkillane: Change and Decline in the West of Ireland*, The Penguin Press, London.

Buchanan, C. (1968) *Regional Studies in Ireland*, Stationery Office, Dublin.

Caldwell, J. and Greer, J. (1984) Physical planning for the rural areas in Northern Ireland, in Jess, P., Greer, J., Buchanan, R. and Armstrong, J. (eds) *Planning and Development in Rural Areas*, Institute of Irish Studies, Queen's University Belfast, 63-86.

Cawley, M. (1979) Rural industrialisation and social change in Western Ireland, *Sociologia Ruralis*, 19, 43-57.

Cawley, M. (1986) Disadvantaged groups and areas: problems of rural service provision, in Breathnach, P. and Cawley, M. (eds) *Change and Development in Rural Ireland*, Geographical Society of Ireland, Special Publication No 1, Maynooth, 48-59.

Cockcroft, W. (1978) *Review of Rural Planning Policy*, HMSO, Belfast.

Commission on Emigration and other Population Problems 1948-54 Reports, Stationery Office, Dublin.

Compton, P. (1981) Demographic and geographical aspects of the unemployment differential between Protestants and Roman Catholics in Northern Ireland, in Compton, P. (ed.) *The Contemporary Population of Northern Ireland and Population Related Issues*, Institute of Irish Studies, The Queen's University of Belfast, 127-42.

Compton, P. (1982) The demographic dimension of integration and division in Northern Ireland, in Boal, F. and

Douglas, N. (eds) *Integration and Division: Geographical Perspectives on the Northern Ireland Problem*, Academic Press, London, 49-74.

Compton, P. (1986) Demographic Trends in Northern Ireland, *The Northern Ireland Economic Council, Report No. 57*, Northern Ireland Economic Development Office, Belfast.

Compton, P., Coward, J. and Wilson-Davis, K. (1985) Family size and religious denomination in Northern Ireland, *Journal of Biosocial Science*, 17(2), 137-45.

Compton, P., Coward, J. and Power, J. (1986) Regional differences in attitude to abortion in Northern Ireland, *Irish Geography*, 19(2), 59-68.

Courtney, D. (1986) Demographic structure and change, in Clancy, P., Drudy, S., Lynch K., and O'Dowd, L. (eds) *Ireland: a Sociological Profile*, Institute of Public Administration, Dublin, 22-46.

Courtney, D. and McCashin, A. (1983) Social Welfare: the Implications of Demographic Change, *The National Economic and Social Council Report No. 72*, The National Economic and Social Council, Dublin.

Coward, J. (1980) Recent characteristics of Roman Catholic fertility in Northern and Southern Ireland, *Population Studies*, 34(1), 31-44.

Department of the Environment, Northern Ireland (1977) *Northern Ireland Regional Physical Development Strategy*, HMSO, Belfast.

Drudy, P. (1986) Regional strategy, in *A Report on the Dublin Crisis Conference*, Dublin, 62-63.

Duffy, P. (1978) Population change in the Irish countryside, *Geographical Viewpoint*, 7, 20-33.

Duffy, P. (1986) Planning problems in the countryside, in

Breathnach, P. and Cawley, M. (eds) *Change and Development in Rural Ireland*, Geographical Society of Ireland, Special Publication No 1, Maynooth, 60-68.

Elwood, J., McIlwaine, W., McCrum, E. and Evans, A. (1986) Trends in mortality from cardiovascular diseases in Northern Ireland, *Irish Journal of Medical Science*, 155, 39-44.

ERDO (1985) *The Eastern Region Settlement Strategy 2011*, ERDO, Dublin.

Gillmor, D. (1985) *Economic Activity in the Republic of Ireland: A Geographical Perspective*, Gill and Macmillan, Dublin.

Gillmor, D. (1986) Rural industrialisation, in Breathnach, P. and Cawley, M. (eds) *Change and Development in Rural Ireland*, Geographical Society of Ireland, Special Publication No 1, Maynooth, 25-33.

Haughton, J. (1983) The demographic resources of Ireland, in Clinch, P. and Mollan, R. (eds) *A Profit and Loss Account of Science in Ireland*, Royal Dublin Society, Dublin, 67-78.

Horner, A. (1985) The Dublin region 1880-1982: an overview on its development and planning, in Bannon, M. (ed.), *A Hundred Years of Irish Planning, Volume 1: The Emergence of Irish Planning 1880-1920*, Turoe Press, Dublin, 21-75.

Horner, A. (1986) Rural population change in Ireland, in Breathnach, P. and Cawley, M. (eds) *Change and Development in Rural Ireland*, Geographical Society of Ireland, Special Publication No 1, Maynooth, 34-47.

Horner, A. and Daultrey, S. (1980) Recent population changes in the Republic of Ireland, *Area*, 12(2), 129-135.

Hourihan, K. (1982) Urban population density patterns and change in Ireland, 1901-79, *The Economic and Social Review*, 13(2), 125-47.

Hourihan, K. (1983) Population redistribution in Irish cities: Dublin, Cork and Limerick, 1971-1981, *Irish Geography*, 16, 113-20.

Howe, M. (1986) Does it matter where I live? *Transactions of the Institute of British Geographers*, New Series, 11(4), 387-414.

Kennedy, R. (1973) *The Irish: Emigration, Marriage and Fertility*, University of California Press, London.

Matthew, R. (1964) *Belfast Regional Survey and Plan 1962*, HMSO, Belfast.

McCarthy, P. and Reid, N. (1986) Mortality in Northern Ireland, *Public Health*, 100, 286-92.

Mulvihill, R. and Van der Kamp, H. (1985) Rapporteurs Report, in *Ireland in the Year 2000: Urbanisation*, An Foras Forbartha, Dublin, 1-3.

OPCS (1985) *Population Projections 1983-2023*, HMSO, London.

Osborne, R. and Cormack, R. (1986) Unemployment and religion in Northern Ireland, *Economic and Social Review*, 17(3), 215-25.

Pringle, D. (1982) Regional disparities in the quantity of life: the Republic of Ireland 1971-77, *Irish Geography*, 15, 22-34.

Pringle, D. (1983) Mortality cause of death and social class in the Belfast Urban Area, 1970, *Ecology of Disease*, 1, 1-8.

Pringle, D. (1986) Premature mortality in the Republic of Ireland, 1971-81, *Irish Geography*, 19(1), 33-40.

Raftery, J. (1983) Health Services: the Implications of Demographic Change, *The National Economic and Social Council Report No. 73*, The National Economic and Social Council, Dublin.

Singleton, D. (1981) Planning implications of population trends in Northern Ireland, in Compton, P. (ed.) *The Contemporary Population of Northern Ireland and Population Related Issues*, Institute of Irish Studies, The Queen's University of Belfast, 102-14.

Slattery, D. (1978) Some aspects of (net) emigration from Northern Ireland, *Journal of the Statistical and Social Inquiry Society of Ireland*, XXIV(2), 133-47.

Sweeney, J. (1982) Air pollution and morbidity in Dublin, *Irish Geography*, 15, 1-10.

Walsh, B. (1968) Some Irish Population Problems Reconsidered, *The Economic and Social Research Institute*, Paper No. 42, Dublin.

Walsh, B. (1974) Ireland, in Berelson, B. (ed.) *Population Policy in Developed Countries*, McGraw-Hill, New York, 8-41.

Walsh, B. (1980) Recent demographic changes in the Republic of Ireland, *Population Trends*, 21, 4-9.

Walsh, B. (1985) Marriage in Ireland in the twentieth century, in Cosgrove, A. (ed.) *Marriage in Ireland*, College Press, Dublin, 132-50.

Walsh, J. (1984) *To go or not to go - the Migration Intentions of Leaving Certificate Students*, Discussion Paper No. 2, Department of Geography, Carysfort College, Co. Dublin.

Ward, J., Healy, C. and Dean, G. (1978) Urban and rural mortality in the Republic of Ireland, *Journal of the Irish Medical Association*, 71(3) 73-80.

Wilson, T. (1965) *Economic Development in Northern Ireland*, HMSO, Belfast.

Wright, M. (1967) *The Dublin Region: Advisory Plan and Final Report*, Stationery Office, Dublin.

5 CRIME IN GEOGRAPHICAL PERSPECTIVE

David Rottman

INTRODUCTION

A single scenario summarises the main changes in crime
experienced by most Western European countries over the
last 150 or so years. Crime changed from being primarily
rural in location and assaultive in nature to overwhelmingly
urban and aimed at acquiring property. Urbanisation initially
had a beneficent impact on crime levels: they declined from
the middle of the nineteenth century, only rising in earnest
after World War II. The discipline of industrial work, education, and professional law enforcement made urban living less
rather than more conducive to crime, despite the anxieties
evoked by the 'dangerous classes' of London, Paris and other
major European cities (Lane, 1979, 1986; Brantingham and
Brantingham, 1984). It was only with the rising economic
prosperity of the 1950s that European cities truly experienced
the consequences of crime patterns that were strongly urban
based. Rates of property crime, and to a lesser extent assaultive crime, grew and grew inexorably. Clearly, the factors
that had contained the level of crime in Europe's cities for
decades had been replaced by others tending to promote criminal behaviour.

GEOGRAPHY AND CRIME

This chapter examines the applicability of this scenario to Ireland. It views crime as a social phenomenon that follows the flow of people, goods, employment, and ideas in a society; as these alter with general processes of economic and social change, so does crime. Societal change generates consequences that differentially affect regions, subregions and neighbourhoods. These consequences impinge most directly on *legal* rather than on criminal behaviours. But in doing so, they affect the opportunities to commit crime and the motivations to commit crime present in various sections of a society. 'Legal activities take place in space and time and thus set the stage for spatial and temporal patterns of crime as well. Certain parts of cities, regions, nations, hours of the day, days of the week, and months of the year have a much higher crime incidence than others. Such patterns in space and time are structured in part by the settings and rhythms of legal activities. Because technological and organisational changes influence the patterns of daily life, crime patterns and rates might be expected to change accordingly' (Felson, 1983, p.665). Economic and social change in Ireland, north and south, did not correspond with the 'typical' European experience, and thus the geography of Irish crime, does not fit precisely the scenario described at the start of the chapter. Irish crime patterns are distinctive, diverging in important respects from those of other countries.

CRIME BEFORE PARTITION

Though this chapter focuses on Irish crime patterns since 1960, those patterns have strong historical roots. Until the conclusion of the Land War of the 1880s, Irish crime was decisively agrarian in origin and location. Faction fighting, meleés involving hundreds of participants, were 'ubiquitous in the 1820s' (Townshend, 1983, p.11). A one hundred year long struggle, marked by rural secret societies and 'agrarian outrages', established security of tenure for most tenant farmers, greatly reducing rural tensions. With the 'Wyndham Act' of 1903 and amendments, purchase of land became feasible for many of these tenant farmers, further reducing the main traditional source of conflict - and crime - in Ireland (Fitzpatrick, 1977, 1982). Though the potential for conflict in this 'land-obsessed society' (O'Tuathaigh, 1982, p.181) remained between owners and those without or awaiting ownership (intergenerational conflict made patricide the classic Irish crime of the era), emigration and an inherent conservatism blunted its impact. Crime instead followed the route toward the more dynamic sections of Irish society in the late nineteenth century: its cities, especially Dublin.

Irish urbanisation in the late nineteenth century was accompanied by a decline in the overall crime rate, a shift toward property crime like theft and away from violent crime like murder and assault. Yet Dublin in this period (roughly 1870 to 1895) was the scene of one half of all serious (indictable) crime in the country, except during the Land War years, when the capital's share fell to one third. The area under the jurisdiction of the Dublin Metropolitan Police at that time contained about one fifteenth of the country's population (O'Brien, 1982). Belfast's prosperity was then peaking and its population approaching the size of Dublin, yet it had but one tenth the number of serious offences. From the 1890s, Belfast and other cities began to leave their mark on the crime statistics and Dublin's contribution waned to between a quarter and a third of the total.

Still, by international standards this was an extraordinary concentration of crime in one urban centre: 'relatively speaking, Dublin could be described as a foremost centre of crime in the United Kingdom, not only during the late Victorian era but also for the early years of the present century ....' (O'Brien, 1982, p.184). 'Relatively speaking' refers both to the usual procedure of comparing crime rates per 100,000 population and the derisory level of crime by contemporary levels of crimes like burglary. A sense of the crime problem in Irish cities around 1910 is shown by the following rates of indictable offences per 100,000 population: Dublin, 852; Belfast, 541; Cork, 215; London, 253; Birmingham, 301; and Manchester, 361; only Liverpool, with a rate of 1,715 exceeded the Dublin level for indictable crime (O'Brien, 1982, Tables 19 and 20).

Since indictable crime covers a wide range of misdeeds, from murder to petty theft, it is plausible to view urban crime in Ireland as part of a general problem of disorderliness. There were more arrests for drunkenness in Dublin early in this century than in London; and Irish cities like Waterford and Limerick appear to have had higher arrest rates for that offence than Dublin (O'Brien, 1982). Prostitution was a flourishing service industry in Dublin at that time, with the most notorious red light district in the British Isles centered around Montgomery Street. Yet by 1916, Dublin was greatly altered, a far more staid capital even before independence. This preceded the loss of the role of garrison town, reflecting instead change processes at work in Ireland in the decades before partition.

CRIME SINCE PARTITION

Change processes north and south after 1922 generated and maintained levels of crime well below those in Britain or in most of Europe, despite periodic resurgences of violence associated with paramilitary activity. World War II prompted a brief upward trend in the crime statistics, but it was not sustained. Crime, such as it was, could be attributed to a small group of not particularly talented repeat offenders, noted more for the frequency of their stays in prison than for the violence or the profit their activities generated.

Given Ireland's isolation from the economic and social changes sweeping Europe in the post-war period and the dearth of criminal opportunities, a low level of crime could be expected. Emigration in the Republic provided the safety valve that permitted the social controls of a traditional, conservative society to persevere; indeed, those with criminal inclinations were often given judicial assistance in leaving the country (Russell, 1964). After 1960, economic prosperity transformed the Republic rapidly and pervasively. Ireland's geographic remoteness was also vastly eroded by new forms of communication and transportation. That change was ultimately to be most starkly registered in the emergence, about 1980, of a substantial heroin abuse problem in Dublin. But even by 1960, the Teddy Boy phenomenon was introducing anxieties among those who associate protest, crime, and social disintegration as parts of a single package.

From 1960, the same forces that had transformed crime elsewhere in Europe were clearly operating in Ireland. But crime in Ireland reflects distinctive local influences, most notably the northern conflict. In political crime, after all, Ireland has been something of an innovator. So crime in Ireland is best understood as a variation on, not a repetition of, a typical European pattern.

It is useful to consider post-1960 crime patterns in the south and north of Ireland separately. That presentation will be preceded by a discussion of the data problems that are shared by students of crime in either jurisdiction and followed by a conclusion that considers similarities and contrasts between the geography of crime in Northern Ireland and the Republic.

CRIME INDICATORS

Crime refers to all actions that are in violation of a jurisdiction's legal code. It cannot therefore simply be equated with either immorality or injurious behaviour. Still, crime covers a wide range of actions, extending from homicide to driving a motor vehicle without insurance. All can potentially result in punishment by the criminal courts; that is their only common element. In practice, we base indicators of crime levels on responses to actions: those instances in which the public or the police decide to intervene. So we really count the number of interventions rather than the number of crimes.

Both the Garda Siochána and the Royal Ulster Constabulary (RUC) retain English police record keeping practices, distinguishing 'indictable offences', those that can or must be tried before a jury, from summary offences, which are automatically dealt with before a justice or magistrate sitting without a jury. Social change and statutory reform have diluted the meaningfulness of that distinction and comparability between the two jurisdictions.

Prior to the 1980s, police statistics provided the only basis for examining variation in Irish crime levels between areas or over time. Indictable offences are measured as offences known - that is, reported to or discovered by the police authorities and accepted as genuine violations of the law. Summary offences, however, are usually tabulated on the basis of persons who are prosecuted. Often, comparisons can only be made based on the number of indictable offences, and thus despite its crudeness, it must be used as an index of crime prevalence. Where possible, comparisons based on selected specific offences, such as homicide or burglary are more revealing. However, all police statistics suffer from limitations. First, they are strongly influenced by public opinion: police forces only know about such crime as the public chooses to divulge. Second, all police forces screen offences of which they are aware when classifying incidents into 'known' crime. Public tolerance of crime and cooperation with the police vary over time and between places, as do police practices. Since Ireland has traditionally been one of the most heavily policed areas in Europe (O'Brien, 1982; Rottman, 1985, p.129), it is easier for the public to report crime to the police than in many other countries, though one must consider the way in which police forces are integrated into the communities they serve.

Crime victimisation surveys offer another measure of crime levels. Here, social scientists go directly to the public and solicit reports of crime victimisations, typically over the preceding year. Such surveys do capture information on incidents not reported to the police because victims felt the offence was too trivial or too personal to be made the subject of official action (Hough and Mayhew, 1983). Generally, incidents are more likely to be recorded by police and victim surveys if they were serious in their consequences and were committed by strangers. The two ways of counting crime agree on which social and geographic locations have the highest rates of various crimes (Nettler, 1984, p.79).

This chapter uses published data on indictable offences from the RUC and Gardai to construct trends for the 1960-85 period. It supplements that information with the results from two crime victim surveys. In the Republic, the Economic and Social Research Institute carried out a nationwide survey of 9,000 households in 1983 and in Northern Ireland data are available on 6,800 households from questions in the 1984 and 1985 Northern Ireland Continuous Household Surveys.

CRIME IN THE REPUBLIC: RECENT TRENDS

The Dublin Metropolitan Area in 1965 housed roughly one quarter of the Republic's population and was the location of one half of all burglaries (48 per cent of shopbreakings and 49 per cent of housebreakings). Dublin's share of other property crime was as generous or higher. Crimes of interpersonal violence were distributed very differently. Rural rates for assault were higher than in Dublin, with the highest rates found in cities like Cork, Limerick, Waterford, or Galway.

1964 marks the point at which the new vibrancy in the Republic, initiated by state-directed policies of industrial development and economic growth dating back to 1958, is reflected in the crime statistics. The initial change was in the sheer frequency of crime. Rates of property crime soared between 1964 and 1975: housebreaking recorded a 4.3-fold increase; shopbreaking, a 3.2-fold increase; these rises were surpassed by motor vehicle theft and robbery, which experienced 7.5-fold and 11.4-fold increases, respectively. Assaults became more prevalent as well, but at the far slower pace of a 2.3-fold increase. If we trace indicators of criminal sophistication or organisation, it is less evident that the mid-1960s represents a watershed. Property crime adapted to the opportunities of new found affluence, whether in the form of cash or commodities like motor vehicles, but until the late 1970s there is little sign of an enhanced professionalisation to crime in the Republic.

Nor is there a clear urban focus to the upward trend. Crime rates were on the increase in all areas. The economic and social changes to which crime was responding were so rapid and so pervasive as to be truly national in their effects. This is contrary to the usual pattern, in which growth in crime originates in the major urban centres and only gradually filters down to other cities, then to towns, and finally to villages. Yet Dublin's early lead and a somewhat steeper upward trend after 1964 tended to concentrate the nation's crime there despite the increases recorded in all areas. Pre-1964 crime rates outside of Dublin were so low as to be negligible.

Dublin's importance was enhanced after 1975 when the main trend ceased to be the rise in numbers but instead in the skill and profit with which crime was being undertaken.

Thus, the late 1970s marked a second basic transformation in the Republic's crime. Crime was becoming more rational, more business-like, and therefore a more serious problem. The second transformation was most apparent in the use of firearms in offences, especially robberies. Virtually unknown in the 1960s, armed robbery reached a level of about 200 annually in the late 1970s. This change underscored the urban concentration of crime in the Republic. Crime was being committed at a level of specialisation and sophistication that required networks of participants, including receivers of stolen property and providers of firearms. Earlier in the century rural areas possessed the dynamic of 'organisational density' that placed them in the forefront of the nationalist struggle (Fitzpatrick, 1978, pp. 132-133). That density, now available for a variety of uses in urban Ireland, had its primary impact on legal activities, but also a clear secondary effect on crime. With the 1980s the upward spiral in the crime statistics resumed and crime became a potent symbol of the nation's problems. The mass media competed in fixing public concern on one manifestation after another, from joyriding to heroin abuse. In this, crime was portrayed as an urban problem. More specifically, certain neighbourhoods of local authority housing became synonymous with the threat posed by crime.

Crime statistics in the mid-1980s suggested that the threat was receding; substantial decreases were recorded in the number of offences 'known' to the Gardai. At the same time, a small number of particularly brutal attacks on elderly residents of households in some of the most rural sections of the West blunted somewhat the association between cities and crime in the public mind. However, in the context of post-1964 crime trends, these developments do not alter the consequences of the transformations to a high level of crime *and* to more organised criminality in the Republic. Those changes appear to be permanent, unfortunate though they may be (Rottman, 1980, 1985).

Some features of the transformed crime pattern are noteworthy in how they differ from what many other countries experienced. First, assaultive crime has not greatly increased over the last quarter century. Homicide, the most reliably measured of all forms of criminality, remains at a low level by British or continental European standards. Interpersonal violence, however, shifted more toward cities. Outside of cities, assaultive offences seem to have retained the traditional pattern of violence within families and between neighbours or friends. In Dublin and other cities, assaults typically involve strangers. Yet infrequency remains the main characteristic of violent crime in the Republic. Second, urban property crime in the 1980s was tied to an addictive drug. Early in the century alcohol had fuelled Dublin's crime rate; now it was heroin. As recently as 1966, the drug abuse problem could be ascribed to a dozen or so individuals. The use of even a 'soft' drug like marijuana was virtually unknown, despite its prevalence elsewhere in Europe. By the mid-1970s, marijuana use and abuse of drugs like barbiturates and amphetamines was sufficiently commonplace to make Dublin less distinctive. But the emergence of heroin abuse certainly was distinctive. Heroin did not gain a foothold and then expand; rather, in late 1979 it was smuggled into the country in large quantities and its use expanded with extraordinary rapidity in Dublin over the next few years (McGroarty, 1986). The impact on crime is not difficult to determine. Addiction causes crime directly through 'transfer offences' carried out to pay for the illegal substance and crimes committed while under the influence of the drug. The profitable business of supplying illegal drugs generates violence among competitors and generally increases the organisational capacity of criminal networks. So heroin exacerbated the crime trends emerging in the late 1970s.

The key to understanding the ease with which heroin became established is the characteristics of the users: 'Garda information and experience from the outset showed that the type of teenagers and young persons involved were mostly from deprived backgrounds' (McGroarty, 1986, p.2). That statement is generally applicable to crime in the Republic (Rottman, 1985, Appendix II). Rapid economic progress was expected to promote greater equality automatically - the rising tide that would float all boats. It did not. Old employment opportunities were eliminated and new ones created that required credentials attesting to substantial educational qualifications. Young people growing up in working class neighbourhoods were marginalised by these changes, left without a realistic point of entry into the post-1958 economy. Social class inequalities in educational participation and the risk of unemployment are greater in the Republic than in England or France (Rottman and O'Connell, 1982; Whelan and Whelan, 1984). A syndrome of early school leaving and high unemployment became characteristic of many Dublin working class neighbourhoods. Local authority housing policies reinforced the impact of such social trends by uprooting old working class neighourhoods and relocating their residents in new estates on the city's outskirts. Dublin is more segregated by social class than other major European cities (Bannon, Eustace and O'Neill, 1981, p.85).

This maximizes the proximity of criminal opportunities and criminal motivations. Areas of privilege and disadvantage crisscross Dublin. Burglary, car theft, and similar crimes can be pursued on a rational basis - targeting those locations with the most valuable property. A generation of young people has been raised without strong ties to conventional activities: schooling, work, church, or community. And they have an abundance of time to pursue illegal activities and little to lose if caught and punished. Here, too, crime is rational. Now high levels of unemployment and poverty did not suddenly afflict the Irish working class in the 1970s. Census data make clear that such conditions have been manifest since the 1920s and earlier (Breen *et al.,* 1988). But the very geography of Dublin, as well as the mass media, made for a new awareness of the standard of living obtaining elsewhere in Irish society. A new, and large, middle class had emerged from the period of economic growth. Its lifestyle set the standard by which others compare their relative circumstances.

The results of crime victimisation surveys support these claims. By the early 1980s, Ireland had a rate of property crime that often exceeded that in Britain. Using the results from the British Crime Survey of 1982, we can obtain the following comparisons: the rate of household burglaries was 450 per 10,000 Irish households and the rate of car theft, 562 per 10,000 vehicle owning households. In England and Wales, the burglary rate was 260 and that for vehicle theft, 232. Scottish rates were virtually identical to those in England (257 and 280). Irish and British rates for other offences were similar, but it is clear that some extraordinary factors were driving Irish property crime levels upwards in the early 1980s.

The dynamic underlying that pressure was Dublin based. Table 1 compares the rates of crime victimisation in four types of areas: Dublin (all of County Dublin and the town of Bray), large cities (Cork, Galway, Limerick and Waterford), all other urban places of more than 10,000 population, and rural areas.

Table 1:   **Victimisation rates according to urban/rural distinctions**

Victimisation rates per 100 households

	Rural	Urban	Cities		
			Dublin	Cities*	Towns**
Burglary	1.0	6.9	8.3	3.5	3.6
Theft around house	2.7	7.4	7.4	7.2	7.7
Vandalism of house	1.9	6.4	6.3	8.1	3.4
Theft from person	1.7	9.8	12.0	5.2	4.1
Theft of car	1.4	8.1	10.4	3.3	3.6
Theft from car	3.1	10.0	10.7	8.1	8.9

Source: Breen and Rottman, 1985, p.49
*Cork, Limerick, Galway, Waterford
**over 10,000 population

A consistent picture of the distribution of major property crime emerges. Rates of victimisation from burglary, car theft, and theft from the person are significantly higher in Dublin than elsewhere. Dublin's standing is also clear when comparing rates of thefts from vehicles, and, indeed, Dublin's rate may be understated since incidents are allocated to areas based on the victim's residence, not the site of the offence; shoppers, businessmen, and tourists visiting Dublin are at risk of victimisation in the capital. Crime rates for urban areas other than Dublin are similar (the differences are rarely statistically significant). Cork's high rate of vandalism is the only exception. The remaining 57 per cent of the population residing in what have been designated rural areas (all places of less than 10,000 population) are exposed to a negligible risk of becoming crime victims. The extent to which that risk is concentrated in Dublin can be illustrated by a simple comparison to London's share in English crime. A family living in Dublin's inner city has 12 times the risk of becoming a burglary victim than someone living in rural Ireland; the typical family in inner city London has a 5-fold greater risk than their rural counterpart.

Fear of crime is more widespread today than the actual geographic distribution of crime risk justifies. That fear and its ubiquity is recent. A 1977 survey found urban residents to be far more concerned over crime than residents of rural areas, though the difference was greatest for assaultive offences. But the actual levels of concern were insubstantial: even in urban areas, half the respondents perceived 'no problem' in their own areas (Whelan and Vaughan, 1982, pp. 64-71). Both the perceived seriousness of the crime problem and the location of fear had altered by the early 1980s. A 1983 study found levels of fear of 'street crime' in Dublin comparable to large American and British cities (Irish Marketing Surveys, 1983). A substantial rural/urban differential existed in the level of fear, but by the 1980s, fear of crime was far more geographically dispersed than any objective measures of degree of risk.

The failure of fear of crime and objective risk to coincide geographically within urban areas offers a clue to the causes of crime in the Republic. A survey of residents of nine Cork City wards found that those living in high risk areas were the least likely to be willing to participate in crime prevention efforts or to believe that their neighbours would be so willing (Hourihan, 1986, p.22). So there appears to be a spatial correlation between crime victimisation rates and the willingness (and perhaps the ability) to mobilise an effective community response. It is likely that the neighbourhood characteristics that generate criminal motivations are the same ones that produce instability and disorganisation. American and British research consistently finds that residents of low income, minority group neighbourhoods bear the highest levels of crime victimisation (Conklin, 1986; Lea and Young, 1984). Dublin offers a variation on that relationship. The distinctive spatial pattern of the city makes it less likely that offenders and victims live in the same localities; generally, high income households are more at risk from crime than residents of low income areas. But there is an important exception to this. The ESRI survey found that where the head of a household is unemployed, the risk of crime victimisation was similar to that observed for high status households (Breen and Rottman, 1985). Observers of crime in Edwardian Dublin recognised the connection between the misery in the city's slums and its relatively high crime rate (O'Brien, 1982, pp.183-84). A different sort of misery, real in its low quality of living but especially so when compared to the standard of the rest of society, generates crime in the slums of urban Ireland today.

CRIME IN NORTHERN IRELAND: RECENT TRENDS

Industrialisation, prosperity, and sectarian conflict traditionally fostered different crime patterns in the north than in the south of Ireland. While the south moved from civil war into a stability which some might call stagnation, the new Northern state experienced high levels of violence, which peaked in 1935 (Darby, 1983, pp. 22-23). Twenty years of relative peace followed, leading into a period of economic expansion fueled first by World War II and then by state-induced industrialisation through external investment. After 1962 a 'belated modernisation of outlooks both north and south' (Townshend, 1983, p.388) offered great promise for the future. North and south raced to produce jobs at a rate faster than losses from declining sectors; at least in the 1970s, the south was swifter and its economic performance higher (Simpson, 1983, p.109). Acute economic distress in the 1970s exacerbated the deprivation of traditional enclaves of disadvantage and created new ones.

The violence that marked the Northern state was sectarian, continuing an Irish crime pattern noted by Cornewall Lewis in 1836, in which crime is comitted with the intent of creating a general effect on a large audience of other people not directly affected but who are deterred from or compelled to certain actions as a result. This contrasts with the crime characterising most countries in which the intent is limited to a specific purpose, like acquiring money (Lewis, 1836, p.78).

'Ordinary crime' was a constant but usually unobtrusive presence in the North. This section attempts to disentangle trends and distributions for the two types of crime, north and south: Does the geographic distribution of 'ordinary' crime in Northern Ireland parallel that in the south, or does it coincide with the distribution of paramilitary activities?

By the early 1960s there was a clear upward trend in Northern Ireland's crime statistics. That trend can be traced and compared to that in the Republic through Table 2, which also provides rates for Belfast and Dublin. Throughout the period considered, the overall rate of indictable crime has been higher in the North than in the Republic, with the gap between the two growing even prior to 1969 and diverging most substantially in the early 1970s. Between 1960 and 1968, the rate for Northern Ireland rose from 5.96 per 1,000 population to 10.84; the Republic experienced a change from a 5.46 to a 8.83 rate over the same years. Significant increases are recorded for most years in both, but the upward trends were dissimilar, especially recently, with Northern crime rates essentially level in the early years of this decade, when they were rising dramatically in the Republic.

Indictable crimes cover too wide a range of conduct to be particularly revealing. However, some broad patterns do emerge. First, crime in Northern Ireland is less concentrated in urban areas. Belfast's contribution to the crime statistics is more in line with its share of the North's population than Dublin's is with its share of the population of the Republic. Over time, Belfast contributes a diminishing share of the total: 56 per cent in 1964 and about 52 per cent in the 1980s. Second, the crime rate in Belfast was traditionally lower than that in Dublin but since the early 1970s the reverse has been true. This shift occurred despite a consistently higher overall Northern Ireland rate for indictable offences. Third, there are interesting similarities between the Belfast and Dublin trends. In both cities, the rate of crime doubled over the 1970s, though it took the influx of heroin to propel such a change in Dublin.

Some commentators (Heskin, 1985; Craig, 1986) cite the similarity in trends north and south - and indeed to those observed for England and Scotland - as evidence that the northern conflict is not generating a disintegration of the

Table 2: Rates of indictable crime, 1960-85 (per 1,000 population)

	Belfast	Northern Ireland	Dublin	Republic Ireland
1960	n.a.	5.96	n.a.	5.46
1961	n.a.	6.90	n.a.	5.24
1962	n.a.	7.16	n.a.	5.39
1964	9.74	7.15	12.84	6.18
1965	11.94	8.75	11.50	5.83
1966	12.80	9.94	13.91	6.60
1967	13.61	10.35	15.08	7.08
1968	14.09	10.84	17.24	7.91
1969	15.28	13.41	18.74	8.83
1970	22.98	16.25	23.37	10.39
1971	27.31	20.04	28.00	12.69
1972	32.03	23.23	27.54	12.96
1973	27.91	20.72	25.38	12.36
1974	29.84	21.54	25.46	12.83
1975	33.34	24.23	29.56	15.25
1976	35.69	25.95	32.20	16.88
1977	41.54	29.50	37.85	19.24
1978	41.87	29.46	36.13	18.68
1979	55.71	35.17	36.51	19.02
1980	55.48	36.40	41.29	21.44
1981	62.26	39.96	51.92	26.00
1982	62.98	39.58	56.02	28.03
1983	65.15	40.78	58.36	29.19
1984	76.57	42.30	55.11	28.21
1985	69.50	41.18	52.66	26.51

Source: Heskin, 1985 (1960-83); NI Office, 1985, p.68 (1985); 1984 estimates based on Heskin, 1985.

north's social fabric: the rise in crime is what would largely have been anticipated given the economic and social changes the province experienced since 1960. Based on trends in rates of indictable crimes, this is plausible but not convincing. Certainly there is contrary evidence, such as the apparent deurbanisation of crime in the North. Statistics on burglary offer another basis for making a judgment on the forces underlying the rise in Northern Ireland's crime rate. Table 3 examines the rise over the 1970s in the number of burglaries known to the police in England and Wales, Scotland, Northern Ireland and the Republic. Rates were highest in Scotland at all four points of comparison, similar in England and Northern Ireland, and significantly lower in the Republic. The results of crime victimisation surveys suggest that these comparisons are largely artifacts of differences in (a) public reporting to the police and (b) police recording practices (Smith, 1983; Breen and Rottman, 1985). Burglary rates based on police statistics rose dramatically in both Northern Ireland and the Republic over the 1970s and crime victimisation surveys suggest that by the early 1980s such offences were far more prevalent in Ireland than in England or Scotland. A burglary was reported by 4.2 per cent of Northern Ireland households in the 1984 Continuous Household Survey, and 3.6 per cent of the Republic's households did so in a 1983 survey; these are significantly higher than rates in Britain (Breen and Rottman, 1985; Craig, 1986).

Belfast's crime rate recorded its most substantial proportionate increase over 1969/70, strongly suggesting that the locus of the conflict shifted there; the 1968/69 rise in the north's crime rate is more evident in the national than in the Belfast statistics (see Table 2). Poole's (1983) study of the geography of 'fatal incidents' (sectarian violence resulting in a death) offers a useful counterpoint to what we can determine about the distribution of crime. Of 1,715 fatal incidents over the 1969-81 period, virtually three quarters (73.5 per cent) were in urban locations (defined as any town or city with a population of over 5,000). Though Belfast had the second highest rate *per capita* population of the twenty seven urban places (Derry had the highest), it was the scene of 57.1 per cent of all fatal incidents and 77.7 per cent of all urban incidents; that 56.9 per cent of the urban population lived in

Table 3: Internal comparisons: burglary

	Burglary		
	Known number	Number per 100 Adults	Detection (%)
England and Wales			
1971	451,000	12.1	37
1976	515,500	13.6	34
1980	622,600	16.0	31
1981	723,700	18.6	30
Scotland			
1971	59,200	15.3	26
1976	79,800	20.2	20
1980	78,400	19.5	22
1981	95,700	23.8	20
Northern Ireland			
1971	10,600	9.8	27
1976	15,900	14.5	16
1980	19,700	17.4	20
1981	20,500	18.1	22
Rep. of Ireland			
1971	9,971	4.9	51
1976	19,336	8.7	44
1980	22,175	9.4	41
1981	25,300	10.7	38

Offences known in 1980 and 1981 are not precisely
comparable with earlier years because of changes in
the counting rules 'to improve the consistency by
the police of the recording of multiple, continuous
and repetitive offences'.
1980 and 1981 population bases: 1980 estimates
of population age 15 and older.

Belfast offers a useful standard by which to gauge the degree of concentration. There was a similar concentration of rural incidents, nearly two thirds of which (62 per cent) occurred in six localities adjoining the border (Poole, 1983, p.175). Fatal incidents tended to polarise into two geographical areas: Belfast and several of the 'largest' towns, on the one hand, and certain rural areas adjacent to the border. Belfast police statistics can only be allocated among four divisions, each covering a wide range of territory, both geographically and in terms of socio-economic status. The differences between those areas, however, suggests a pattern of urban crime that is more typical than that described for Dublin, in which the highest rates of victimisation risk are concentrated in low income areas and in the commerical sections of the city centre. Data from crime victim surveys offer a basis for testing this assertion.

In 1971, only one town apart from Belfast had a population exceeding 50,000. There is a significant gap between the size of Belfast, with its 554,000 residents, to Derry, the next largest, and then to the next largest town, Bangor, with a population of 35,000. The north's legacy as the industrial base of Ireland was translated into a single urban centre; in practice, like Dublin, an urban region. Other urban places are on a far smaller scale. Unemployment in the north has retained its traditional geographic pattern, high in the inner Belfast area and then rising with distance from Belfast. It is the level of unemployment that has varied over the last 30 or so years. In the 1960s, Belfast benefited disproportionately from efforts to reduce unemployment, a record that was reversed in the 1980s (Simpson, 1983).

If we look at the crime rates in the north today, we find that the lowest crime rates are in the RUC divisions that cover predominantly rural areas, including South Armagh, and stretch from the border into the centre of the province. The rate of indictable crime in that area is roughly 10 per cent of that found in South Belfast Division (which includes the city centre). The next lowest levels are found in the southwestern part of the province, including much of Fermanagh and parts of Tyrone. Similar levels are found in the border areas southeast of Belfast. The primarily rural division that has Derry, one of the most depressed cities in Europe, at its centre, has a crime rate similar to that of the more affluent areas of the Belfast region. South Belfast, which includes the city centre and the province's most prosperous neighbourhoods, has a 1985 rate of indictable crime (192 per 1,000 population) that is nearly five times the Northern Ireland average (Craig, 1986, p.6).

In part, it is the absence of police or victim survey-generated estimates of urban crime rates (apart from the total number of indictable offences in Belfast) that makes regional differences more important in Northern Ireland than in the Republic. In the North, it is not possible to disaggregate an area's crime rate into rural and urban contributions. There is substance to that focus, however. This is evident in the persistence of strong regional differences in the 'crime mix', the proportion of the total that various crime categories represent. The regional effect is strong along a general violence vs. property crime dimension, with interpersonal violence forming a larger share of the total in rural areas generally and especially in those to the south of Belfast. Yet the lowest percentage of such offences is in the RUC Division which covers Co. Antrim. Burglary formed a major component of the crime totals in two of the four Belfast Divisions: North Belfast and East Belfast; its percentage elsewhere in the urban region is not atypical. The lowest proportions of theft and burglary are recorded in Eniskillen, which, however, has the highest proportion of 'fraud' offences. Regions also differ in the extent to which offences fall into a broad 'other' category comprised mainly of public order offences whose enforcement is subject to considerable police discretion. Such offences are an unusually large component in the division with Derry as its centre. Here, however we are examining the composition of the crime recorded in a region, not the actual level of offences. As in the Republic, the most prominent feature of crime in the north is its concentration in a single urban centre. The second most prominent, again shared with the Republic, is a high rate of incidence. Despite official commentaries offering reassuring international comparisons (Northern Ireland Policy, Planning and Research Unit, 1984; Northern Ireland Office, 1985) it appears that the level of major property offences that are unlikely to be tied to 'The Troubles' is atypical for these islands and for much of Europe.

It is unnecessary to view this transformation as evidence of 'societal disintegration'. Instead, it reflects the change to a different type of society, one whose pattern and distribution of ordinary activities - commerce, leisure, consumption - generates a high rate of property crime. Of course, conflict in the north affects that pattern and distribution; indeed, 'violence is a factor affecting urban spatial structure' (Poole, 1983, p.153), a cause as well as a consequence. Violence certainly is a causal factor in the shaping of crime in Ireland, and its most durable effects may be indirect. Groups not directly participating in the conflict imitate the methods and weaponry of those which are and networks of contacts and markets for criminal activities are fostered. Irish crime statistics reflect these altered features of society.

CONCLUSION

Economic and social change indisputably alter the level and nature of crime in a society. It is not the case, however, that their impact is inevitably to generate inexorably rising levels of crime. Historically, this was not the case in Europe during the last century and today it does not appear to be true in Switzerland (Clinard, 1978) or in Japan (Rutherford, 1984). Even in the United States the recent trend has been for crime rates to decline (the evidence from police statistics) or to remain stable (the evidence from crime victimisation surveys). The impact may instead by registered mainly in the geography of crime. For example, in the United States, the regional pattern of crime has been transformed in recent decades. American television portrays crime accurately in one respect: it is leaving the old commerical centres of the east coast and midwest and moving to the sunbelt. Murder and burglary are now most prevalent in cities like Miami, Las Vegas and Atlanta; Chicago and Detroit do not even rank in the top 25 cities for such crimes, measured as rates per 100,000 population (Brantingham and Brantingham, 1984). The old metropolitan centres hold their own only in levels of robbery. This has occurred during a period of continued economic decline in the industrial/commerical heartland and of expansion in the newly crime-prone sunbelt.

A map which indicates the distribution of unemployment, low income households, or discrimination in Ireland today would only partially coincide with one depicting the distribution of crime. Crime is likely to occur where ordinary activities are such as to bring together motivated offenders and opportunities for crime. The location of opportunities is decisively urban, perhaps more so in Ireland than in other parts of Europe. For opportunities to turn into crimes, potential offenders must first come into contact with them. Research suggests that the same regularities that govern ordinary human interactions apply to criminal activities: interactions diminish in frequency as distance increases, though the extent of the distance decay process is greater for violent than for property offences (Brantingham and Brantingham, 1984). Generally, this represents another basis for expecting high levels of urban property crime. Those residing outside urban areas lack suitable targets for crime within an acceptable search area. It is also likely that differences between Dublin and Belfast reflect the greater restrictions in the latter city on movement. Different local authority housing policies also contribute, though in the north their impact is probably as much overshadowed by, as it is expressed in, religious segregation.

Such a perspective assumes a roughly equal level of motivation to commit crime across a society's social groups. If we restrict our consideration to conventional property crime (burglary, robbery, shoplifting or cheque fraud), this is manifestly untrue. Police bias cannot explain the social class differentials in arrests for such offences. But why should Ireland over the last twenty years have generated so substantial a proportion of its population prepared to avail of criminal opportunities. The rapidity and incompleteness of socio-economic change offer partial answers. Change has brought marginalisation to many segments of the urban working class, along with exposure to expected standards of living based on consumerism. It is incorrect to regard the resulting situation as social disorganisation or disintegration. The change is toward new forms of behaviour, new rules for conduct. If we think that the change has been limited to the less privileged sections of either north or south, it would be instructive to view - if it were possible - a map indicating the distribution of white collar crimes, such as tax fraud, embezzlement, and insurance fraud (Rottman and Tormey, 1986). Such crimes are far more costly financially than the conventional crimes about which politicans and the public are obsessed (Conklin, 1986). They also inflict substantial social costs, by diminishing trust and confidence in societal institutions. A focus on such offences might clarify in our minds the extent to which the geography of crime in Ireland resembles that of ordinary, legal activities. Crime and employment, crime and commerce, crime and demography share a common geography in Ireland.

REFERENCES

Bannon, M.J., Eustace, J. and O'Neill, M. (1981) *Urbanisation: Problems of Growth and Decay*. Stationery Office, Dublin.

Brantingham, Paul and Brantingham, Patricia (1984) *Patterns in Crime*, Macmillan, New York.

Breen, R. and Rottman, D. (1985) *Crime Victimisation in the Republic of Ireland*, Economic and Social Research Institute Paper No. 121.

Breen, R., Hannan, D., Rottman, D. and Whelan, C. (1988) *Class and State in the Evolution of Modern Ireland*, Macmillan.

Clinard, M. (1978) *Cities With Little Crime: The Case of Switzerland*, Cambridge University Press.

Conklin, J. (1986) *Criminology* Second Edition. Macmillan, New York.

Craig, J. (1986) Crime in Northern Ireland, paper read to a conference Crime in Ireland: Crisis or Manageable Problems? National Association of Probation Officers and Probation Officers' Branch of the Union of Professional and Technical Civil Servants, Belfast.

Darby, J. (1983) The historical background, in Darby, J. (ed.), *Northern Ireland: The Background to the Conflict*, Appletree Press, Belfast, 13-31.

Felson, M. (1983) Ecology of crime, in Kadish, S. (ed.) *Encyclopedia of Crime and Justice*, The Free Press, New York, 665-70.

Fitzpatrick, D. (1977) *Politics and Irish Life: 1913-21*, Gill and Macmillan, Dublin.

Fitzpatrick, D. (1978) The geography of Irish nationalism 1910-1921, *Past and Present*, 77, 113-44.

Fitzpatrick, D. (1982) Class, family and rural unrest in nineteenth century Ireland, in Drudy, P. (ed.) *Ireland: Land, Politics and People, Irish Studies*, Vol. 2, Cambridge University Press, Cambridge, 37-75.

Heskin, K. (1985) Societal disintegration in Northern Ireland: a five-year update, *Economic and Social Review*, 16(3), 187-99.

Hough, M. and Mayhew, P.(1983) *The British Crime Survey*, First Report, HMSO, London.

Hourihan, K. (1986) Community policing in Cork: awareness, attitudes and correlates, *Economic and Social Review*, 18(1), 17-26.

Irish Marketing Surveys (1983) *The I.M.S. Poll*, 1(4).

Lane, R. (1979) *Violent Death in the City: Suicide, Accident and Murder in 19th Century Philadelphia*, Harvard University Press, Cambridge, Mass.

Lane, R. (1986) *Roots of Violence in Black Philadelphia 1860-1900*, Harvard University Press, Cambridge, Mass.

Lea, J. and Young, J. (1984) *What is to be Done About Law and Order?*, Penguin, Harmondsworth.

Lewis, G. C. (1977 reprint) *Local Disturbances in Ireland*, Tower Press, 1836, Cork.

McGroarty, J. (1986) 'Drugs and Crime', paper presented at the Forensic Science Symposium, St Patrick's College, Dublin.

National Economic and Social Council (1985) *The Criminal Justice System: Policy and Performance*, NESC Report 77, Stationery Office, Dublin.

Nettler, G. (1984) *Explaining Crime*, Third Edition, McGraw Hill, New York.

Northern Ireland Office (1984) *Policy, Planning and Research Unit A Commentary on Northern Ireland Crime Statistics: 1969-1982*, PPRU Occasional Paper 5, Belfast.

Northern Ireland Office (1985) *A Commentary on Northern Ireland Crime Statistics: 1985*, HMSO, Belfast.

O'Brien, J. (1982) *'Dear, Dirty Dublin': A City in Distress, 1899-1916*, University of California Press, Berkeley.

O'Tuathaigh, M.A.G. (1982) The land question, politics, and Irish society, 1922-1960, in Drudy, P. (ed.), *Ireland: Land, Politics and People*, Irish Studies Vol. 2. Cambridge University Press, Cambridge.

Poole, M. (1983) The demography of violence, in Darby J. (ed.), *Northern Ireland: The Background to the Conflict*, Appletree Press, Belfast, 151-80.

Rottman, D. (1980) *Crime in the Republic of Ireland*, Economic and Social Research Institute Paper No. 102, Dublin.

Rottman, D. (1985) *The Criminal Justice System: Policy and Performance*, National Economic and Social Council Report 77, Stationery Office, Dublin.

Rottman, D. and O'Connell, P. (1982) The changing social

structure of Ireland, in Litton, F. (ed.), *Unequal Achievement: The Irish Experience 1957-1982*, Institute of Public Administration, Dublin, 63-88.

Rottman, D. and Tormey, P. (1986) Respectable crime: occupational and professional crime in the Republic of Ireland, *Studies*, 75 (297), 43-55.

Russell, M. (1964) The Irish delinquent in England, *Studies*, 53 (summer), 136-148.

Rutherford, A. (1984) *Prisons and the Process of Justice*, Heinemann, London.

Simpson, J. (1983) Economic development: cause or effect in the Northern Ireland conflict, in Darby, J. (ed.), *Northern Ireland: The Background to the Conflict*, Appletree Press, Belfast, 79-109.

Smith, L. (1983) *Criminal Justice Comparisons: The Case of Scotland and England and Wales*, Home Office Research and Planning Unit Paper No. 17, London.

Townshend, C. (1983) *Political Violence in Ireland: Government and Resistance since 1848*, Clarendon Press, Oxford.

Whelan, B. and Vaughan, R. (1982) *The Economic and Social Circumstances of the Elderly in Ireland*, Economic and Social Research Institute Paper No. 102, Dublin.

Whelan, C. and Whelan, B. (1984) *Social Mobility in the Republic of Ireland: A Comparative Perspective*, Economic and Social Research Institute Paper No. 116, Dublin.

6 THE HISTORICAL LEGACY IN MODERN IRELAND

Stephen A. Royle

In 1988 the Bushmills whiskey distillery ran an advertising campaign, the slogan of which was '380 years practice makes perfect', a reference to the distillery's original licence to operate having been granted in 1608. In Belfast and other parts of Northern Ireland some of the posters referring to 1608 were close by graffiti slogans exhorting the reader to 'remember 1690' and the Battle of the Boyne. History and tradition are very important in Ireland. Her peoples have long memories for certain aspects of their past, some would say too long given the legacy of contemporary conflict arising out of past events. Here, more than most other places, the past conditions the present in terms of the attitudes of the people as well as acting in the usual fashion upon the human and physical geography of the area. The effect upon the physical landscape is clear, relicts from former climatic and geological conditions upon Ireland are everywhere from the distribution of different rock types to the legacy from the last glaciation. In the human landscape, too, the past is all pervasive.

THE NON-MATERIAL HISTORICAL LEGACY

One obvious legacy is in the naming of places and features. Most place names in Ireland date back centuries to an Irish language root. For example, Ballymena, one of the staunchest

113

bastions of loyalism in the north, doubtless to the chagrin of some of its citizens derives its name from the Irish Baile-Meadhonach, middle town (Joyce 1870). Sometimes the name remains appropriate: 'Inishmore' means big island and it is still the big island of Aran; elsewhere the name that lives on in the present looks back to something since forgotten or now unimportant. There are enough crossings over the Lagan now for Belfast's root, Belfierste, the ford of the sandbank, to be anachronistic today but Joyce (1870) thought that Dublin's Duibh-linn, black pool, root remained a good description of that part of the Liffey on which the city is built. However Dublin's Irish name, Baile-atha-cliath, the town of the hurdle-ford, is now an equivalent anachronism to Belfast's sandbank ford of centuries gone by - but the names live on, the past feeds through to the present. The history of Irish place names also has a political-cultural side to it that has sometimes constrained which aspect of the past is to be preserved. The regal names of two Midland counties in the Republic, Kings and Queens, have been replaced by the more politically acceptable Offaly and Laois respectively. Similarly, Maryborough, Queenstown, and Kingstown have become Portlaoise, Cobh and Dun Laoghaire. In Northern Ireland the name of the second city remains in dispute. The official name and that favoured by unionists is Londonderry, the second part from the Irish, doire, or oak grove, the first acknowledging the role of the London companies in the settling of that part of Ulster in the seventeenth century. Nationalists, wanting to downgrade the British influence, would like to see the name revert to just Derry and as the local authority has a nationalist majority it has renamed itself Derry City Council although the official name of both city and county remains Londonderry. Ironically, the city is called 'Derry' on Orange arches erected annually before the 12th of July celebrations. The BBC in Belfast trying to be even handed, but probably pleasing neither side, says 'Londonderry', then 'Derry', alternately. Such disputes over place names and the history/culture/language they represent of course are not confined to Ireland. Bilingual Welsh/ English signs are now the norm in Wales after a nationalist campaign. Even in the peaceful island of Majorca the author has seen Spanish language signs changed by islanders into their native Mallorquin. The English version of most Irish placenames are anglicized phonetic renderings of Irish language names, a result of the work of the Ordnance Survey in the early nineteenth century who, after local research, provided standardized and usually acceptable versions of names for the first Ordnance Survey maps, though many places still retain alternative spellings (Andrews, 1975). On the Republic's bilingual signposts one normally sees an obvious root and a phonetic derivative as - Tralee, Tra Li, for example. However, sometimes no linguistic link is present as in Wexford, Loch Garman, for the anglicized version here is not of the Irish root but of Waesfjord, the harbour of the mudflats, the name given to the place by the Vikings, another era of Ireland's past coming out in contemporary placenames. More recent changes, probably not unconnected with political feelings, include the transformation of anglicized Navan back to An Uaimh and, in the same county of Meath, Kells back to Ceanannas Mor though both retain their anglicized version in brackets in the Atlas of Ireland produced by Dublin's Royal Irish Academy - itself an institution whose name is redolant of a vanished era of Ireland's past.

Like placenames, boundaries represent an historical legacy. As Nolan has recently emphasized, 'many of Ireland's county boundaries date from the 16th and 17th centuries when they were superimposed by outside - i.e. British - interests . Yet this territorial division which, above all others, represents conquest and assimilation owes much of its vitality to its adoption by the Gaelic Athletic Association as its basic organizational unit!' (Nolan, 1986, p.74). This is another paradox of Irish cultural history as the GAA was set up in the late nineteenth century as an overtly anti-British sporting organization with its 'uncompromising hostility to all foreign sports - the games of the protestant incomers in effect' (Magnusson, 1978, p.115). The international border across the island isolating six of the nine counties of Ulster is, of course, the most significant boundary legacy left to the present by Irish history. 'Political boundaries do not exist in nature, they are cultural features which result from human decisions and are found only where man establishes them. The Irish border.... was established in 1920 to allow two sets of Irishmen with opposing national aspirations to live apart because they could not live together' (Douglas, 1976, p.162). But the border still put together within Northern Ireland those who 'could not live together' and the political historical legacy of Ireland and her irreconcilable peoples remains a very tragic contemporary problem.

Ireland's non-material historical legacy is not all tainted by political controversy. In music, sport, literature, drama, broadcasting and many other fields, representatives of the few million people from this small island have played a disproportionately significant role and Ireland has a splendid cultural life stemming from its ancient traditions. The Chieftans, traditional Irish musicians, have entertained the Pope and much of the rest of the world; another paradox is that writing in the imported language of English has been dominated for generations by writers from Ireland, giving the island an unrivalled literary legacy; at times it seems as if the BBC in London never mind RTE in Dublin or the BBC in Belfast is staffed almost entirely by men and women from Ireland peddling their inherited gift for words. Cultural as well as political history matters to the residents of Ireland and to the Irish abroad. St Patrick's Day is celebrated in North America still with cards and, in New York, green beer; the Scotch-Irish, Ulster, heritage, is found in places like Toronto where the union jack can be seen still, flying alongside the maple leaf outside Orange Halls although the Canadian Orange Movements have lost a lot of their former power recently (Houston and Smyth, 1980).

THE MATERIAL HISTORICAL LEGACY

In addition to the past's effects on such phenomena as names and boundaries, its role in conditioning the appearance of the human landscape should not be overlooked. The siting of towns, the arrangement of fields, the location of industry and farms - all these and many other landscape phenomena reflect decisions taken in the light of conditions and constraints operative in times gone by. Aalen's important book, *Man and the Landscape in Ireland*, (1978) devotes much discussion to this theme - for example 'it is the planned elimination of rundale which has produced the present landscape of many areas. The bulk of the formerly fragmented holdings have now been

wholly or partially consolidated, communal practices abandoned and fences erected to define the fields' (p.223). In some
areas, despite such reorganization, earlier practices have left
their scars on the landscape too (Figure 1).

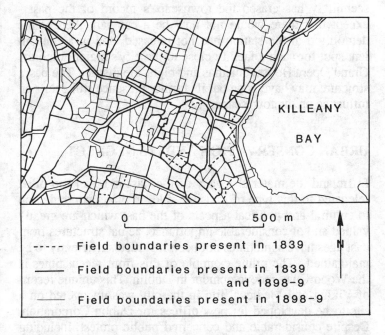

**Figure 1. The effects of past agricultural constraints upon
the present landscape - tiny walled fields, Inishmore,
Aran, Co. Galway from the 1839 OS Map.**

Slievemore, a deserted village on Achill Island, is only
one of many places where, in this case, the stone walls and
field banks delimiting the boundaries of a reorganized agricultural system cut across the ribs of the older lazy beds. It is
not always easy to obliterate or cancel out the marks of earlier
events even in the cities. Belfast's first railway station opened
in 1839 on what is now Great Victoria Street. In 1848 the
station was rebuilt on the same site but by the 1970s this location was no longer convenient and the station was pulled
down to be replaced by a new terminus at East Bridge Street

where now sits the city's futuristic, inappropriately named Central Station at a fair hike from the city centre. At the time of writing, the site of the old Great Victoria Street station is a car park with little to show from its previous life. Contemporary decision making in the light of contemporary pressures seemingly has erased the townscape's record of the past - except that there would not now be a car park there if the demolition of the station had not released land in a convenient spot for such a facility close to the city's major hotel, the Grand Opera House, cinemas and night life district. The past's structure may have gone but its former presence still exerts an influence on the townscape.

URBAN CONSERVATION AND ITS NEGLECT

In Ireland the material legacy of the past is rather too often relegated to this type of invisible role. In direct contradiction to cultural and political aspects of the past which are greatly valued and of considerable importance, actual structures from past ages have sometimes not been regarded sufficiently to be maintained. The prime example of this from recent times is the Wood Quay redevelopment in Dublin. This unique record of Viking and Medieval life in the city was uncovered on a site to be developed for new offices for Dublin Corporation. Despite considerable and concerted public protest, including the occupation of the site (Martin, 1984) the Corporation was still able to go ahead and build its offices. Wood Quay was lost forever. Simms noted that despite this 'irresponsible destruction' (1984, p.155) 'we have increased our knowledge about the origins of Ireland 's capital city' (p.157) so there has been a cultural historical benefit but the material legacy - as so often in Ireland - was destroyed. One can point to the actual site, but one can no longer see the remains there, *in situ*, just the office building. Similarly, one can walk along Great Victoria Street in Belfast, point to the car park and say 'there stood the station', one cannot point to a building and say 'there is "a modern mercantile revival of a Renaissance Italian *palazzo* with a heavily rusticated lower floor and a refined *piano nobile* above" (Walker and Dixon, 1984 p.8) once the station, now put to alternative use'.

Residents of the largely Georgian English city of Bath would almost certainly tend to have a smaller knowledge of and certainly less interest in the seventeenth century political events or the nineteenth century agricultural affairs in these islands than residents of Dublin, Cork or Belfast - nobody 'remembers 1690' or 1845, for that matter, in Avon. But Ford could write in the *Geographical Review* in 1978 a fairly critical piece on Georgian Bath's conservation yet still find that about half of the city is in Conservation Areas and two-thirds are subject to some kind of special control on its development whereas, in the same journal, for Dublin, Kearns had to report that in the twenty years up to his writing 'the Georgian structures of the city continued to be altered and demolished. Governmental neglect, developers' greed, and public apathy have persisted' (1982, p.290). Kearns, in fact, identified more than just simple apathy. The Irish, he says, have an anti-urban bias that leads many to care little about cities, others are just ill-informed about the merit of the architecture but there is also a more sinister reason for the lack of care (1983a). To the Anglo-Irish period that is represented materially by Georgian Dublin, 'the sentiments of most Irish have ranged from benign insouciance to overt hostility' (1983b, p.81). The hostility comes about, again, because of cultural and political legacies. The Anglo-Irish period was the time of the Protestant ascendancy, that 'valley of humiliation' (Beckett, 1976, p.63) for nationalistic Ireland, and 'the perpetuation of perjorative stereotypes of the Anglo-Irish has been injurious to the objective appraisal of eighteenth century architecture' (Kearns, 1982, p.273). Hence, much of Georgian Dublin has been allowed to decay or has been demolished - 40% gone in the two decades to 1982 - the past's material legacy sacrificed, partly at least, because it was of less importance than a selective cultural legacy, a folk memory of Ireland's 'humiliation'.

The greatest force for conservation is inertia - most Irish small towns and villages have very similar centres because, since the rush of building in the eighteenth and early nineteenth century, there has been little pressure for change : 'the charm of the Irish village is that it still retains much of the original layout and that it is often graced by buildings . . . acquired at an early stage in its history' (Cullen, 1979 p.20). But, as Cullen goes on to point out, this charm, largely the product of inertia, is now 'imperilled by the replacement of its ageing buildings by residential and commercial interests alike. ' (p.20). If this is the case for a village, it is even more problematical in a great city when pressures for change can be almost irresistible. When there is insufficient check on these pressures and the people involved in redevelopment place a low value on the existing structure, as was the case with Dublin's Georgian terraces, much of the old townscape may be lost. In a similar way, ancient monuments or neolithic burial sites in agricultural areas may be destroyed to allow for increased production. A low public esteem for conservation plus financial analysis which dictates that the function of a site and thus the infrastructure on it appropriate to that function should be that which produces the maximum return work against conservation. However, if legislators have imposed conservation practices, great economic pressures for change can be ameliorated. This can be seen most notably within the British Isles in the Scottish city of Aberdeen where the major oil-related development pressures of the 1970s were countered by a local authority with Conservation Area legislation in place that was keen to keep the 'Granite City' intact. The new developments were restricted largely to part of the harbour and to the outskirts of the city outside its historic granite areas (Royle, 1982). In Ireland, by contrast, there has been neither the will, nor sufficiently powerful legislation to resist change in favour of conservation.

Legislative Framework

Newcomb in his *Planning the Past* is in no doubt of the importance of a legislative framework: 'needed is a set of promulgated regulations governing relics and sites so that their discoverers or developers may find guidelines for their legitimate actions in order to prevent plunder or blunder' (1979, pp.80-81). For Aberdeen there was a wide range of powers available to the council. For England, mention has already been made of the two-thirds of Bath being subject to some legislative protection. In much of the rest of Europe conservation has also been given a high priority (Contacuzino, 1975), but in the Republic of Ireland 'noble government proclamations about preservation have not been buttressed by affirmative legislation' (Kearns, 1983b, p.85). Despite the Planning Act of 1973, the conservation policy for Dublin was really formulated only in the Dublin City Development Plan of 1976 and the city earlier was 'unprepared for and vulnerable to redevelopment' (Kearns, 1982, p.278). Even in 1980 Dublin was described in conservation terms as 'the most vulnerable great historic city in Europe' (Binney, 1980, p.208). The Republic's other cities and towns were also vulnerable but at least did not have the extreme pressures placed upon the rapidly growing capital city.

Urban Conservation in Northern Ireland

In Northern Ireland as late as 1968 the Ulster Architectural Heritage Society, 'claimed with some justice that the approach to conservation of the built environment was fifty years behind the times (Hendry, 1977, p.373). However, the Planning (Northern Ireland) Order of 1972 enabled Northern Ireland to catch up with British practices whilst the Historic Monuments Act (Northern Ireland) 1971 dealt with structures in state care. In 1974 under the 1972 order, the Historic Buildings Council was set up to advise what is now the Department of the Environment for Northern Ireland and the Council started to prepare listings of buildings of architectural and historic interest - 6,500 were so designated by 1984 (Pierce and Coey, 1984). Under the order, Northern Ireland now also has conservation area legislation (Twenty-two such areas were declared by 1984 with perhaps another eight to come (Pierce and Coey, 1984), compared to about 5,500 in England and Wales where the architectural heritage is rather richer. Amongst areas protected are town centres (e.g. Enniskillen, Moira), villages (Cushendall), areas with an industrial heritage (Sion Mills), and areas of good Victorian residential property (some of the streets around Queen's University Belfast). The new Belfast Urban Area Plan 2001 (Belfast Divisional Planning Office, 1987) proposes 'the protection and enhancement of the most important areas of special townscape value' (p.11) and lists again more areas of the city for conservation area status, mostly in and around the city centre. The National Trust, though most noted for their protection of coastal areas (e.g. much of north Co. Antrim) and country houses (e.g. Castle Coole, Co. Fermanagh, Castle Ward and Mount Stewart, Co. Down amongst others), also have holdings in settlements including the villages of Kearney Co. Down and Cushendun, Co. Antrim with its late nineteenth century whitewashed houses and the 1925 Cornish style cottages by Clough Williams-Ellis. Additionally, the National Trust holds a Belfast public house, The Crown Liquor Saloon on Great Victoria Street (opening hours are licensing hours and 'refreshments' mentioned in the Trust's literature are the full bar facilities). This is one of Belfast's finest high Victorian, buildings and receives the highest 'Baedeker' rating of three stars in Brett's (1967) guide to the city's historic buildings. The Crown received a major grant from the Historic Buildings Council towards repair and restoration.

Most property in Northern Ireland neither receives nor deserves any specific protection outside that given by the normal planning permission regulations and many outworn structures in city and towns alike have been and are being replaced. Belfast for long had the reputation of having some of the worst housing in Western Europe, usually being bracketed with Naples, Italy in this regard. Because of Belfast's rapid industrialization in the second half of the nineteenth century, a rather neglectful and uncaring local authority in the inter-war period and the usual delay in making progress here compared to the rest of the UK in the post-war period (Boal and Royle, 1986), Belfast entered the 1970s with much very poor housing (Birrel *et al.*, 1971). Since then many areas have been subject to comprehensive redevelopment and although Belfast as well as Londonderry has had its share of planning disasters in housing (most notoriously Belfast's truly awful Divis Flats thankfully again being considered for total not just partial demolition *(Belfast Telegraph,* 25 September 1986), much development under the Northern Ireland Housing Executive and various housing associations has been of acceptable standards. In fact, although some of the old residential property has been saved by rehabilitation and some has yet to be dealt with, the mean terraced streets so redolent of Belfast's past are becoming a lot less plentiful, so much so that, that ultimate safety net of architectural preservation in Ulster, the Ulster Folk and Transport Museum, has rebuilt a row of Belfast terraces at its park in Cultra, Co. Down.

Another factor affecting building conservation in many towns in Northern Ireland are the troubles, the local euphemism for the bloody inter-ethnic conflict, itself a cultural/ political legacy from the past. Much property has been abandoned, particularly in interface zones where met the two sides always largely separate residential areas (e.g. Boal, 1969), much of it has been cleared. Residential property has suffered as a result of attacks upon its occupants, bomb attacks upon private and, especially, commercial and public buildings have caused dreadful damage. In recent years the use of the appallingly destructive car bomb has diminished somewhat - although Enniskillen is an exception - but many

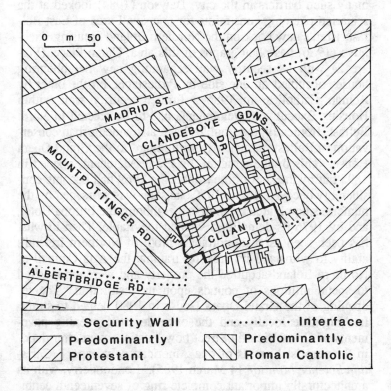

Figure 2. Defensive planning in Belfast (Source: Dawson, 1984)

police stations continue to be attacked whilst Belfast Town Hall (designed 1869 and 'a sturdy brick essay in mixed historical styles (Walker and Dixon, 1983, p.30) and not to be confused with the stone City Hall of 1906) was badly damaged in 1985 when a magistrate's court was bombed. Bombed buildings of any architectural merit are usually restored but sometimes when the damage has been extensive a pastiche is erected instead. Near Queen's University some Georgian style facades hang anachronistically upon steel girder frameworks for this reason. Another effect of the troubles upon the townscape has been the need to plan for separate communities sharing the urban space. In Belfast the fence marking the Peace Line between the Falls and the Shankill is only one of many such barriers in the city. Dawson (1984) looked at the sectarian planning constraints for one small area of East Belfast; Brett's new book on the Northern Ireland Housing Executive (1986) has photographs of this area, Cluan Place, (Figure 2) with its security wall separating new neighbouring houses occupied on one side by Protestants, on the other by Roman Catholics. In 1986 a further security wall within Housing Executive developments had to be erected in North Belfast. Ironically, the troubles have also aided urban conservation in Belfast, particularly with regard to the Grand Opera House. This ornate structure by Frank Matcham, opened in 1895 but by the start of the troubles in 1969 had become considerably decayed. In the 1970s its state was worsened by the numerous bomb attacks upon its neighbour, the Europa Hotel, and by 1976 its demolition seemed inevitable. Then in what has been seen as an important symbol of Northern Ireland's ability to withstand its dreadful trauma, the Arts Council of Northern Ireland acquired the property and were empowered to spend millions of pounds upon its restoration. The reopening of the Grand Opera House in 1980 helped to stimulate nightlife locally and the once deserted evening pavements of Great Victoria Street now 'bustle - even after dark!' in the surprised words of one American newspaper (*Christian Science Monitor* 14 March 1985). Londonderry, with its architecturally important complete ring of seventeenth century walls, has also had conservation successes recently. The most notable was the Derry Inner City Project involving the restoration and return to use of old buildings which in June

1986, from an entry of 184, won the national Community Enterprise Awards competition.

Despite recent progress in both the public awareness of the importance of conservation and the legislative powers to enforce it open to the authorities, losses of good buildings in Northern Ireland continue. A recent example was the Old Music Hall of 1840 in May Street Belfast which despite being a listed building was allowed to decay into an unsaveable condition after its last users moved out in 1973. It was demolished to make way for yet another car-park. Another loss was the demolition of Belfast's main post office, a century after its erection in 1886. An unlisted building and rather plain considering its High Victorian date, nonetheless conservationists were disappointed that its facade at least could not be retained when its site and that of the old Smithfield Market - a victim of the troubles in 1974 - was required for a new shopping centre (Brown, 1985). (Dublin's main post office of 1815 remains, presumably, inviolate, not just because of its architectural merit though it is a much finer building than Belfast's but because of political historical associations as the site of the Easter Rising of 1916). Other dangers in Northern Ireland are the ever present forces of time and decay, too often helped by local pollution including acid rain (McGreevy *et al.*, 1983).

Urban Conservation in the Republic of Ireland

The quality of the atmosphere in cities in the Republic has also been subject to considerable criticism recently despite Binney's praise for 'Dublin's clean air' (1980, p.208) as a positive factor in its conservation. However, others have said that because of atmospheric pollution 'every major building in Dublin is damaged to the point where things are literally falling off such as the balustrade of Trinity College (McIvor, 1986, p.220). Over 140,000 households in Dublin (58%) use solid fuel - coal and/or turf - as their principal method of domestic heating (Brady, 1986) but the role of these fires in causing pollution and the extent of the pollution itself has been challenged by the Coal Information Service (Linehan, 1986) though even this body supported some form of clean air legislation for the Republic and a measure was under discussion in 1986. This is an example of progress towards a more pro-environmentalist stance in the Republic, another being the £7m p.a. now spent on cleaning and restoration work in Dublin (McIvor, 1986) to which the newly revealed splendours of the Bank of Ireland is a fitting tribute. Even Kearns in his 'Dublin's crumbling fair city' conceded that there are now 'some encouraging signs' of a change in attitude towards conservation (1983b, p.85). An Taisce in 1980 also got 'a sense . . . that the tide is turning' (p.16). Pressures to turn this tide had been building up for some time. In 1966 Flora Mitchell published her *Vanishing Dublin* and the attractive drawings therein are in too many cases an historical record of what used to be ; The Royal Institute of the Architects of Ireland in 1975 published *Dublin: a city in crisis*, ('renewal and expansion need not be carried out at the expense of our architectural heritage' (p.ii)); An Taisce's *Dublin's Future* of 1980 helped, as did Kearns' works (1982, 1983a, 1983b); for areas outside Dublin one can point to Shaffrey's important *The Irish town: a guide to survival* (1975). European Architectural Heritage Year, 1975, also served to stimulate public awareness of Ireland's fine architectural heritage and the dangers it faced (e.g. Architectural Association of Ireland, 1975).

Problems remain of course. Many otherwise attractive small towns are in desperate need of a coat of paint. More seriously, Dublin and to a lesser extent the other towns remain troubled by dereliction and are tormented by vandalism and crime, which as Cabot (1985) points out, cannot be disassociated from an unpleasant physical environment. However, a recent summary on Irish conservation matters, Cabot's *The State of the Environment* (1985) is not entirely gloomy about urban areas. It notes that their body has produced a series of evaluations of important buildings in urban areas, many written by Garner (e.g. 1980, 1981); Dublin's neglected Georgian interiors have been inventoried; its property of architectural merit has been listed; mention is made of Cork's architectural advice service; of the Bord Fáilte's Tidy Towns Competition (there are similar types of competition in Northern Ireland); and of Wexford's Beautiful Wexford campaign as one example of a number of urban environmental schemes. Conservation and amenity bodies such as the Irish Georgian Society, An Taisce, and a number of Dublin based groups keep up the pressure. For example, the Dublin Architectural Study Group (1982) recommend that a supervisory body along the lines of the Edinburgh New Town Conservation Committee be set up for the Irish capital. One can only speculate, but it may be that a different result may have occurred had the Wood Quay development/conservation battle been fought a few years later. However, the ability of Dublin Corporation to demolish what was a National Monument does lead one to wonder what force, even now, urban conservation could command in situations where a lot of money was at stake. But Wood Quay and the extraordinary protests its redevelopment generated (Martin, 1984) at least did have the effect of gaining considerable public notice for the conservation lobby and 'it is hoped that the lessons of Wood Quay will be learned' (Friends of Medieval Dublin, 1984, p.168). There are financial benefits from maintaining Ireland's historic urban fabric. One of 'the best preserved . . . of Europe 's capital cities' (Binney, 1980, p.208) to say nothing of the splendours of many of the country towns such as Trim and Cashel, the fine commercial and residential property of Cork and Limerick - all these things are attractive to tourists. Knock all the old stuff down and the tourists are less keen to come. In Ireland, where tourism is marketed on history as much as landscape - 'a unique and little known (sic) land that has preserved the inheritance of over 3000 years of civilization' (Automobile Association, 1972, p.577) - it is as well to retain some material evidence of this history.

THE RURAL LANDSCAPE

Mitchell considers the tourist in the context of the retention of the traditional Irish rural landscape: 'the image dear to the city man and to the still more important tourist is one of happy peasants living in thatched cottages from which they sally forth now and again to dig in patchwork fields or fish from rocks. Are those who like the image prepared to pay to preserve their illusion and employ the surviving inhabitants as landscape gardeners to keep the countryside reasonably free of weeds and bushes with a few property cows standing around here and there and picturesque but uneconomic sheep wandering on the hillsides' (1976, p.218). Mitchell was particularly concerned here about the agricultural and landscape future of the west of Ireland but he serves to point out the major dilemma facing all traditional Irish agriculture - whether to change, to commercialise, even at a cost to the landscape, even at the cost to the rural society in areas, particularly in the west, where fully commercial agriculture within a competitive system could not flourish given the inherent constraints. Mitchell feared further abandonments and depopulation. The date of his extract, 1973, is significant because it marked the accession of Ireland and the UK into the EEC. As an agricultural region in the mid 1970s, Ireland achieved considerable benefit from EEC membership and a short-lived economic boom helped in the process already identified by Mitchell, the transformation of Irish agriculture at a cost to the historical legacy on the Irish rural landscape. For example, Mitchell said that for the more favoured east of the island 'larger units and larger machines will necessitate still larger fields and we may expect to see many of the hedgerows and their trees disappear, just as they are disappearing in England.' (1976, p.220). By 1985, as Duffy (1986) demonstrated, this had happened. He presents a map of field boundary changes in Co. Wexford to prove it, as part of a piece in which he states his concern over the effects of the radical rearrangement of the Irish countryside (p.63). The boom may be over but the effects of this era upon the landscape together with that of other eras, lingers on.

Much of Ireland's agricultural past can still be read on the ground but as circumstances change, so may the landscape. Many great estates have undergone alteration through time for example. Many have long since been broken up, much estate land was redistributed by the Congested Districts Board for instance. Some have been taken into state care, Glenveagh, estate of the Adair family, is now a National Park with strict controls imposed on the use of its land. In Northern Ireland the Belmore estate of Castlecoole in Co. Fermanagh and the Londonderry's establishment at Mountstewart, Co. Down are amongst others now belonging to the National Trust (Figure 3).

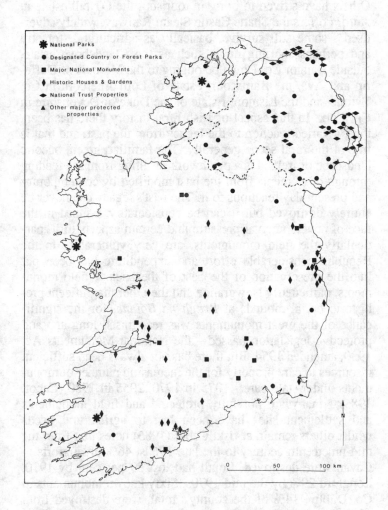

Figure 3. Protected parks and property in Ireland

Other estates have rented out or sold, the land and the houses where they remain are country mansions, often now occupied by business people or institutions, the former landscape and agricultural unity of the estate having been lost. The Ulster Folk Museum at Cultra is an example here. Others have striven to bring in tourists - the O'Neill estate in Antrim offers the Shane Castle Steam Railway, widely advertised - some still survive, basically as agricultural, forestry and perhaps sporting units, such as the Clandeboye estate, outside Bangor Co. Down belonging to the Marquis of Dufferin and Ava, the Barnscourt estate of Lord Abercorn in Co. Tyrone, and the Lismore Estate of the Duke of Devonshire in Co. Cork. In this aspect of Irish rural history there has been, thus, a varied reaction to the legacy from the past, and that is true of the rural scene generally. The familiar, green, bucolic landscape of Ireland is a patchwork resulting from the feeding through of elements from the past modified by contemporary and present day reactions to it. The past's legacy can never be entirely destroyed but it can be considerably reduced in the face of developmental pressure and certain aspects of it, particularly the field monuments, are very vulnerable. In the Republic considerable effort and expenditure has been put into the preservation of the best of the ancient field monuments, particularly Newgrange and the other magnificent prehistoric burial mounds at *Brugh na Boinne*, for the significance of the great monuments was recognised long ago and protective legislation passed - the National Monuments Act 1930, amended 1954. But there has not always been sufficient resources to care properly for the increasing number of monuments under state care - 1075 in 1970, 2055 in 1980 (Cabot, 1985) - many are not fully protected and field monuments and settlement sites have been lost to agricultural needs whilst others remain at risk. Cabot (1985) notes that from the mid nineteenth century to the late 1960s, 46% of ringforts in Cavan were destroyed, Louth had lost 39 per cent by 1970, Longford 29% by the mid 1970s. Sixty known monuments in Co. Dublin, 14% of the county's total, were destroyed from 1930, the year of the National Monuments Act, to 1975. The Dublin losses were largely to urban expansion but 80% of ringfort destruction has been to agricultural activity. Moated sites and standing stones have also been destroyed, or greatly modified; 52% of the former in Carlow, Kilkenny, Tipperary, and Wexford have been lost since the 1840s (Barry, 1978). Barry (1979) points out that even in Northern Ireland which has greater legal powers of protection for monuments than the Republic, under the Historical Monuments (Northern Ireland) Act 1971 only 500-600 out of c.15,000 field monuments are protected and he compares Ireland as a whole very unfavourably to Denmark, half its size, where c.2,500 prehistoric monuments are in state care.

'Had economic growth continued . . . I tremble at what that might have meant in impact upon the scenic, natural and historic heritage' (Dower, 1984, pp 17-18). In Ireland, there remains, then, a conflict between development and conservation in the rural as well as the urban areas. In the former, as in Dower's quote, this conflict is usually widened to include scenic beauty and pollution as well as the material heritage from past time, ancient monuments and country houses alike. An Taisce (1977) has produced a report on the Republic's historic houses with the alarming title, *Heritage at risk*. Development is no longer seen as being without cost. Even the example of Scotland's Highlands and Islands Development Board, in particular its EC Integrated Rural Development Scheme for the Western Isles, often seen as a worthwhile development model at scales ranging from rural Ireland in general (Conway and O'Hara, 1984) to the Irish offshore islands (Royle, 1986) recently has been criticized in Ireland because of 'an excessive promotion of land reclamation and the very low priority given to spending on the protection of the environment.' (Armstrong, 1986, p.73). This fear about the threats to the environment as a whole is becoming increasingly a positive force for conservation.

Rural Conservation in Northern Ireland

As recently as 1970 Eggeling could present a paper at a conservation conference entitled 'Why conserve in Northern Ireland?' , but there are now what can only be described as a plethora of conservation bodies in the province, so much so that there has been a need to coordinate their efforts and an umbrella organization, the Northern Ireland Environment Group has been set up. One of the most important of the independent groups is the Ulster Society for the Preservation of the Countryside founded in 1937 on the model of the Council for the Preservation of Rural England of 1926 after the usual time lag for diffusion across the North Channel. At that time there were worries about urban expansion ('a gimcrack civilization crawls like a gigantic slug over the country leaving a foul trail of slime behind it' (Marshall, 1938)); crude afforestation; thoughtless roadbuilding, and power line erections, all of which put the countryside under pressure. It was a period of revival for outdoor living: the Ramblers' Association and the Youth Hostel Association were founded in the 1930s. The Kinder Scout trespass took place then, when deliberate, mass trespass by ramblers onto private land helped publicise the need for freer access to the countryside. Pressure for National Parks in Great Britain began then and they were proposed for Northern Ireland in the report, *The Ulster Countryside* in 1947, as were areas of special control, nature reserves and field museums. As Wilcock and Guyer have summarized recently (1986, also Buchanan 1982), little happened until the Abercorn committee of 1962 reiterated the plans and the Amenity Lands Act (Northern Ireland) of 1965 put some of them into practice. Northern Ireland now has several designated Areas of Outstanding Natural Beauty but still no National Parks. AONBs are under the control of local authorities with the Ulster Countryside Committee set up also under the 1965 Act as an advisor, but nationalist politicians especially feared that the designation of National Parks would reduce local powers and might hinder economic development by imposing upon their areas unrealistically high environmental safeguards (Wilcock and Guyer, 1986). The beautiful Erne Basin in Co. Fermanagh is only recently an Area of Special Control though the 1947 report recommended it become a National Park. A Fermanagh District Council officer stated at a conference on rural planning that official policy in this area 'was seen as an imposition by an insensitive government located several hundred miles away completely immune to the way of life and to the wishes of the local rural population' (Burns, 1984, pp.200-1). However, two of the county's finest houses, Florencecourt and Castlecoole are National Trust properties, Castle Archdale is one of the province's Department of the Environment Country Parks and Fermanagh also has two of the Department of Agriculture's Forest Parks one of them around Florencecourt and running up to the spectacular summit of Cuilcagh on the Irish border.

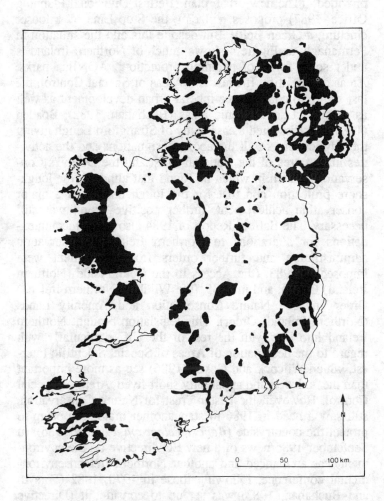

Figure 4.   Protected land in Ireland

After the Cockcroft Report of 1978, more building in the countryside has been allowed and from the mid-1970s the planning restriction around Belfast, the Stop Line, imposed by the Matthew Plan of 1964 has been breached, most notably with the Poleglass estate on the southwest edge of the city which allowed further housing for Catholic families to be provided. (The new urban plan (Belfast Divisional Planning Office, 1987) proposes replacing the Stop Line by a looser cincture, a Green Belt). But despite this and the situation in Fermanagh, as Figure 4 shows, much of Northern Ireland is under some form of planning protection, AONBs, parks, Areas of Scientific Interest and Areas of Special Control, the last set up to constrain peripheral urban development as well as to protect some scenic areas (Buchanan, 1982). Boaden (1984) has presented a case study of Strangford Lough giving the implications of all the above designations and the activities in that area of the National Trust and the DoENI's Conservation Branch in which he found that although the loughshore protection and that for the lough itself was a major conservation achievement, further positive action was still necessary. The Balfour Report of 1984 also made recommendations for alterations to Northern Ireland's conservation administration and further orders from Parliament were imposed in 1983, (the Access to the Countryside (Northern Ireland) Order), and in 1985, (the Wildlife (Northern Ireland) Order and the Nature Conservation and Amenity Lands (Northern Ireland) Order). This legislation brought Northern Ireland into line with the rest of the UK, particularly with regard to the designation of Areas of Special Scientific Interest, which Wilcock and Guyer (1986) see as more important than the 'unloved and apparently short-lived' Areas of Special Control. However, the Ulster Trust for Nature Conservation, still saw a need in 1986 to start another major campaign to protect the countryside (*Belfast Telegraph* 19 June 1986). In September 1986 news of a new EC directive on the environment was announced and another Northern Ireland environmental conference, following those in 1970, 1982 (Forsyth and Buchanan, 1983) was set up to consider it (Directive (85/337/EEC) - how it will change the way we live (*Belfast Telegraph* 10 September 1986). Furthermore 1987 was designated European Year of the Environment.

Rural Conservation in the Republic of Ireland

Under its different government, the Republic of Ireland, of course, has developed independent rural conservation practices (Convery, 1978) although EC directives will affect both parts of the island and Cabot (1984) recently called upon the Republic to develop some aspects of conservation legislation along similar lines to that in the North. There has been the same halting progress towards greater protection of the environment. An Foras Forbartha in its report, *The Protection of the National Heritage* (1969), pointed out the continued need for conservation although the requirements of the Local Government (Planning and Development) Act of 1963 had already instructed each of the county authorities to prepare development plans taking into account areas of outstanding landscape based on scenic, historic, architectural or scientific grounds. Some authorities so designated a large part of their areas - Wicklow 51%, Mayo 36%, Galway 29% etc. - and the national figure was 17.6%, not dissimilar from the proportion of specially protected land in Great Britain which was 16. 7% (An Foras Forbartha, 1977). This designation did not provide complete protection, leaving the local authority open to a claim for compensation if it turned down a planning application on amenity grounds alone. It was possible to declare a Special Amenity Area Order to overcome this problem but though Dublin Bay was so protected in 1964 and 'several planning authorities (were) considering others' in 1977 (An Foras Forbartha, 1977, p.12) no others had been set up by 1984. Many of the areas designated by the counties featured again in An Foras Forbartha's *Inventory of Outstanding Landscapes in Ireland* (1977) though this work, which was in response to a request from the International Union for the Conservation of Nature and Natural Resources in 1972, could also benefit from the Conservation and Amenity Advice Service set up in 1971 and the Conservation orders and Tree Preservation Orders of the 1976 Local Government (Planning and Development) Act. A conservation weakness remains in that the use of land for agricultural and forestry purposes is exempt from planning control, but a strength compared to the situation in the North is that the Republic has been able to set up several national parks. These areas containing fine landscapes and in some cases important houses and monuments, are strictly conserved. They consist of parks in Killarney (8,038 ha established 1932), Connemara (2,699 ha, established 1980), Glenveagh (9,667 ha established 1975 with the addition of Glenveagh Castle and its gardens in 1981) and the Burren (410 ha, under preparation). The National Parks and Monuments Service also maintains monuments and other properties. Additionally, Phoenix Park in Dublin, (709 ha) has been under public management since 1829 and the Commissioner of Public Works also has a role in the management of the Curragh grasslands of Co. Kildare (Craig, 1984). Also, by 1981 about 1,000 areas amounting to 231,500 ha had been categorised as Areas of Scientific Interest, many already protected by the counties or under the nature reserves set up by the Wildlife Act, 1976 (An Foras Forbartha, 1981). However, listing is not always a guarantee of protection in the field and a spot survey of 26 of these sites in 1983 in Donegal, Wexford, Galway and Cork found 16 damaged or adversely affected whilst a more detailed survey of Donegal found over a fifth damaged and a number threatened by drainage, construction, bog development, water pollution and the provision of sports facilities (Cabot, 1984), Mawhinney's conclusion about the Republic of Ireland's physical planning process was that it 'has had a noticeable though circumscribed amount of success in environmental control functions' (1984, p.98). An Foras Forbartha remains concerned about capitalist pressures: 'the market may possibly be relied upon to generate economic growth; unaided it certainly will not regulate environmental quality' (Cabot, 1985 p.2).

CONCLUSION

Ireland remains a land that looks to its past. Certain aspects of this past are not attractive and some would say that Ireland has become a land haunted by history. During the course of the preparation of this chapter the newspapers were full of the latest three killings of the troubles, carried out in the city in the northwest of Ulster whose very name cannot be written or spoken without the choice from the two alternatives used becoming a statement of the historico-political beliefs and traditions of the person speaking or writing. That type of legacy from Ireland's non-material history will not disappear; more pleasingly nor will other aspects such as Ireland's literature, culture, or music, which will continue to be cherished. With regard to the important material legacy of this ancient land, destruction of landscape elements from hedgerows to houses continues. The pressures leading to change or potential change do not diminish. Population redistribution and in some areas, growth, development, conflict, industrial and commercial activity and change, the ever present forces of time and decay - all these factors and others can have a deleterious affect upon the structures of the past. However, there are signs that the considerable recent efforts by both legislatures and pressure groups affecting both parts of the island are having some constraining effects.

REFERENCES

Aalen, F.H.A. (1978) *Man and the Landscape in Ireland*, Academic Press, London.

Andrews, J.H. (1975) *A Paper Landscape : the Ordnance Survey in Nineteenth Century Ireland*, Clarendon Press, Oxford

An Foras Forbartha (1969) *The Protection of the National Heritage.* An Foras Forbartha, Dublin.

An Foras Forbartha (1977) *Inventory of Outstanding Landscapes in Ireland*, An Foras Forbartha, Dublin.

An Foras Forbartha (1981) *Areas of Scientific Interest in Ireland*, An Foras Forbartha, Dublin.

Architectural Association of Ireland (1975) *Architectural Conservation: An Irish Viewpoint*, The Architectural Association of Ireland, Dublin.

Armstrong, W.J. (1986) Integrated rural development in Western Europe - theory and practice, in Breathnach, P. and Cawley, M.E. (eds), *Change and Development in Rural Ireland*, Geographical Society of Ireland Special Publications No 1, Maynooth, 69-76.

Automobile Association (1972) *Treasures of Britain and Treasures of Ireland*, Drive Publications, London.

Balfour, J. (1984) *A New Look at the Northern Ireland Countryside: A Report prepared for the Parliamentary Under-Secretary*, Department of the Environment Northern Ireland, HMSO, Belfast.

Barry, T.B. (1978) The Destruction of Moated Sites in Southeast Ireland, *Moated Sites Research Group Report*, 5.

Barry, T. B. (1979) The destruction of Irish archaeological monuments, *Irish Geography*, 12, 111-13.

Beckett, J.C. (1976) *The Anglo-Irish Tradition*, Faber and Faber, London.

Belfast Divisional Planning Office (1987) *Belfast Urban Area Plan 2001:Preliminary Proposals*, HMSO, Belfast.

Binney, M. (1980) Dublin's vulnerable beauty, *Country Life*, 7 July, 208-9.

Birrel, W.D., Hillyard, F.A.R., Murie, A. and Roche, D.J.D. (1971) *Housing in Northern Ireland*, Centre for Environmental Studies, London.

Boaden, P.J.S. (1984) Strangford Lough - a conservation history, in Jeffrey, D.W. (ed.), *Nature Conservation in Ireland : Progress and Problems*, Royal Irish Academy, Dublin, 57-66.

Boal, F.W. (1969) Territoriality on the Shankill-Falls Divide, Belfast, *Irish Geography*, 6, pp 30-50.

Boal, F.W. and Royle, S.A. (1986) Belfast: boom, blitz and bureaucracy, in Gordon, G. (ed.) *Regional Cities in the U.K. 1890-1980*, Harper and Row, London, 191-216.

Brady, J.E. (1986) The impact of clean air legislation on Dublin households, *Irish Geography*, 19, 41-4.

Brett, C.E.B. (1967) *Buildings of Belfast 1700-1914*, Weidenfeld and Nicolson, London.

Brett, C.E.B. (1986) *Housing a Divided Community*, Institute of Public Administration, Dublin.

Brown, S. (1985) Smithfield Shopping Centre, Belfast, *Irish Geography*, 18, 67-9.

Buchanan, R.H. (1982) Landscape, in Cruickshank, J.G. and Wilcock, D.N. (eds) *Northern Ireland : Environment and Natural Resources*, Queen's University, Belfast and The New University of Ulster, Coleraine, 265-89.

Burns, G. (1984) A commentary on a local situation in Northern Ireland, in Jess, P.M., Greer, J.V., Buchanan, R.H. and Armstrong, W.J. (eds) *Planning and Development in Rural Areas*, Queen's University, Belfast, 199-208.

Cabot, D. (1984) The future of nature conservation in Ireland, in Jeffrey D.W. (ed.) *Nature Conservation in Ireland : Progress and Problems*, Royal Irish Academy, Dublin, 161-74.

Cabot, D. (1985) *The State of the Environment* An Foras Forbartha, Dublin.

Contacuzino, S. (1975) *Architectural Conservation in Europe*, Watson-Guptill Publications, New York.

Convery, F.J. (1978) Some environmental policies - reviews and outlook, in Dowling, B.R. and Durkian, J. (eds) *Irish Economic Policy: a Review of Major Issues*, Economic and Social Research Institute, Dublin, 355-88.

Conway, A. and O'Hara, F. (1984) Integrated rural development in the West of Ireland, in Jess, P.M., Greer, J.V., Buchanan, R.H. and Armstrong, W.J. (eds) *Planning and Development in Rural Areas*, Queen's University, Belfast, 103-54.

Craig, A.J. (1984) National Parks and other conservation

areas, in Jeffrey, D.W. (ed.) *Nature Conservation in Ireland : Progress and Problems*, Royal Irish Academy, Dublin, 122-37.

Cullen, L. (1979) *Irish Towns and Villages*, Eason, Dublin.

Dawson, G.M. (1984) Defensive planning in Belfast, *Irish Geography*,17, 27-41.

Department of the Environment for Northern Ireland (1978) *Review of Rural Planning Policy. Report of a committee under the chairmanship of Dr W.H. Cockcroft*, HMSO, Belfast.

Douglas, J.N.H. (1976) The irreconcilable border, *Geographical Magazine*, XLIX, 162-8.

Dower, M. (1984) Rural planning in the British Isles, in Jess, P.M., Greer, J.V., Buchanan, R.H. and Armstrong, W.J. (eds) *Planning and Development in Rural Areas*, Queen's University, Belfast, 13-30.

Dublin Architectural Study Group (1982) *Urbana - a Study of Dublin* An Taisce and the Heritage Trust, Dublin.

Duffy, P.J. (1986) Planning problems in the countryside, in Breathnach, P. and M.E. Cawley, M.E. (eds) *Change and Development in Rural Ireland*, Geographical Society of Ireland Special Publications No. 1, Maynooth, 60-9.

Eggeling, W.J. (1970) Why conserve in Northern Ireland, in Forsyth, J. and Boyd, D.E.K. (eds) *Conservation in the Development of Northern Ireland*, Queen's University, Belfast, 84-92.

Ford, L.R. (1978) Continuity and change in historic cities : Bath, Chester and Norwich,*The Geographical Review*, 68, 253-73.

Forsyth, J. and Boyd, D.E.K. (1971) *Conservation in the Development of Northern Ireland*, Queen's University,

Belfast

Forsyth J. and Buchanan, R.H. (1983) *The Ulster Countryside in the 1980s*, Queen's University, Belfast.

Friends of Medieval Dublin (1984) A policy for Medieval Dublin, in Bradley, J. (ed.) *Viking Dublin Exposed : the Wood Quay Saga*, O'Brien Press, Dublin, 168-77.

Garner, W. (1980) *Bray: Architectural Heritage*, An Foras Forbartha, Dublin.

Garner, W. (1981) *Ennis: Architectural Heritage*, An Foras Forbartha, Dublin.

Hendry, J. (1977) Conservation in Northern Ireland, *Town Planning Review*, 48, 373-88.

Houston, C.J. and Smyth, W.J. (1980) *The Sash Canada Wore: A Historical Geography of the Orange Order in Canada*, University of Toronto Press, Toronto.

Joyce, P.W. (1870) *Irish Local Names Explained*, (republished as *Pocket Guide to Irish Place Names*, Appletree, Belfast 1984).

Kearns, K.C. (1982) Preservation and transformation of Georgian Dublin, *The Geographical Review*, 72, 270-89.

Kearns, K.C. (1983a) *Georgian Dublin: Ireland's Imperilled Architectural Heritage*, David and Charles, Newton Abbot.

Kearns, K.C. (1983b) Dublin's crumbling fair city, *The Geographical Magazine*, LV, 80-5.

Linehan, S. (1986) Facts about Dublin's air, *Irish Times*, 3 July 1986.

Magnusson, M. (1978) *Landlord or Tenant: A View of Irish History*, The Bodley Head, London.

Marshall, H. (1938) The Rake's Progress, in Williams-Ellis, C. (ed.) *Britain and the Beast*, J.M. Dent, London.

Martin, F.X. (1984) Politics, public protest and the Law, in Bradley, J. (ed.) *Viking Dublin Exposed : The Wood Quay Saga*, O'Brien Press, Dublin, 8-16.

Matthew, Sir Robert H. (1964) *Belfast Regional Survey and Plan 1962*, HMSO, Belfast

Mawhinney, K. (1984) Physical planning for rural areas in the Repubic of Ireland, in Jess, F.M., Greer, J.V., Buchanan, R.H. and Armstrong, W.J. (eds) *Planning and Development in Rural Areas*, Queen's University, Belfast, 87-102.

McGreevy, J.P., Smith, B.J. and McAlister, J.J. (1983) Stone decay in an urban environment: observations from South Belfast, *Ulster Journal of Archaeology*, 46, 167-71.

McIvor, C. (1986) Dublin under a pall, *The Geographical Magazine*, LVIII, p.220.

Mitchell, F. (1966) *Vanishing Dublin*, Allen Figgis, Dublin.

Mitchell, G.F. (1976) *The Irish Landscape*, Collins, London.

Ministry of Home Affairs (1962) *Nature Conservation in Northern Ireland*, Report of a Committee under the Chairmanship of the Duke of Abercorn, HMSO, Belfast.

Newcomb, R.M. (1979) *Planning the Past: Historical Landscape Resources and Recreation*, Dawson, Folkstone.

Nolan, W. (1986) Some civil and ecclesiastical territorial divisions and their geographical significance, in Nolan, W. (ed.) *The Shaping of Ireland : The Geographical Perspective*, Mercier Press, Cork, 66-83.

Pierce, R. and Coey, A. (1984) *Taken for Granted : A Celebration of Ten Years of Historic Buildings Conservation*, Royal Society of Ulster Architects and Historic Buildings

Council, Belfast.

Royal Institute of the Architects of Ireland (1975) *Dublin: a City in Crisis*, Royal Institute of the Architects of Ireland, Dublin.

Royal Irish Academy (1979) *Atlas of Ireland*, Royal Irish Academy, Dublin.

Royle, S.A. (1982) Urban conservation in Aberdeen: a personal view, *The Planner*, 68, 47-9.

Royle, S.A. (1986) A dispersed pressure group: Comhdhail na nOilean, The Federation of the Islands of Ireland, *Irish Geography*, 19 92-5.

Shaffrey, P. (1975) *The Irish Town: An Approach to Survival*, O'Brien Press, Dublin.

Simms, A. (1984) A key place for Dublin, past and present, in Bradley, J. (ed.), *Viking Dublin Exposed: The Wood Quay Saga*, O'Brien Press, Dublin, 154-63.

An Taisce (1977) *Heritage at Risk*, An Taisce, Dublin.

An Taisce (1980) *Dublin's Future : The European Challenge*, Country Life , London.

The Ulster Countryside (1947) HMSO, Belfast.

Walker, B.M. and Dixon, M. (1983) *No Mean City: Belfast 1880-1914*, Friar's Bush Press, Belfast.

Walker, B.M. and Dixon, M. (1984) *In Belfast Town 1864-1880*, Friar's Bush Press, Belfast.

Wilcock, D.N. and Guyer, C.F. (1986) Conservation gains momentum in Northern Ireland, *Area*, 18, 123-9.

7 THE PROBLEMS OF RURAL IRELAND

Mary Cawley

INTRODUCTION

A general concern with rural areas as lived-in environments, in addition to their function as sources of employment and income, occupies much of the planning-related literature in Ireland as in many other Western European countries in recent years (Hodge, 1986; Cullingford and Openshaw, 1982; Commins *et al.*, 1978). This concern relates in part to the realisation that despite the widespread application of regional development policies in the era since World War II, many areas of the countryside contain sizeable populations which lag significantly behind both national norms and large proportions of their urban counterparts along a range of social and economic indicators of well-being (Curry, 1981; Orlando and Antonelli, 1981). Such indicators include employment opportunities and income, but educational attainment, housing standards, transport availability and general demographic structures, which in turn influence the quality of life experienced, have received increased attention from researchers (Joyce and McCashin, 1982). The reallocation of resources that has taken place from central sources to areas characterised by a poor agricultural resource base and lacking alternative sources of employment, often with a view to encouraging industrial development, has been inadequate in many instances to enable rural economies to keep pace with national growth rates (Keeble *et al.*, 1981, 1982). This failure to take

145

part in national growth trends is particularly the case in areas such as rural Ireland where farm structures and systems have inhibited full participation in the benefits of the Common Agricultural Policy (Cuddy 1981; Tracy, 1982).

Low farm incomes and inadequate levels of rural employment opportunities outside agriculture inevitably affect levels of disposable income within local economies with implications for housing standards and general living conditions (Commins *et al.*, 1978; Shaw, 1979). The absence in most peripheral areas of employment opportunities offering prospects of career advancement means that many of the younger better-educated members of communities continue to emigrate. The rural brain drain is a well-established phenomenon and is widely acknowledged to have severe inhibitory effects on local development in Ireland (O'Cinneide, 1985). A cycle of cumulative decline has become established in many instances. Where this cycle has been interrupted in recent years through community development efforts, it is of interest to note that the organisational abilities of innovative individuals and development-orientated educational programmes have been of central importance (Keane and O'Cinneide, 1986).

Apart from the evidence that disparities between urban and rural living conditions persist at an aggregate level, there is recognition also, in Ireland as elsewhere in Western Europe, that significant spatial variations exist within the countryside in terms of structural planning needs and social service provision (Cloke, 1983; Curry, 1980; Caldwell and Greer, 1984). Broad contrasts are apparent between 'pressured' areas of rapid population growth in the immediate environs of expanding urban centres and 'remote' locations which remain outside the effective zone of influence of large towns and cities (Duffy, 1978). In the pressured areas, controlling new residential and non-residential development to reduce negative visual and environmental effects and to maximise on economic efficiency is a prime concern (Shaffrey, 1985). In remote locations, by contrast, low density and dispersed populations prevail, some still experiencing decline, and the provision of social and consumer services assumes major significance for state, private and voluntary agencies (Cawley, 1986). Population thresholds are frequently so low that only a skeletal network of fixed service points remains thereby giving rise to long journeys to basic educational and medical facilities (Thomas, 1986).

Some of the problematic dimensions of life in contemporary rural Ireland are of recent origin, associated with increasing urban encroachment on rural land, a phenomenon that has been particularly marked in the Republic since the late 1960s. The continuing contraction of resident populations in upland, coastal and agriculturally difficult environments inland is part of a long-protracted process but has now reached a critical stage in many areas where the dominant elderly and low-income groups have greatest need of the services which have gradually been centralised in towns and villages. Agricultural improvement and farm structure reform have absorbed the efforts of rural planners in Ireland since at least the late nineteenth century. While the problems of rural Ireland are not necessarily all new, novel approaches are being adopted towards their alleviation in pursuance of broad trends within the Western European periphery. Such is the case in particular in relation to agriculture (Armstrong, 1986).

This chapter describes the principal dimensions of economic and social underdevelopment which characterise rural Ireland as identified in recent research publications. Some contrasts in the definition of problems and in the design of remedial strategies are apparent between Northern Ireland and the Republic associated with basic differences in government policy between the two jurisdictions. The distinctive features of each area are recognised but an attempt is made here to identify the common difficulties shared by both and to set them within a broad European reference context. The discussion deals first with issues of rural social and economic deprivation and then with specific aspects of structural and functional planning.

RURAL DEPRIVATION IN IRELAND

The social and economic problems of rural areas have for long occupied the attention of planners in Ireland, particularly in the Republic where employment opportunities outside agriculture have traditionally been fewer than in Northern Ireland. Thus, the Underdeveloped Areas Act of 1952 designated twelve western counties as meriting preferential industrial development grants as a means of attracting investment and so offsetting continued outmigration from agriculture and reconstituting imbalanced population structures. Sectoral issues have dominated in the definition of rural underdevelopment and in the design of strategies for its alleviation until very recently in both Northern Ireland and the Republic. Farm structure reform, increasing agricultural production and establishing an industrial base in the countryside have figured largely in the remedial policies pursued which have also included afforestation, fisheries development and tourism. The initiative and finance for these policies have come from central government and European Community sources thereby reflecting a strong 'top down' approach to planning.

Sectoral policies still dominate regional planning in Ireland. Nevertheless a growing awareness has emerged of the interrelated nature of the factors contributing to rural underdevelopment and of the need for multivariate strategies for its alleviation (Commins *et al.,* 1978; Conway and O'Hara, 1986). The involvement of community groups, including community councils and co-operative organisations, in overcoming local problems is also a novel dimension of emerging approaches to rural development (Commins, 1985).

A number of studies have documented the geographical distribution of deprivation thereby providing insights into the interrelationships between various social and economic dimensions of rural life and also permitting areas of need to be identified with some precision. The spatial pattern of levels of living in Northern Ireland, as revealed by Goodyear and Eastwood (1978), illustrates a broad east-west division, at the level of the 26 District Council and 83 local government areas. Using a summary index of four combined measures relating to residential overcrowding, household amenities, underemployment and the incidence of elderly people in the population a broad centre-periphery pattern was identified (Table 1, Figure 1). The most favourable living conditions occurred within 20 miles of Belfast. The worst conditions were present west of the river Bann, in south Down and in parts of North Antrim, reflecting a deterioration with distance from Belfast (Figure 1). Within the periphery, urban centres appeared as islands experiencing superior levels of living. The overall pattern led the authors of the study to conclude that 'a pattern of rural deprivation' exists in Northern Ireland. Deprivation was not exclusively a rural problem, however. Comparably low living conditions were present within Belfast, notably in the inner city and in some western wards, where further deterioration has taken place in recent years (Pringle, 1983).

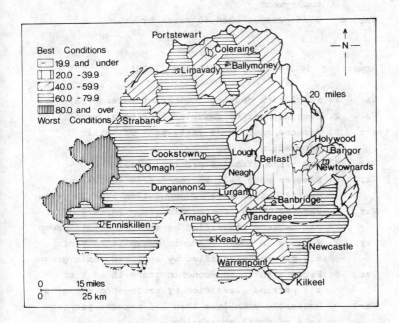

Figure 1. Northern Ireland: Levels of living by local government area (Source: Goodyear and Eastwood, 1978)

The broad spatial pattern of variation in living conditions identified by Goodyear and Eastwood (1978) was confirmed by Armstrong *et al.*, in 1980, using a larger number of measures (Table 1). With the electoral ward as the basic unit of analysis rural problem areas emerged which were characterised by a marked bias towards agricultural employment, high levels of overcrowding, low levels of educational attainment,

Table 1: Variables used in studies of rural deprivation

Goodyear and Eastwood (1978)

Var. 1   Average number of persons per room
Var. 2   % of private households without exclusive use of a fixed bath
Var. 3   % of economically active persons out of employment
Var. 4   % of persons aged 60 years or over

Armstrong, McClelland and O'Brien (1980)

Var. 1. Population density (number of persons/ha)
Var. 2. Persons over 65 years as % of total population
Var. 3. Persons aged 15-34 years as % of total population
Var. 4. Sex ratio (females per 100 males)
Var. 5. Persons with HNC, HND or degree equivalent qualification as % of population over 20 years
Var. 6. Persons in socio-economic group 13 as % of all persons classified by socio-economic group
Var. 7. Persons in socio-economic group 14 as % of all persons classified by socio-economic group
Var. 8. Persons in socio-economic group 15 as % of all persons classified by socio-economic group
Var. 9. Household with less than 1.5 persons per room as % of total households
Var. 10. Households with hot water, shower or bath and inside toilet as % of all households
Var. 11. Distance from Belfast

Cawley (1986)

Var. 1. Population change (1979 population as %
of 1971 population)

Var. 2. Population density (persons per km, 1979)

Var. 3. Migration (population aged 25-29 in 1971
as % of population aged 5-9 in 1951)

Var. 4. Age structure (0-14 age group as a
% of total, 1979)

Var. 5. (20-44 age group as a
% of total, 1979)

Var. 6. (65 and over age group as a
% of total, 1979)

Var. 7. Distribution of single males in population
(single males aged 45-54 as % of total
population, 1979)

Var. 8. Isolation (% of population aged 65 years or
over and living alone, 1971)

Var. 9. Socio-economic index (% of dwellings with
bathrooms, 1971)

Var. 10. (% of dwellings without
piped water, 1971)

Var. 11. (Cars per 1,000
population, 1971)

Var. 12. (Land valuation per ha
in IR£, 1971)

low levels of provision of household amenities and high unemployment. The locations involved accounted for approximately two-thirds of the land area and 15% of the population of Northern Ireland, embracing an extensive region stretching from northeast Antrim southwest to incorporate almost all of counties Fermanagh and Tyrone (Figure 2). Two compact areas in south Armagh and south Down displayed similar characteristics of underdevelopment as did a zone extending from the Tyrone-Armagh-Monaghan border to the south western end of Lough Neagh. A strong but not identical relationship with the distribution of 'less-favoured areas' designated as eligible for special farm aid under Directive 268 of the European Community of 1975, emerged, illustrating that deprivation is not solely an agricultural problem.

The interrelated nature of the social and economic structural problems of rural areas is apparent in the Republic of Ireland also (Cawley, 1986). Of 155 combined Urban and Rural Districts used as a base in a multivariate analysis, forty emerged as displaying strong features of deprivation as reflected in continuing population decline or very low rates of increase during the 1970s, low population densities, above average proportions of elderly people present, a depleted 25-29 age year cohort, below average housing conditions as reflected in levels of bathroom and piped water provision, and below average levels of motor car ownership (Table 1, Figure 2).

The deprived areas have a marked western distribution and include many coastal districts from Donegal to west Cork together with a sizeable inland belt extending from east Mayo through south Sligo, north and west Roscommon into Leitrim and Longford and with an upland rural district outlier in County Carlow (Figure 2). In general the areas involved are those where small farm structures dominate and the urban network is weakly developed. Nevertheless, the distribution is considerably less extensive than the 'less-favoured areas' classified under Directive 268 which include expanses of east Connacht, east Kerry and mountainous districts elsewhere. Differences in the areas used as a base in both instances, in the measures of living conditions applied, as well as the more stringent methods of measurement used in the study of deprivation explain these discrepancies.

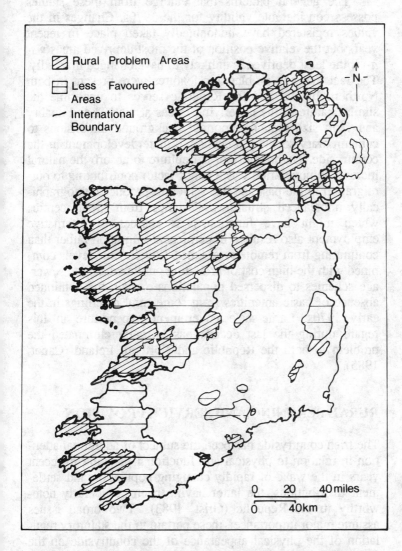

Figure 2. Northern Ireland and Republic of Ireland: rural problem areas and less favoured areas (Sources: Armstrong *et al.*, 1980; Cawley, 1986)

The general patterns that emerge from these studies possess considerable validity for the 1970s. Changes in the values registered have undoubtedly taken place in recent years but the relative position of the most deprived areas *vis-à-vis* the least deprived is unlikely to have changed markedly. The fact that key problematic features were present in both Northern Ireland and the Republic serves to underline the similar nature of the underlying forces at work. Land quality and farm size remain important background dimensions to contemporary social and economic underdevelopment in the countryside. The inability of agriculture to absorb the natural increase in population was a major factor contributing to out-migration in the past and has served to produce demographically imbalanced structures. Distance from urban centres which might have functioned as sources of alternative employment also resulted in permanent migration rather than commuting from remote rural areas. Low income levels combined with the high cost of providing public water and sewerage schemes to dispersed populations explain the continued absence of basic amenities from some rural dwellings in the early 1970s. Large-scale government expenditure in this regard during the last decade has virtually eliminated the problem in both the Republic and Northern Ireland (Cabot, 1985).

RURAL PLANNING AND SERVICE PROVISION

The Irish countryside has been the subject of additional attention in relation to physical and functional planning in recent years in the wake of rapidly changing population and settlement distributions, the latter having been particularly noteworthy in the Republic (Grist, 1983). Two broad issues assume major importance: these pertain to the statutory regulation of the physical appearance of the countryside on the one hand, and to the distribution of basic social and consumer services on the other (An Taisce, 1983; Mawhinney, 1984). Patterns of demographic change and distribution dictate that regulatory planning has hitherto attracted greatest attention in the vicinity of urban centres where growing populations create increased demand for residential land. In remote areas of diminishing population, lower income households, including the elderly, dominate and the enhancement of physical access to basic social and consumer services assumes major importance. The need for statutory regulation of the physical appearance of the landscape is no less important here, however (Buchanan, 1982; Shaffrey, 1985).

Countryside Planning

Statutory planning provisions relating to the location and design of physical structures in the countryside date to the 1960s in both the Republic and Northern Ireland (Caldwell and Greer, 1984; Mawhinney, 1984). The restriction of new residential development to the environs of existing settlements was inherent in the initial policy objectives formulated in both areas. The application of policy has, however, been much more stringent in Northern Ireland than in the Republic, with the result that notable contrasts exist in the physical appearance of the rural landscape in the two parts of Ireland. A key settlement strategy was applied in Northern Ireland until the late 1970s leading to the concentration of most new housing in the environs of existing towns and villages. In the Republic, by contrast, a stated commitment to containment failed to be upheld for a variety of reasons, and suburbanisation, including ribbonisation along access routes, proliferated during the late 1960s and the 1970s (Duffy, 1983).

The factors which guided planning policy in Northern Ireland from 1964 until 1979 included the need to preserve high quality agricultural land and natural amenities, to control development costs and to provide for access to communication networks, places of employment, services and commercial facilities (Ministry of Health and Local Government, 1964). Where permission for constructing a residence in a rural area was sought, evidence of the 'need' for such a residence had to be provided. This requirement effectively limited planning permission for new dwelling construction to farmers, part-time farmers, persons involved in other types of agricultural work or those who for poor health or some other exceptional circumstance warranted a home outside a nucleated settlement. These restrictions remained in force until the late 1970s when a major review of rural planning policy by the Department of the Environment (1978) pointed out that the provisions were more appropriate to densely populated areas of lowland Britain where control of urban sprawl rather than the preservation of rural communities was of prime concern. In 1978, it was decided that outside Designated Areas of Special Control, the restrictions would no longer apply. Areas of Special Control were to be defined in consultation with the District Councils and were to include parts, but not all, of the Areas of Outstanding Natural Beauty, Areas of Scientific Interest, the Belfast sub-region and areas close to all other urban centres (Caldwell and Greer, 1984). Future levels of development in the countryside were, however, to be guided closely by the size and proximity of the adjoining built-up areas so that the viability of existing towns would not be negatively affected.

In the Republic of Ireland, provision has been made for the regulation of the physical development of rural areas (places outside settlements of 1,500 population or more) since the early 1960s under the terms of the Planning and Development Acts of 1963, 1976 and 1982. The desirability of concentrating non-farm housing in the environs of existing settlements has been acknowledged since the beginning. Nevertheless, there has been a consistent failure to implement this policy. This failure has been attributed in part to the strong tradition of the isolated dwelling in rural Ireland (Mawhinney, 1984). The sense of relief, bordering on euphoria, engendered by the return of population to areas of countryside where outmigration had been the prevailing trend for generations is also forwarded as an explanation for the ease with which planning permission was granted for single dwellings, often located on half-acre sites, during the 1960s and the 1970s. The absence of a proper appreciation by the public at large of the negative aesthetic consequences of ribbon-type development must be recognised as contributing to the widespread roadside sprawl that took place.

The achievement of personal preferences for individual rural dwellings whether in isolated locations in the countryside or as a series on half-acre plots along approach roads to towns and villages was made possible mainly through two circumstances: a ministerial directive of 1973 which instructed planning authorities to facilitate such preferences

(Department of Local Government, 1973); and the operation of Section Four of the City and County Management Act of 1955 which permits local representatives to override the decisions of planning officials in advising county managers on development. This was designed initially for use in relation to council house allocation but it has been applied by some county councils in a discretionary way in relation to private housing. It is estimated that 11,000 detached dwellings are built each year (40% of all new houses in the early 1980s) and that most are located in rural areas (McCarthy, 1982). The numbers of such dwellings are greatest in Counties Cork, Donegal, Galway, Mayo and Wexford, which all contain areas of great scenic interest. Cork, Donegal and Galway, together with County Kerry have a notorious record of over-use of the Section Four instrument to override the planning control regulations (An Taisce, 1983). Since 1980, the desirability of concentrating new residential development in the vicinity of existing settlements has received new emphasis. A ministerial circular of that year urged the channelling of new public and private housing into smaller towns and villages which offer scope for the utilisation of existing physical and infrastructural facilities (Department of the Environment, 1980). In this way, it is hoped to encourage the maintenance of existing communities, to counteract the higher energy demands generated by isolated housing and to offset the imbalances created by the over-rapid development of large urban concentrations.

Rural Accessibility

While urban-fringe locations have become subject to increasing population pressure with its attendant implications for housing and ancillary services, areas of countryside outside an acceptable commuting distance of urban-based employment have continued to experience population decline through outmigration. Where stabilisation or growth occurred during the late 1970s, the numerical increases tended to be small (Horner, 1986). As a result, dispersed low density settlement patterns now dominate thereby contributing to substantial costs being associated with gaining access to the basic fixed-point service facilities that exist. A

157

combination of demographic and economic factors together with state centralisation strategies designed towards achieving improved provision and increased economies in public expenditure have served over the past twenty years to further deplete the network of social and consumer service provision points throughout the island of Ireland. One of the major effects of this process has been to place rural dwellers in remote areas at a severe disadvantage by comparison with their urban counterparts in terms of proximity to basic welfare facilities. Some of the major dimensions of this phenomenon include education, health, retail provision and public transport facilities.

The systematic closure of small one- and two-teacher schools in areas of depleted population structures became a feature of both Northern Ireland and the Republic during the 1960s, and in the former area even three-teacher units were subject to rationalisation. One-teacher schools declined from constituting 9% of all primary education units in Northern Ireland in 1958 to under 7% in 1968 and a further 300 small schools have closed since the 1960s. The phased closure of one- and two-teacher schools in the Republic was recommended in a report on Investment in Education published in 1965, and by 1977 1,800 such units had closed. The introduction of a free transport system for children living more than two miles from a primary school in 1967-68, a measure already in operation in Northern Ireland, facilitated the concentration of primary education in larger units in the Republic. Curriculum policy changes in both parts of the island involving the introduction to the primary school syllabus of environmental studies, art, and physical education, all of which required specially trained teachers, favoured larger school units. Difficulties in attracting and retaining highly trained teachers also presented problems in rural Northern Ireland.

Since the late 1970s, a re-evaluation of the policy of closure has taken place partly because of the cost of school transport systems. (In the Republic, travel charges have been in operation since 1983 for all but the children of low income families.) Social considerations have also received attention with special recognition being given to the role of the primary school in maintaining cohesion within local communities and its beneficial effects in the socialisation of young children within a familiar environment (Northern Ireland Economic Council, 1982).

Rationalisation within the primary health sector has been taking place in Northern Ireland since the 1950s, in line with general developments within the National Health Service in Great Britain. The majority of rural practices were still single-handed in the early 1970s, but it was becoming increasingly difficult to attract young doctors to sparsely populated areas and vacancies were emerging. As part of a major reorganisation of the health sector in 1973, community health services were integrated with primary care provision, and health centres, usually with a group practice, assumed new importance (Department of Health and Social Services, 1983). Because of the centralisation of facilities that has occurred, however, urban residents have benefitted disproportionately.

Centralisation of health care facilities in the Republic of Ireland date to 1971 when the dispensary system established in 1851 was replaced by a 'choice of doctor scheme' (Hensey, 1979). Many rural dispensaries closed during the early 1970s although difficulties had existed for some time in attracting young doctors to practise in remote rural localities. As a result of closure, access to medical facilities decreased for some patients who now had to travel several miles to attend their general practitioner's surgery in a town. Telephone ownership which would facilitate contacting GPs and ambulance services in emergency circumstances is notoriously low in remote western areas, particularly among the elderly.

The local shop, frequently with a public house attached, provided a very important service in rural areas in the past and often functioned as a social focus as well (Curran, 1972). The number of rural shops has declined continuously over the past two decades, although the closure of small outlets is a general feature of the retail grocery trade in Ireland (Parker, 1985). The demise of the 'country general shop' was noted by Dawson for Northern Ireland in 1970. Since then the main developments in retailing in the North reflect the trends noted for the Republic, with the emergence of major shopping centres in the vicinity of large centres of population dominated by multiple chains being the predominant pattern (Parker,

1984). The closure of the local grocery shop invariably causes deprivation for some members of its former clientele including the elderly, low-income households and mothers with young children (Standing Conference of Rural Community Councils, 1978).

The reduced levels of access to fixed services that result from closure and centralisation tend to be exacerbated by the generally low density route networks and infrequent public transport services that it is possible to provide in rural areas of sparsely distributed population. In both the Republic of Ireland and Northern Ireland, rail services serve mainly as a link between major centres of population. Many branch rail lines have been taken out of use since the 1950s resulting in a 50% decline in the total length of the network in the Republic by 1980. The absence of rail services is compensated for only in part by public bus provision where high operation costs mean that low route densities and infrequent schedules are the norm. Some villages have only a once weekly bus connection with a larger town and substantial numbers of people live many miles from a bus stop. Additional bus routes were introduced to replace withdrawn rail services during the 1960s but in many instances these serve as a social rather than an economic service as the McKinsey Report (1979) commissioned by Coras Iompar Éireann points out. As a result, heavy subsidisation of rural public transport has become a feature of the Republic and Northern Ireland in recent years (Department of Communications, 1985; Hewitt *et al.*, 1982).

OFFSETTING RURAL DEPRIVATION

Sectoral policies, notably in agriculture, still dominate approaches to rural development in Ireland as in many other countries of Western Europe. Changes in national policy are therefore viewed as an initial prerequisite for socio-economic advance. The analysis of structural problems within a regional context is particularly strong in the Republic and one of the outcomes is an increased level of awareness of the need for coordinated effort between various agencies at regional level in applying policy. Integrated Development Programmes which have been initiated in a number of marginal locations elsewhere within the European Community are increasingly being looked to as providing the optimum path to rural advance in Ireland also. At the same time considerable confusion exists as to what the term 'integrated development' means in practice (Armstrong, 1986).

Coordination of local and central decision making is recommended by Conway and O'Hara (1984, 1985) as the optimum method of overcoming the negative aspects of resource endowment, low levels of income and employment, and the imbalanced population structures that prevail in rural areas of the Republic. The basic objective of the suggested strategy is development rather than growth with an emphasis on unlocking the cycle of persistent disadvantage by giving the residents of underdeveloped areas access to occupations which would enable them to achieve a reasonable standard of living. Frequently this would involve providing off-farm employment to supplement inadequate agricultural incomes. The need for greater mobility of land, through sale or renting is stressed (Conway, 1985). The adoption of minimum standards of land use is also advocated, to be enforced through the imposition of a tax where such standards are not met. In this way it is expected that the supply of underutilised land for sale and renting would increase. Paid work on public schemes is envisaged as an alternative to the current system of payment of unemployment benefits to those out of work.

Agriculture, forestry, industry and tourism are all allocated roles in the development strategy proposed for Northern Ireland by Armstrong *et al.* (1980). Within the agricultural sector, increased price support from the EC is viewed as necessary as is rationalisation of farm structures and layout, improved productive capacity and reduced costs on farms. The promotion of rural industry is recommended as a means of supplementing low incomes on small farms. Forestry is seen to have a special role to play as a supplementary source of income for farmers in upland areas of low agricultural potential. The tourist potential of many problem areas is also very great, especially such designated Areas of Outstanding Natural Beauty as the Glens of Antrim, the Mourne Mountains, the Fermanagh Lakeland, the Sperrin Mountains and North Derry. Apart from their highly-prized scenic attractions, these areas also offer considerable scope for activities such as fishing, boating, sailing, walking, climbing and pony trekking. It is suggested that substantial emphasis should be placed on tourism as a provider of employment opportunities and generator of income, particularly in the remote areas of the countryside.

The case for an integration of effort, resources and of policy in pursuing rural development is supported also by community councils and co-operative organisations working independently or under the aegis of national government or European Community programmes such as that to Combat Poverty. One of the major obstacles that such community-based organisations face in obtaining information, advice and financial assistance relates to the need to consult a multiplicity of agencies at state and regional level (O'Cinneide, 1985). Such consultation is expensive both in terms of financial expenditure and time. A nationally coordinated policy for rural areas administered by a single agency would simplify the consultative process considerably.

OFFSETTING THE PROBLEMS OF RURAL ACCESSIBILITY

In both the Republic and Northern Ireland, many measures have been introduced by public, private and voluntary agencies to compensate for the low levels of fixed service facilities that exist in remote areas of the countryside through the provision of mobile services or through the adoption of innovative transport strategies. Much remains to be done, however. Mobile libraries and mobile banks are familiar features in many areas of the countryside although their value to those without easy access to transport to towns is not always fully appreciated by the public at large. The provision of mobile libraries is largely dependent on local authority priorities in expenditure in the Republic and the service is therefore not as widespread as in Northern Ireland. The postal delivery service is not designed as a social facility, although it frequently functions in that way with individual postmen and women selling stamps to and posting mail for people who live many miles from the nearest post office or mail box. Mobile shops survive in many remote rural areas, but decreasing densities of population are gradually increasing travel costs to a point where the maintenance and running of vehicles is no longer economically possible.

Among statutory agencies, the regional Health Boards in the Republic and Northern Ireland provide a range of home-based services for the elderly and infirm. These include regular visits by nurses to the chronically ill and the more general assistance with household chores provided by home helps. Social Service Councils in the Republic, which are partially funded from central sources but which also have a voluntary component frequently provide additional services for the elderly including Meals on Wheels in village locations.

The housing conditions of the rural elderly are a matter for very serious concern in the Republic of Ireland in particular because of the dominance of old structures in a very poor state of repair, subject to dampness and frequently lacking piped water and indoor sanitation. Public authorities provide remedial housing strategies in the form of essential repairs or more usually a pre-fabricated structure where an elderly person lives alone. Purpose-built village-based housing is also being provided on an experimental basis in a number of locations to bring the elderly closer to fixed services but strong attachments to the family home and land are often difficult to overcome. In theory, the elderly are well provided for in relation to public transport because of the availability of free travel passes in the Republic and on payment of a small fee in Northern Ireland. In practice, however, because of low route densities in remote rural areas and occasionally because of physical disabilities, public transport for the aged is generally insufficient. Experimental remedial transport services in the Republic include community buses operated in association with the health services and a postbus in County Clare. In Northern Ireland, taxis have recently been shown to function as an efficient suburban transport service in the Londonderry area.

It is rapidly becoming apparent that a comprehensive approach to the issue of rural service provision is necessary in the Republic and in Northern Ireland if major inequities between urban and rural populations are to avoided. Part of the solution in this regard lies in a more flexible approach to rural transport provision. This may involve the introduction of non-conventional vehicles such as minibuses on rural routes where capital and running costs are lower than for conventional buses, the use of school buses out of school hours, and greater flexibility with regard to private individuals obtaining insurance coverage to carry passengers for a fee, in areas where an adequate taxi or hackney service is not available. Such issues are currently under consideration in the Republic. A more imaginative use of mobile facilities that already exist, such as postal, milk and coal delivery services, is also desirable and is indeed a recommended strategy for remote rural locations in Great Britain (Moseley and Packman, 1984).

SUMMARY AND CONCLUSIONS

Rural Ireland is experiencing a series of problems which relate in part to the underlying resource structure, but which are influenced also by recent patterns of population change and the continuing decline of remote areas of the countryside. In both Northern Ireland and the Republic, the spatial distribution of rural deprivation reflects the underlying influence of the agricultural resource base and distance from urban-based employment opportunities. Partly for income reasons, but due also to a combination of other factors, the whole fabric of social and economic life for many rural populations is characterised by relative disadvantage *vis-à-vis* national norms. Many decades of experience illustrate that the pursuit of agricultural reform and industrial development by conventional measures is not adequate to reverse the downward spiral that is in train. Integrated approaches which will take account of the multidimensional nature of the problem are now recommended as providing the optimum path to development in rural Ireland. Such measures will require an innovative approach not only to designing remedial strategies but also in relation to their implementation. This will demand close co-operation between agencies and institutions which have hitherto functioned in virtual independence of each other.

A second series of problems relating to residential development, on one hand, and aspects of access to social and consumer services in remoter areas of the countryside, on the other, has emerged as also requiring urgent attention. Recent changes in the statutory planning regulations in Northern Ireland and the Republic as well as changing public attitudes with regard to environmental conservation, should help to offset the major issues that pertain to housing provision. The accessibility problem requires yet other solutions. Remedial strategies already in operation designed to offset the effects of distance include mobile services and innovative transport options. Progress in the latter context has been minimal to date, focusing mainly on pilot schemes. Yet, the experience of other European countries suggests that transport is the area in which innovative options must be considered because few opportunities exist for extending the network of fixed service points in the countryside.

Any remedial measures that are introduced to meet the development and planning needs of rural Ireland will require a level of co-operation between European Community, state, regional, local, private and voluntary agencies which has not hitherto taken place and it is here that the greatest challenge lies. As yet, no coordinated rural policy exists within the European Community as Clout (1984) points out and until such is achieved it is likely that the problems of rural Ireland will continue to be tackled in a piecemeal and ineffective manner. A new emphasis on Integrated Development Programmes in marginal areas of the EC to be funded by a doubling of structural funds over the years 1988-92 offers new hope, however, for the less-favoured areas of Ireland.

REFERENCES

An Taisce (1983), *Planning; Use and Mis-use of Section 4 Resolutions*, An Taisce, Dublin.

Armstrong, J.A., McClelland, D. and O'Brien, T. (1980) *A Policy for Rural Problem Areas in Northern Ireland*, Ulster Polytechnic, Belfast.

Armstrong, W.J. (1986) Integrated rural development in Western Europe: theory and practice, in Breathnach, P. and Cawley, M.E. (eds) *Change and Development in Rural Ireland*, Geographical Society of Ireland, Special Publications No 1, Maynooth, 69-76.

Buchanan, R.H. (1982) Landscape: the recreational use of the countryside, in Cruickshank, J.G. and Wilcock, D.N. (eds) *Northern Ireland; Environment and Natural Resources*, The Queen's University and New University of Ulster, Belfast and Coleraine, 265-89.

Cabot, D. (1985) *The State of the Environment*, An Foras Forbartha, Dublin.

Caldwell, J. and Greer, J. (1984) Physical planning for rural areas in Northern Ireland, in Jess, P.M., Greer, J.V., Buchanan, R.H. and Armstrong, W.J. (eds) *Planning and Development in Rural Areas*, The Queen's University, Belfast, 63-86.

Cawley, M.E. (1986) Disadvantaged areas and groups; problems of rural service provision, in Breathnach, P. and Cawley, M.E. (eds) *Change and Development in Rural Ireland*, Geographical Society of Ireland, Special Publications No. 1, Maynooth, 48-59.

Cloke, P.J. (1983) *An Introduction to Rural Settlement Planning*, Methuen, London.

Clout, H. (1984) *A Rural Policy for the EEC*, Methuen, London.

Commins, P. (1985) Rural social change, in Clancy, P., Drudy, S., Lynch, K. and O'Dowd, L. (eds) *Ireland a Sociological Profile*, Institute of Public Administration, Dublin, 47-69.

Commins, P., Curry, J. and Cox, P. (1978) *Rural Areas: Change and Development*, National Economic and Social Council, Report No. 41, Dublin.

Conway, A.G. (1985) Development and disadvantage in rural

areas; some issues for farm development, in Conway, A.G. (ed.) *Agricultural Economics Society of Ireland Proceedings 1984-85*, 1-27.

Conway, A.G. and O'Hara, P. (1984) Integrated Rural Development in the West of Ireland, theoretical and methodological issues, in Jess, P.M., Greer, J.V., Buchanan, R.H. and Armstrong, W.J. (eds) *Planning and Development in Rural Areas*, The Queen's University, Belfast, 103-50.

Conway, A.G. and O'Hara, P. (1985) *Integrated rural development: an Irish Case Study to Identify Strategies for the West Region*, An Foras Taluntais, Dublin.

Conway, A.G. and O'Hara, P. (1986) Education of farm children, *Economic and Social Review*, 17(4), 253-76.

Cuddy, M. (1981) European agricultural policy: the regional dimension, *Built Environment*, 7(3/4), 200-10.

Cullingford, D. and Openshaw, S. (1982) Identifying areas of rural deprivation using social area analysis, *Regional Studies*, 16(6), 409-47.

Curran, R.J. (1972) Ireland, in Boddewyn, J.J. and Hollander, S.C. (eds) *Public Policy towards Retailing; an International Symposium*, Lexington Books, Mass., 167-90.

Curry, J. (1980) *The Irish Social Services*, Institute of Public Administration, Dublin.

Curry, J. (1981) Rural poverty, in Kennedy, S. (ed.) *One Million Poor*, Turoe Press, Dublin, 93-110.

Dawson, J.A. (1970) The development of self-service and supermarket retailing in Ireland, *Irish Geography*, 6(2), 194-99.

Department of Communications (1985) *Transport Policy; a Green Paper*, Stationery Office, Dublin.

Department of Education (1965) *Investment in Education*, Stationery Office, Dublin.

Department of Health and Social Services (1983) *Regional Strategic Plan for the Health and Personal Social Services in Northern Ireland 1983-85*, HMSO, Belfast.

Department of Local Government (1973) *Planning Control Problems*, Circular PL 210/8, Stationery Office, Dublin.

Department of the Environment (1980) *Circular PD 2/80*, Stationery Office, Dublin.

Department of the Environment (Northern Ireland) (1978) *Review of Rural Planning Policy; Report of the Committee under the chairmanship of Dr W.H. Cockroft*, HMSO, Belfast.

Duffy, P.J. (1978) Population change in the Irish countryside, *Geographical Viewpoint*, 7, 20-33.

Duffy, P.J. (1983) Rural settlement change in the Republic of Ireland: a preliminary discussion, *Geoforum*, 14(2), 185-91.

Goodyear, P.M. and Eastwood, D.A. (1978) Spatial variations in level of living in Northern Ireland, *Irish Geography*, 11, 54-67.

Grist, B. (1983) *Twenty Years of Planning; a Review of the System since 1963*, An Foras Forbartha, Dublin.

Hensey, B. (1979) *The Health Services of Ireland*, Institute of Public Administration, Dublin.

Hewitt, V.N., McCarthy, C. and Crowley, J. (1982) *Transport in the Republic of Ireland and Northern Ireland*, Co-operation North, Belfast and Dublin.

Hodge, L. (1986) Rural development and the environment, *Town Planning Review*, 57(2), 175-86.

Horner, A.A. (1986) Rural population change in Ireland, in Breathnach, P. and Cawley, M.E. (eds) *Change and Development in Rural Ireland*, Geographical Society of Ireland, Special Publications No. 1, Maynooth, 34-47.

Joyce, L. and McCashin, A. (1982) *Poverty and Social Policy*, Institute of Public Administration, Dublin.

Keane M.J. and O'Cinneide, M.S. (1986) Promoting economic development amongst rural communities, *Journal of Rural Studies*, 2(4), 281-89.

Keeble, D., Owens, P.L. and Thompson, C. (1981) EEC, Regional disparities and trends in the 1970s, *Built Environment*, 3/4, 154-61.

Keeble, D., Owens, P.L. and Thompson, C. (1982) Regional accessibility and economic potential in the European Community, *Regional Studies*, 16(6), 419-31.

McCarthy, J.A. (1982) *Private Housebuilding in Ireland, 1976-81*, An Foras Forbartha, Dublin.

McKinsey International Inc. (1979) *The Transport Challenge*, Stationery Office, Dublin.

Mawhinney, K.A. (1984) Physical planning for rural areas in the Republic of Ireland, in Jess, P.M., Greer, J.V., Buchanan, R.H. and Armstrong, W.J. (eds) *Planning and Development in Rural Areas*, The Queen's University, Belfast, 87-99.

Ministry of Health and Local Government (1964) *Proposals for the Erection of Subsidy Houses in Rural Areas*, Circular No. 56/64, HMSO, Belfast.

Moseley, M.J. and Packman, J. (1984) *Mobile Services in Rural Areas*, Geo Publications Ltd, Norwich.

Northern Ireland Economic Council (1982) *Public Expenditure Priorities:Education*, NIEC, Belfast.

O'Cinneide, S. (1985) Community response to unemployment, *Administration*, 33(2), 231-57.

Orlando, G. and Antonelli, G. (1981) Regional policy in EEC countries and Community Regional Policy; a note on problems and perspectives in developing depressed rural regions, *European Review of Agricultural Economics*, 8, 213-46.

Parker, A.J. (1984) Northern Ireland; Planned Shopping Developments, *Estates Gazette*, 272, 131-36.

Parker, A.J. (1985) Small shops in Ireland: the Government takes a hand, *Retail and Distribution Management*, 13,(4), 22-26.

Pringle, D.G. (1983) Mortality, cause of death and social class in the Belfast Urban Area, *Ecology of Disease*, 2(1), 1-8.

Scully, J.J. (1971) *Agricultural Development in the West of Ireland*, Stationery Office, Dublin.

Shaffrey, P. (1985) Settlement patterns: rural housing, villages, small towns, in Allen, F.H.A. (ed.) *The Future of the Irish Rural Landscape*, Department of Geography, Trinity College Dublin, 56-79.

Shaw, J.M. (1979) *Rural Deprivation and Social Planning, An Overview*, Geo Books, Norwich.

Standing Conference of Rural Community Councils (1978) *The Decline of Rural Services; A Report*, National Council of Social Services, London.

Thomas, C. (1986) Distance, time and rural communities in Ireland, in Thomas, C. (ed.) *Rural Landscapes and Communities*, Irish Academic Press, Dublin, 145-78.

Tracy, M. (1982) *Agriculture in Western Europe; Challenge and Response 1880-1980*, Granada, London.

8 AGRICULTURAL DEVELOPMENT

Desmond A. Gillmor

INTRODUCTION

Agriculture in developed countries is characterised by certain common trends and problems. With increasing affluence, a diminishing proportion of national expenditure is spent on food, and as more of this expenditure is absorbed in marketing and processing, the farmers' share declines even more. Governments may intervene to regulate and financially assist agriculture for social and political reasons and to promote the sector's contribution to national economic development. Stimulated by such aid, agricultural output expands by application of technology and other innovations, even though the farm labour force is diminishing. These processes lead to surplus production and to social and environmental repercussions. The escalating cost of the farm support, combined with general budgetary difficulties, often results in government curtailment of agricultural expenditure and production.

Irish farming has shared in these trends and problems but shows additional facets which make the process of agricultural development even more vital, complex and interesting. Although the role of farming has diminished as in other countries, it remains crucial to the Irish economy. Agriculture accounts directly for 16% of employment in the Republic of Ireland and 5% in Northern Ireland, with substantial additional part-time farming and ancillary processing and servicing. There is a regional dimension in that agriculture's share of

171

direct employment in the less-developed western areas is over one-quarter. The contribution of farming and processing to gross domestic product is about 18% in the Irish Republic and 10% in Northern Ireland. Irish agriculture is strongly export-oriented, with more than half of production being sold outside the country, so that external trading environments and international linkages are of great importance. It is clear that the political and economic contexts within which farming operates are critical. In this respect, Irish agricultural development in the twentieth century is of particular interest, developing initially within one political unit and then within two separate sovereign states from partition in 1922. However, since accession of both the Republic of Ireland and the United Kingdom (UK) to the European Communities (EC) in 1973, Irish agriculture has fallen within the framework of the Common Agricultural Policy (CAP).

Agriculture in the Republic of Ireland and Northern Ireland is basically similar, comprising predominantly livestock production on medium and small owner-occupied farms. This resemblance is to be expected, given the shared island location, largely common heritage and similar physical environment. The mild moist climate and the nature of the soils, combine with traditional practices and market outlets, to favour pastoral farming. Thus 80% of agricultural output in the Republic of Ireland in 1985, and 73% in Northern Ireland, was based on grass production. In the same year, total agricultural land covered over four-fifths of the island, and of this area, 51% was improved pasture for grazing, 22% was grassland that was mowed for conserved fodder and 18% was rough grazing which was mainly on upland and bog. Only 10% of the improved agricultural land was under arable crops. The limited amount of land with a wide use range places a major constraint on the choices available to farmers.

POLITICAL AND ECONOMIC FRAMEWORK

The commercial evolution of Irish agriculture occurred within the political and economic context of the UK, with rural Ireland being a supplier of primary produce to the British market (Crotty, 1966; Baillie and Sheehy, 1971). The trading environment for the Irish Free State did not alter abruptly on partition, for it continued to have preferential access to the UK market. The divergence between the two parts of Ireland came mainly in the early 1930s, when the Irish Free State adopted a policy of maximum self-sufficiency in food production with protection of the home market and there was an economic dispute with the UK. This dispute was ended by an Anglo-Irish trade agreement in 1938 which granted duty-free access to the British market for agricultural produce. However, further difficulties were encountered in the early 1960s over import quota controls introduced in the UK because of increasing supplies from British and foreign farms. The Irish Republic's disadvantageous trading position was redressed, in part, under the Anglo-Irish Free Trade Agreement in 1965. There have been no trade barriers for Northern Ireland farmers as they retained their UK status.

Agricultural policies in the two parts of Ireland have differed because of contrasting state interests and objectives. The UK favoured consumers by following a cheap food policy because of its role as an industrialised country and major food importer. It allowed market forces to determine product prices, usually at international levels, and it made deficiency payments to bring farmers' receipts up to target levels. This system could be financed from the national budget due to the small share of agriculture in the overall economy. In contrast, the Irish Republic used trade restrictions and a limited price guarantee system to stabilise and raise prices for its producers. Commensurate with such expenditure on price support, the levels of production and development support, together with the range of measures, were much greater in Northern Ireland than in the Irish Republic. Total state supports, as a percentage of the value added in agriculture for 1970/71, were 68 in Northern Ireland and 28 in the Republic, and expenditure per male engaged in farming was more than three times greater in Northern Ireland (Cuddy and Doherty, 1984). These differentials had narrowed considerably from earlier years, mainly as a result of increased price supports in the Republic of Ireland, but this cost put a considerable burden on the exchequer. Differences in the prices received by farmers fell also, those in Northern Ireland being 19% higher in the 1940s and 7% in the 1960s (Sheehy *et al.*, 1981).

Much of the attraction of EC membership to the Republic of Ireland derived from the perceived benefits which would accrue to the agricultural sector under the CAP. These advantages were mainly in access to guaranteed markets at relatively high prices and the transfer of much of the burden of agricultural support from the Irish exchequer to the Community. Prior to 1973 there were major increases in the prices received by farmers, because of upward movement in anticipation of membership, and afterwards through harmonisation with EC levels over the five year transitional period. Increases up to the late 1970s were also associated with the annual farm price reviews of the EC and favourable movements in financial exchange rates. The gains to Northern Ireland farmers through access to markets, prices and government support were much less (Stanier, 1985). Also the UK policy of favouring consumers meant that its agricultural currency exchange rates were detrimental to farmers in Northern Ireland. Reversal of this policy since 1979, together with less advantageous currency movement in the Irish Republic, have switched the relative advantages between the Republic and Northern Ireland. Conversely, the differing emphases of both governments and the farmers' organisations have allowed some relative advantages to the Republic of Ireland under the CAP, although various beneficial measures instigated by the Republic were extended subsequently to Northern Ireland.

Although there are still some differences in policies and prices, CAP developments in the 1980s have been unfavourable to the agriculture of both territories. Some reorientation was needed because of the extent to which the EC conditions had promoted large production surpluses, huge increases in expenditure on agricultural support in the context of slow economic growth, and consumer prices above world levels. Curtailment by the EC is achieved mainly by minimising price increases to farmers and imposing limits to production. One measure, important in the Irish context, is the milk superlevy introduced in 1984, whereby milk production above certain levels is penalised heavily.

The bulk of expenditure under the CAP has been on prices and markets through its guarantee fund, rather than on structural, social and processing improvement through its guidance fund. Socio-structural policy in Ireland came to focus

on measures under two EC directives to promote farm modernisation and investment and to assist farming in less-favoured
areas, and much less on directives relating to socio-economic
advice and to retirement of farmers (Matthews, 1984). The
Farm Modernisation Scheme provided aid for capital
improvements and herd expansion, in the context of farm
development plans. The fact that low-income farmers did not
qualify for the highest levels of assistance was particularly
unfavourable to the west of Ireland. This restriction ceased
under the replacement Farm Improvement Programme introduced in 1985, which has upper income ceilings on qualification.

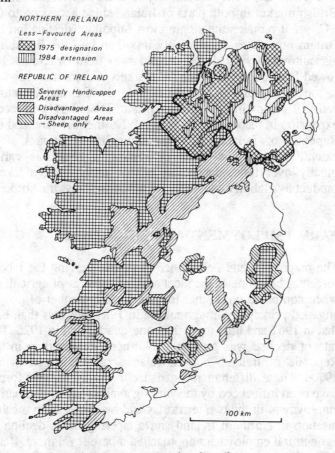

Figure 1 Less-favoured Areas for farming.

175

The Less-Favoured Farming Areas Scheme provides compensatory headage payments on cattle and sheep to farmers in areas of physical difficulty. Different levels of difficulty were distinguished from the outset in the Republic of Ireland, with the combined areas now covering over 60% of the state (Figure 1). The designated area had comprised 44% of Northern Ireland but it was extended to 75% under redefinition of the UK areas in 1984. Lower levels of assistance are given in the areas of lesser difficulty. EC concern for problem regions led to a drainage scheme and a development programme for the west of Ireland.

There has been substantially reduced dependence on the British market in both parts of Ireland since accession to the EC; yet it remains the single most important destination but Britain now takes considerably less than half the Republic's agricultural exports. Reorientation away from Britain has followed from access to other EC markets, Community subsidisation of exports to markets outside the EC and increased self-sufficiency in food within Britain. There has been recent expansion in recorded trade between Northern Ireland and the Republic of Ireland, although smuggling continues. The extent, direction and form of such smuggling has varied greatly over time, depending on the relative price levels, product availability and financial supports across the border.

FARM EMPLOYMENT AND STRUCTURE

The magnitude and persistence of people leaving the labour force has been one of the most striking features of agriculture when compared with other industries. The number of people engaged principally in agriculture in 1981 was less than half that in 1961 and little more than one-quarter that in 1926. The rate of decline was greater in Northern Ireland than in the Republic of Ireland, but the relative rates changed in the 1970s. These different rates of decline in the labour forces have been influenced by the varying availability of alternative employment in the two territories and by earlier farm mechanisation in Northern Ireland and a tapering of its decline as agricultural employment approached a base-level there. Rates of decline have differed according to occupational status; for example the fall in the number of farmers has been much less than the losses in agricultural labourers and in relatives assisting on the farm. Thus the agricultural labour force has become increasingly dominated by farmers, so that of people working full-time on farms in Northern Ireland in 1985, 80% were owners, 13% were family workers and 7% were hired workers.

Adoption of part-time farming has become a distinct feature of the agricultural labour force. This has been related to the growth in other employment opportunities and especially to the desire of small holders for alternative sources of income. There are major difficulties in assessing the role of part-time farming, but in 1985 30% of farm owners in Northern Ireland were classified as part-time, while Higgins (1983) estimated the proportion at 25% in the Republic. Cattle production is particularly favoured on part-time farms because of its low labour requirements. Its prevalence results in comparatively low productivity of the land. Part-time farmers, however, rely principally on their off-farm income, which affords living standards far above those obtainable from farming alone.

A defective tenancy system was abolished in the late nineteenth and early twentieth centuries, so that Irish land tenure today is based predominantly on owner-occupancy, mainly in family farms of small to medium size. There has been a trend towards larger farm holdings but it has been very slow. Reasons why this trend has been much less than suggested by the large fall in the agricultural labour force, include the lower rate of decline among farmers than family members and farm labourers, the growth in the number of part-time farmers (recorded under other industries in the census) and the separate enumeration of multiple holdings belonging to one person. The comparative immobility in Irish land tenure is, at least, in part, attributable to strong personal attachments to individual tracts of land and to the related continuity of family farming, including the tendency for those in other occupations to retain their land. The decrease in the number of farm holdings has been greater in Northern Ireland; reasons for this difference include the more commercial orientation of its farming, the greater opportunity and incentive to leave agriculture because of its smaller role in the economy and the fact that in the Republic many large estates were divided into small holdings by the Irish Land Commission. In the Republic of Ireland, the number of holdings over 2 hectares (ha) in 1980 amounted to 90% of that in 1960 and 80% of the number in 1931. The size of holdings had been much smaller in Northern Ireland than in the Irish Free State at the time of partition, but the size structure is now similar in both territories. However, there is more short-term letting of land on the conacre system in Northern Ireland, 18% as compared with 7% in the Republic, serving to increase the operational size of holdings. The mean size of holding on this basis is 25.4 ha in Northern Ireland and 22.6 ha in the Republic of Ireland. Although only 24% of operational holdings exceed 30 ha in the Republic, these occupy 56% of the farmland. The number of farm businesses in Northern Ireland declined by one-fifth between 1970 and 1985.

Major spatial variations in farm size occur within Ireland, the average area of improved land per holding in County Kildare being nearly three times that in County Mayo. Size tends to diminish northwestwards and small farms are a feature of the west, although local variations are widespread. The problems of small farms are compounded by their tendency to occupy poorer land, so limiting resources in terms of both quantity and quality.

AGRICULTURAL PRODUCTION

Because of the extent to which livestock dominate Irish farming, trends in their numbers may be used as a consistent indicator of the changing intensity of agricultural production (Figure 2). The different types of livestock were compared by using livestock unit equivalents (Gillmor, 1970). Horses were omitted from the computations because appropriate data were not available for all years and in any case their direct contribution to output was small. The base was taken as 1926, the year of the first complete and reliable agricultural census for the whole country after partition.

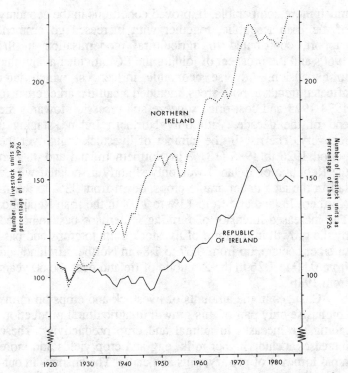

Figure 2 Trends in the numbers of livestock units, 1922-85

The numbers of livestock units in the two territories followed similar trends under the generally common circumstances of the 1920s, but there was marked divergence thereafter (Figure 2). Livestock production in Northern Ireland expanded in the 1930s and 1940s as part of UK agriculture, favoured by market growth and protection, and by state support and wartime conditions. Production in the Irish Free State suffered severely under the policies and the economic war of the 1930s, so that by the late 1940s the number of livestock units was less than in 1926. Growth in the 1950s was mainly on land released by decline in the number of working horses, so that only in 1960 did the total livestock units exceed those at partition. The Republic of Ireland had lost considerable ground, but subsequent trends in the two territories were

179

much more comparable. Improved conditions in the economy of the Irish Republic, together with increased government support, contributed to agricultural modernisation in the 1960s and the prospect of joining the EC provided a stimulus to expansion. Adverse economic influences, which later affected trends in both areas, included a cattle market crisis in 1974-1975 and cost-price squeeze and recession towards the end of the decade, with only Northern Ireland staging a recovery. The rise in the number of livestock units over the period 1926 to 1985 (131% in Northern Ireland and 48% in the Republic of Ireland) was only slightly associated with a fall in the area under arable crops, down from 23% to 9% in Northern Ireland and from 13% to 11% in the Irish Republic. The increased intensity of farming is perhaps best indicated by the growth in the ratio of livestock units to crops and pasture per hectare, up from 0.67 to 1.84 in Northern Ireland and from 0.80 to 1.26 in the Republic of Ireland over the 60 years from 1926.

Changes in the amounts of livestock and crops on farms emphasise only part of this growth in agricultural production, hiding the inceases in animal and crop productivity. These increases include higher milk, egg and crop yields and more rapid turnover of poultry, pigs and cattle. The changes in output of the principal commodities between the years 1960 and 1985 are shown in Table 1. For the individual categories, for example milk and cereals, the differences between the rates of change in output and in the corresponding numbers or areas indicate the relative magnitudes of productivity increase in the livestock and crops. However, these may be distorted by unrecorded cross-border movement, as with respect to sheep. In assessing the full extent of growth in agricultural output, individual commodities may be compared in monetary terms. This is done in the computation of the gross agricultural output, which is the value of production sold by farms or consumed in farm households. Gross output in real terms doubled in the Republic of Ireland between the years 1960 and 1985. Differences in data hinder comparisons with Northern Ireland but the output growth rate there was lower. However, gross output per hectare of crops and pasture in the Irish Republic, expressed as a proportion of that in Northern Ireland, was 75% in 1925, 32% in 1960 and 55% in 1985.

Table 1: Changes in livestock and crops and in output, 1960-85

	Percentage change in livestock numbers and crop areas	
	Northern Ireland	Republic of Ireland
Cows	+76	+62
Other cattle	+41	+38
Sheep	+45	-8
Pigs	-37	+6
Poultry	-3	-32
Cereals	-50	-12
Potatoes	-63	-65
Sugar beet		+23

	Percentage change in output of commodities	
	Northern Ireland	Republic of Ireland
Milk	+122	+116
Cattle	+66	+69
Sheep	-2	+39
Wool	+30	-16
Pigs	-33	-44
Poultry	+493	+312
Eggs	-21	-14
Cereals	+10	+106
Potatoes	-21	-21
Sugar beet		+38

Relationships in output per person were similar. These trends suggest that growth in production began earlier in Northern Ireland, but that since 1960 the differential in output has narrowed.

The expansion in agricultural production has been sustained from an almost constant amount of farm land by a reduced labour force with greatly increased investment. Extensive farm mechanisation has been the most obvious feature of the substitution of capital for labour and horses. The number of tractors more than trebled over the period 1960 to 1980 in the Republic of Ireland and increased by two-thirds in Northern Ireland. At the same time, machines became larger. The level of mechanisation, as indicated by the number of tractors per 1,000 hectares of improved land, remained higher in Northern Ireland in 1980, at 57 as compared with 31 in the Irish Republic. The rate of fertiliser application in 1985 was higher by one-half in Northern Ireland, even though the Northern Ireland increase since 1960 (74%) has been far below that in the Republic of Ireland (about 400%). Similar trends and differences may be discerned in the use of purchased feeding stuffs for livestock and in the construction and modernisation of farm buildings. The earlier capital intensification of Northern Ireland agriculture was facilitated by its more favourable market context and greater state support, but with the improvement in the environment for farming in the Republic of Ireland since 1960 the rate of growth has been greater there. However, capitalisation remains higher in Northern Ireland; in 1985 input costs amounted to 64% of gross output value compared to 47% in the Irish Republic.

Associated with greater purchase of inputs, there was huge expansion in the use of credit by Irish farmers. Because of increased expenditure on inputs, growth in net agricultural output has been less than in gross output, so that in the Republic of Ireland the rates of real increase over the period 1960 to 1985 were equivalent to 100% in gross output, 283% in certain farm material purchases and 67% in net output.

Trends in agricultural incomes have varied with levels and prices of inputs and outputs. Comparisons between Northern Ireland and the Republic of Ireland have been hindered by variations in data between the two areas. The results of investigation differ by the particular time period studied and with the individual authors' interpretations (Smith, 1948-49; Attwood, 1966-67; Symons, 1970; Furness and Stanier, 1981; Sheehy *et al.*, 1981; O'Sullivan, 1983-84; Cuddy and Doherty, 1984; Whittaker and Spencer, 1986). However, the conclusion must be drawn that the comparative advantages of farming in the two territories have altered with the internal and external relationships of the changing political and economic framework. From the 1930s to the 1960s, agricultural development and incomes were appreciably higher in Northern Ireland, being favoured by membership of the UK. Farming in the Irish Republic had been almost static in many respects until the 1960s but then expansion of other sectors of the economy, anticipation and realisation of EC accession and substantial real increases in agricultural prices in the 1970s brought rates of growth which were far in excess of those in Northern Ireland. With lower input expenditure, the income level rose above that in Northern Ireland. From 1979, farming in both areas suffered through difficult circumstances which included falling real prices for agricultural products, escalation in the costs of inputs and high interest rates. Farm incomes fell in real terms to less than those on accession to the EC and far below the level in other sectors of the economy. The situation was less severe in Northern Ireland than in the Republic because of (i) the more favourable exchange rate movements in relation to product prices, (ii) lower input cost inflation and (iii) greater control exercised by its more capitalised farmers over inputs, which, by 1985, were still lower than in 1978. Poor weather affected incomes in 1985 and 1986, emphasising the continuing role of the physical environment in Irish farming. By 1984-1985, mean agricultural incomes, both per hectare and per person, were almost identical in Northern Ireland and the Irish Republic.

CHANGES IN FARM ENTERPRISES

The changing political and economic framework has affected trends in the mixture of farm enterprises in the two territories. The agricultural consequences of partition were most evident in the divergent trends in arable crops and in pig and poultry production over the period 1925 to 1960 (Table 2).

Arable farming was protected and encouraged in the Irish Free State and this led in particular to the development of wheat and sugar beet production. While the area of tillage increased there by 8%, it declined by 34% in Northern Ireland. Imported cereals were available in Northern Ireland at world prices and this was a major asset to the pig and poultry industries; the purchases of feeding stuffs for them contributed to the high level of agricultural inputs. These farmyard enterprises were also favoured by subsidies and market access within the UK, by small farm size, and by organisation of marketing within Northern Ireland. Pigs and poultry accounted for more than 50% of the production value in 1950 and, although the contribution of poultry had fallen, the combined figure of 47% of output in 1960 contrasted with the decline to 19% in the Irish Republic (Table 2).

***Table 2*:** **The percentage of gross agricultural output, 1925, 1960 and 1985.**

	1925	1960	1985
Northern Ireland			
Arable crops	21.6	9.1	6.6
Milk	19.4	17.0	27.8
Cattle	23.2	21.1	40.8
Sheep	4.3	5.8	4.1
Pigs	9.3	28.3	10.5
Poultry	22.3	18.7	10.2
Republic of Ireland			
Arable crops	14.6	20.4	11.3
Milk	24.1	23.1	36.5
Cattle	24.3	30.3	38.8
Sheep	5.0	7.0	4.0
Pigs	16.0	11.4	5.7
Poultry	15.9	7.7	3.7

Over the period 1926 to 1960, pig and poultry numbers in Northern Ireland increased by 524% and 31% respectively, while the corresponding changes in the Republic were +8% and -39%. Enterprise expansion in the Republic of Ireland, apart from arable crops, was mainly in cattle. Production benefitted indirectly from UK subsidies, as store cattle exported to Northern Ireland and Great Britain qualified for payments, after a period of further feeding.

The striking features of enterprise change since 1960 have been the distinct trend towards convergence in the production of the two territories coupled with a joint tendency towards specialisation in cattle and milk production (Table 2). Although pig and poultry production remain more important in Northern Ireland, their roles have diminished greatly. In 1985, there were 42% less pigs and 28% less poultry than in 1970. These activities suffered from increased costs of feeding stuffs, at first in the 1960s through the establishment of a UK floor price for grain and more recently to a greater extent from the protected cereal markets of the EC. Within the UK, there was a shift in competitive advantage towards British producers because of their lower feed costs. Diminished profitability in the pigs and poultry enterprises within Northern Ireland accelerated the shift of farmers to cattle and milk production. As in the Republic, the dairy and beef industries were favoured by high levels of state support and by market opportunities, especially within the EC. Although relative growth was greater in Northern Ireland, especially in cattle production, the combined level of dominance by the beef and dairy industries in 1985 was higher in the Irish Republic, at 75% of total agricultural output, representing a high degree of product specialisation. Sheep tended to be a neglected enterprise until Irish lambs were given preferential access to the French market in 1977, and, more importantly, an EC common sheep policy was adopted in 1980. Sheep numbers increased over the period 1980 to 1985 by 50% in Northern Ireland and by 21% in the Irish Republic.

The trend towards fewer products in total agricultural output suggests that there might also have been increased specialisation at farm level. In fact, enterprise concentration has been much greater than would be expected from the overall trend in output or from the limited change in farm structure alone. Development over the period 1960 to 1980 in the Irish Republic together with a comparison for Northern Ireland in 1980 is indicated in Table 3, although all the data are not strictly comparable. For almost all enterprises, whether expanding or contracting, incidence was restricted to fewer holdings in 1980 than in 1960, while the scale of operation had increased on those holdings where an enterprise was practised.

Table 3: **The structure of farm enterprises, 1960 and 1980**

	Republic of Ireland, 1960	Republic of Ireland, 1980	Northern Ireland, 1980
Percentage of holdings with livestock or crop type			
Dairy cows	80	47	24
All cattle	88	88	85
Sheep	30	20	20
Pigs	38	5	21
Poultry	78	39	22
Cereals	69	33	26
Root and tuber crops	78	48	21
Average herd size or crop area (ha) per holding			
Dairy cows	5	15	27
All cattle	19	35	45
Sheep	49	76	122
Pigs	8	84	85
Poultry	54	110	1,269
Cereals	2.3	5.6	5.6
Root and tuber crops	0.9	0.8	1.6

Uniform data over a long period are not available for Northern Ireland, but Edwards (1980) showed that much concentration and specialisation had occurred in the mid-1970s. Over the decade 1970 to 1980, the size of most farm enterprises increased, with the mean number of dairy cows, pigs and poultry per holding approximately doubling. Between 1970 and 1984, the number of mixed farms declined by two-thirds, while the proportion of specialised types increased. It is evident that the overall levels of specialisation and enterprise concentration are more marked in Northern Ireland than they are in the Republic (Table 3). In both areas the degree of concentration varies by enterprise, being least in beef production and greatest in poultry and pigs. In Northern Ireland in 1985, the levels reached in the latter industries were such that 24 farms with over 100,000 broiler chickens accounted for 50% of all birds and 124 farms with over 1,000 pigs had 42% of the total pigs. Farm enterprises have become concentrated into fewer and larger units and it follows that the range at the individual farm level has diminished, allowing the traditionally diversified Irish farming system to undergo substantial simplification.

The processes of concentration and specialisation in farm enterprises result from numerous varying influences but with several major underlying causes. A prime objective for many farmers has been to increase efficiency, and to this end many have concentrated their resources on fewer enterprises in order to achieve economies. This tendency has been reinforced by the reduced labour force, technical advance and increased investment. The development of intensive management systems and cost-price squeezes in poultry and pig production have led to most small producer units being replaced by fewer large ones. Rises in government support and guaranteed prices have reduced the financial risks associated with specialisation. The extent of mixed farming has lessened along with a reduction in subsistence, and with the growth of part-time farming, which tends to be associated with simpler farming systems. Farmers have shown flexibility in adapting their farming patterns to changing circumstances.

CHANGING PATTERNS IN FARM ENTERPRISES

There is a highly complex regional pattern in Irish agriculture within its small land area; individual farm enterprises vary spatially in their distribution, nature and intensity. Such patterns have been mapped and discussed by Symons, (1970), Edwards (1977), Gillmor (1977, 1985), and Horner *et al.* (1984). Investigation of spatial change in the Republic of Ireland's agriculture over the period 1960 to 1980, demonstrated a distinct tendency towards greater areal concentration (Gillmor, 1987). Five of the seven enterprises investigated became more localised in their distributions, while no significant change occurred in the concentration of the greatly expanded milk and beef sectors. The distributions of all the enterprises became spatially more separated from each other. This resulting increase in the regional specialisation of agriculture relates to the process of enterprise concentration at the farm level. The spatial patterns and changes, however, are the outcome of a complex mix of physical and human influences, so that only some of the main features can be mentioned here.

Arable cropping has become more concentrated in the east of Ireland and on good land in places along the south coast and in the north. In these areas, arable farming is favoured by a drier and sunnier climate and lighter soils and by historical associations, as well as by the larger farm size with highly mechanised production. Dairying is principally in Munster and Kilkenny in the south and in Ulster in the north, but has been actively promoted and expanded throughout much of the country. Production is most specialised in areas of grassland on heavy clay soils which hinder alternative enterprises. Cattle production occurs more frequently and is more widespread than any other farm activity and also it displays a quite variable growth pattern. Reliance on cattle farming is greatest in the midlands and west, but cattle fattening is commonest on the large farms and fertile land of north Leinster and eastern Northern Ireland. Sheep are found most commonly in the uplands and adjacent districts, where they have an advantage compared with other enterprises under the difficult physical conditions, and many also occur on the dry limestone lowland of east Connacht. They had become more concentrated in the uplands due to less competition from alternative enterprises but also because of the payment of state subsidies there. However the recent expansion in sheep farming has also occurred on the lowlands. Traditionally poultry were an important supplementary source of income on small farms in the north and west, while pigs were associated with the surpluses of skim milk on dairy farms in the north and southwest. Changes in feeding practices and management systems have led to the pig and poultry industries becoming more concentrated both in area and by farm, the levels of concentration being greater than in any other enterprise. The principal areas in the Republic, Cavan for pigs and Monaghan for poultry, are cross-border extensions of the main production areas in Northern Ireland. Working horses have declined everywhere but last of all on the small farms of the west. Horses are now mainly for the breeding and racing industry, of which the chief centre is in north Kildare, and for recreational purposes, most commonly around urban areas.

DISPARITIES IN DEVELOPMENT

There are major differences in the levels and rates of development between farms and regions in Irish agriculture. Even the mean values for those categories recorded in the Farm Management Survey in the Irish Republic (Heavey *et al.*, 1984) indicate the extent of disparities in income, with the farm by farm variations within categories being even greater. The income per person was five times greater on dairy and tillage farms than on hill and lowland cattle and sheep farms, it was two and a half times higher on farms of over 80ha than on those of 12 to 20ha, while in Munster it was twice the level of that in Connacht. On the three soil groups distinguished in the Farm Management Survey, average incomes per hectare on the third and second categories of soils were only one-half to three-quarters of those on the top quality soils.

The process of modernisation and its impacts have varied considerably from farm to farm. The great extent of price support has been of much more benefit to the large producers and even investment aid has gone proportionately more towards the higher income farmers. In particular milk suppliers benefitted. The high income dairy and tillage enterprises became more concentrated on the larger farms. The scale of production increased, and also a growing proportion of output was contributed by large holdings, with the top farms showing dramatic advances. Half of total production in the Republic of Ireland is now accounted for by about one-sixth of farms. It was to the commercial and developing sector that research, advisory service, farmer organisation and other institutions were predominantly orientated. Younger, and better educated, farmers who were married and had children were more likely to avail of opportunities, to adopt modern technology and management practices, and consequently to increase production.

In contrast, there were many farms which were almost static or even declining, allowing recognition of a duality in Irish agriculture, although there is a gradation between the two types. Kelleher and O'Mahony (1984) estimated that 55% of farm operators had been left behind in the process of modernisation in the Republic of Ireland, where the problem is more acute than in Northern Ireland. Four-tenths of these operators had alternative income from off-farm employment, so that the social problem related mainly to the residue, which amounted to one-third of all farm operators. They were often smallholders, with poor quality land, and most had extensive low-income cattle and sheep farming systems. Many, but by no means all, were old and unmarried, widowed or without children at home. Poor living conditions and ill health were common. These farmers had been unable to adjust to the transformation of agriculture, so that the incidence of change and modern practices in their farming systems was low. This fringe sector was largely bypassed by modernisation and had not shared in its benefits, leaving a large number of farmers who continue to have unacceptably low incomes.

While the social and personal differences between farmers have been important in affecting the extent of adjustment between individual farms (Kelleher and O'Hara, 1978; Commins, 1980). yet the wider spatial influences have still been sufficient to impart distinct regional characteristics to the process of development. The extent of the spatial disparities in growth and distribution of agricultural incomes on a county basis in the Republic of Ireland is shown in Figure 3.

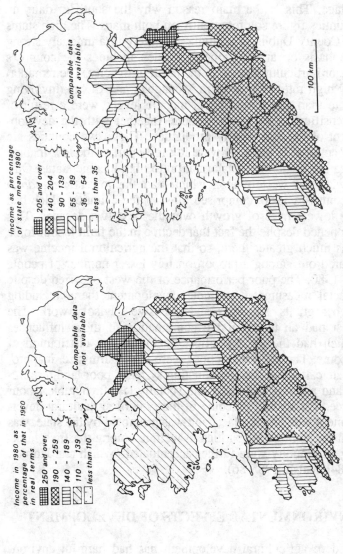

Figure 3 Agricultural income per male member of family, by county.

High growth rates and high levels of income were associated mainly with milk production and to a lesser extent with tillage. This is the main reason why the southern dairying counties figure so prominently on both maps. The high status of County Dublin is attributable largely to a uniquely strong emphasis on arable cropping, which includes the country's major horticultural area, to the north of the city. The remarkably high rates of growth in Monaghan and Cavan, involving more than a trebling of real income levels, were a result of intensification of dairying there, together with the development of pig and poultry production. These two counties are exceptional in that they were formerly part of the low-income area but made the transformation to high income status. Otherwise the western area which had the lowest *per capita* incomes in 1960, comprised most of the counties which had the lowest rates of growth over the next two decades. This happened despite the fact that decline in the farm labour force was much greater there, so that the agricultural income was shared out among a proportionately lesser number of people by 1980. The poor performance of the west occurred despite special measures to promote development there, including higher grants, subsidies and intensive advisory work. The west had an unfavourable combination of those influences which had the greatest effect on the spatial distribution of income. The main restrictions were small farm size, inappropriate cattle and sheep production systems, poor land resources and a demographically and socially less favourable labour force. The overall outcome of the type of agricultural development which took place, and was promoted by the state, was that the regional disparities in welfare among the Irish farming community were further accentuated (Walsh and Horner, 1984; Walsh, 1985-86).

ENVIRONMENTAL EFFECTS OF DEVELOPMENT

Modern agricultural development has had harmful environmental side-effects through the greatly increased levels of activity, the application of modern technology and the generation of large quantities of organic waste. The consequences include water pollution, destruction of wildlife, landscape deterioration, soil degradation and threats to human health. The increased intensification, concentration and specialisation in farming have resulted in an agricultural ecosystem which is much simpler, but consequently less stable, than the traditional mixed farming system. Yet the environmental impacts in Ireland have been less than in places where agricultural development is more advanced and intensive, and the pastoral component much less.

One of the changes in farming with the greatest environmental implication has been the replacement of the traditional haymaking by silage production, to the extent that silage accounts for four-fifths of fodder conservation in Northern Ireland. Silage effluent has a high polluting capacity and the volume of slurry generated by large livestock production units is another major source of contamination. This has followed mainly from the intensification of pig and poultry production and to a lesser extent from dairying and cattle fattening. The modern large buildings associated with these and other farm purposes are often an intrusive element in the rural landscape. The appearance of the countryside has also been affected in places by the removal of hedgerows, which also means a loss of wildlife habitat. Removal allows more efficient use of modern farm machinery, grassland management and a reduction in what is considered waste land. The extent of hedgerow removal and its consequences are less in Ireland than elsewhere in Europe because of the pastoral emphasis in farming and the high initial density of hedgerows. The greatly increased applications of commercial fertilisers have led to soil and water enrichment by nitrate and phosphate. The use of chemical biocides, antibiotics and hormones has serious implications for human health. Arterial land drainage has impaired wildlife habitats and even its economic benefits are now questioned. Hill reclamation has encroached upon the upland environment, with threats in relation to amenity value, wildlife and soil erosion. Apart from drainage and land reclamation, the environmental impacts of modern agriculture have been greatest in the better farming areas and associated more with the intensive farming of Ulster.

CONCLUSION

Agricultural development in Ireland has followed and shared common features with the process in more developed countries, as investigated for the EC by Bowler (1985). A smaller labour force has expanded output substantially by great use of capital and technical innovations. Scale of farm production increased through amalgamation of businesses, intensification and concentration of enterprises and greater specialisation. In Ireland as elsewhere, the benefits of farm modernisation have been unevenly distributed, between both farms and regions, and there have been harmful impacts on the environment. While political division has led to some divergence between Northern Ireland and the Republic of Ireland in policies and the form of agricultural development, the trend since the early 1970s has been towards convergence. This convergence has been related, in part, to increased international linkages, both through the CAP of the EC and also following the greater export dependency resulting from increased levels and specialisation of production (Cox and Kearney, 1983; O'Connor *et al.*, 1983; Sheehy, 1984). Further absorption into the international economic system means that the problems and prospects faced by Irish agriculture are increasingly similar to those of other countries, and the Irish experience becomes more relevant to economies at earlier stages of development.

The prospects for agriculture are uncertain, but not encouraging, and farmers will have to make difficult adjustments to changed circumstances. Much will depend on the prices and markets policies of the EC, which in the short term, at least, are likely to be cast within the contexts of production surpluses and budgetary constraints. Instead of increasing output further, the emphasis will be on reducing costs and improving efficiency. Rather than applying higher levels of expensive fertilisers, for example, many farmers could achieve improvements through use of clover and better grazing management. Some purchased feedstuffs may be replaced by grass and arable crops grown on the farm. There should be some shift of emphasis in production from quantity to quality, for which Ireland has an unrealised potential capable of exploitation. This could include some production by organic and other less intensive means. Maintenance of a high quality environment will be critical in the establishment of a reputation for pure food.

It is unfortunate for Ireland that both its considerable capacity to produce more and its comparative advantages for production were not realised more fully before controls on output became necessary. It is also unfortunate that the products to which it is highly suited and on which it has increasingly specialised, milk and beef, are ones which are in surplus and which may be injurious to health at high levels of consumption. The trends towards specialisation at national and farm levels may now be reversed but alternative enterprise options are limited. Sheep seem to offer the best prospects at present; for dairy and beef farmers there would be the advantage of more effective pasture utilisation through mixed stocking of cattle and sheep because of their differing grazing habits. The keeping of horses, deer and goats may expand. Plants grown for oil, protein, fibre and energy should figure amongst alternative crops. Agricultural production will have to be related more to the realities of market demands and trends and ideally would be cast within a global framework.

It is essential that agricultural development should not be viewed narrowly in isolation from other rural activities. Instead it should be promoted as an essential component of the interrelated spheres of socio-economic development, land use policy and environmental conservation, within the contexts of integrated rural and regional development with employment provision. There is a need for a much greater emphasis on socio-structural policies to counteract the existing disparities in income; more attention to smaller farms, less favoured regions and environmental impacts are welcome indicators of some reorientation of EC policy. Practices which are environmentally beneficial could be supported financially and countryside conservation will receive greater recognition. The need in land use is not to further extend the agricultural area but to develop alternative land uses such as forestry and recreation. Farmers are likely to show increased interest in subsidiary activities. Provision of alternative employment and income opportunities is essential both for the owners of small holdings and for the members of farm families. The indirect potential for agriculture to generate more employment and income through improved processing and marketing of its produce is far from being fully exploited and offers scope for industrial growth. The vital role which agriculture will continue to play as an integral part of development, both regionally within Ireland and in the total economy of the island, must receive due recognition and consideration at the levels of local, national and Community government.

REFERENCES

Attwood, E.A. (1966-67) Agricultural developments in Ireland, North and South, *Journal of the Statistical and Social Inquiry Society of Ireland*, 21, 9-34.

Attwood, E.A. and O'Sullivan, M. (1983-84) Some aspects of agricultural development in Ireland North and South since accession to the EEC, *Agricultural Economics Society of Ireland Proceedings*, 137-172.

Baillie, I.F. and Sheehy, S.J. (eds) (1971) *Irish Agriculture in a Changing World*, Oliver and Boyd, Edinburgh.

Bowler, I. R. (1985) *Agriculture under the Common Agricultural Policy: a Geography*, University Press, Manchester.

Commins, P. (1980) Imbalances in agricultural modernisation - with illustrations from Ireland, *Sociologia Ruralis*, 20, 63-81.

Cox, P.G. and Kearney, B. (1983) The impact of the Common Agricultural Policy, in Coombes, D. (ed.) *Ireland and the European Communities : Ten Years of Membership*, Gill and Macmillan, Dublin, 158-182.

Crotty, R.D. (1966) *Irish Agricultural Production : its Volume and Structure*, University Press, Cork.

Cuddy, M. and Doherty, M. (1984) *An Analysis of Agricultural Developments in the North and South of Ireland and the Effects of Integrated Policy and Planning : a Study Prepared for the New Ireland Forum*, Stationery Office, Dublin.

Edwards, C.J.W. (1977) The spatial distribution of livestock in Northern Ireland, *Irish Geography*, 10, 58-71.

Edwards, C.J.W. (1980) Recent structural changes in Northern Ireland agriculture, *Irish Journal of Agricultural Economics and Rural Sociology*, 8, 45-50.

Furness, G.W. and Stainer, T.F. (1981) Economic performance in agriculture in Northern Ireland and the Irish Republic, *Annual Report on Research and Technical Work*, Department of Agriculture, Belfast, 57-9.

Gillmor, D.A. (1970) Spatial distributions of livestock in the Republic of Ireland, *Economic Geography*, 46, 587-97.

Gillmor, D.A. (1977) *Agriculture in the Republic of Ireland*, Akademiai Kiado, Budapest.

Gillmor, D.A. (1985) *Economic Activities in the Republic of Ireland: A Geographical Perspective*, Gill and Macmillan, Dublin, 169-211.

Gillmor, D.A. (1987) Concentration of enterprises and spatial change in the agriculture of the Republic of Ireland, *Transactions of the Institute of British Geographers*, New Series, 12, 204-16.

Heavey, J.F., Harkin, M.J., Connolly, L. and Roche, M. (1984) *Farm Management Survey 1983*, An Foras Taluntais, Dublin.

Higgins, J. (1983) *A Study of Part-time Farmers in the Republic of Ireland*, An Foras Taluntais, Dublin.

Horner, A.A., Walsh, J.A. and Williams, J.A. (1984) *Agriculture in Ireland A Census Atlas*, Department of Geography, University College, Dublin.

Kelleher, C. and O'Hara, P. (1978) *Adjustment Problems of*

Low Income Farmers, An Foras Taluntais, Dublin.

Kelleher, C. and O'Mahony, A. (1984) *Marginalisation in Irish Agriculture*, An Foras Taluntais, Dublin.

Matthews, A. (1984) Agriculture, in O'Hagan, J.W. (ed.) *The Economy of Ireland : Policy and Performance*, Irish Management Institute, Dublin, 299-321.

O'Connor, R., Guiomard, C. and Devereux, J. (1983) *A Review of the Common Agricultural Policy and the Implications of Modified Systems for Ireland*, The Economic and Social Research Institute, Dublin.

Sheehy, S.J. (1984) The Common Agricultural Policy and Ireland, in Drudy, P.J. and McAleese, D. (eds) *Ireland and the European Community*, University Press, Cambridge, 79-105.

Sheehy, S.J., O'Brien, J.T. and McClelland, S.D. (1981) *Agriculture in Northern Ireland and the Republic of Ireland*, paper 3, Co-operation North, Dublin and Belfast.

Smith, L.P.F. (1948-49) Recent developments in Northern Irish agriculture, *Journal of the Statistical and Social Inquiry Society of Ireland*, 18, 143-160.

Stainer, T.F. (1985) *An Analysis of Economic Trends in Northern Ireland Agriculture since 1970*, Department of Agriculture, Belfast.

Statistical Office of the European Communities (1985) *Community Survey of the Structure of Agricultural Holdings 1979/1980*, Luxembourg.

Symons, L. (1963) *Land Use in Northern Ireland*, University Press, London.

Symons, L. (1970) Rural land utilisation in Ireland, in Stephens, N. and Glasscock, R.E. (eds) *Irish Geographical Studies in Honour of E. Estyn Evans*, The Queen's University of Belfast, 259-73.

Walsh, J.A. (1985-86) Uneven development of agriculture in Ireland, *Geographical Viewpoint*, 14, 37-65.

Walsh, J.A. and Horner, A.A. (1984) Regional aspects of agricultural production in Ireland 1970-1980, *Irish Geography*, 17, 95-101.

Whittaker, J.M. and Spencer, J.E. (1986) *The Northern Ireland Agricultural Industry : Its Past Development and Medium Term Prospects*, Economic and Social Research Council.

9 THE NEW INDUSTRIALISATION OF IRELAND

Barry Brunt

Manufacturing has traditionally been recognised as playing a fundamental role within the development process. The direct creation of employment and wealth by this sector, together with its induced multiplier effects, have made the attraction of manufacturing enterprises a crucial element in most regional and national planning strategies. During the general economic growth of the 1950s and 1960s, manufacturing was still considered a growth sector and provided a significant pool of mobile, or potentially mobile, investment which could be influenced by spatially selective and growth orientated government policy. The more problematic trading environment of the 1970s and 1980s, however, has modified greatly the conditions under which policies of industrial development operate. Manufacturing is no longer considered a sector likely to generate substantial growth in employment while, at the same, time international recession has reduced significantly the amount of footloose industry available for attraction into areas seeking to industrialise. These changing conditions have caused increasing problems for regions and countries that have based their success upon the ability to promote and attract industrial investment and have necessitated a reassessment of development policy (Goddard, 1980).

The purpose of this chapter is to focus attention on the development of manufacturing in the Republic of Ireland and Northern Ireland, essentially for the period following 1960; a date advanced as a turning point for both economies (Moore *et al.,* 1978; Gillmor 1985). Despite a significant degree of differentiation that existed between the two economies at the time of partition, elements of convergence subsequently emerged and were emphasised by a central commitment to attract foreign enterprise as the basis for economic development (Walsh, 1979; O'Malley, 1985). In this, however, the Republic has proved more successful, with the result that a role reversal has occurred between the North and South in terms of the dominance of manufacturing. The Republic is now the dominant manufacturing economy and this position has tended to increase in the more contemporary period. Convergence is therefore far from complete and aspects of divergence that now exist seem to work in favour of the South rather than the North.

The changing patterns of manufacturing within Ireland will be reviewed at three spatial levels; national, regional and urban-rural. Five processes will then be outlined as promoting the changing spatial patterns; government policy, business organisation, technological change, labour market changes and finally the emergence of a more open trading economy, exemplified by membership of the European Community. Some remarks on the implications for economic prospects on both parts of Ireland will conclude the chapter.

INDUSTRIALISATION TO DEINDUSTRIALISATION

The Irish economy evolved essentially within the context of the nineteenth century capitalist industrial system of the United Kingdom. This bequeathed a specialist role for the economy of the north-eastern counties as distinct from the rest of the country. The north-east, centred on Belfast and the Lagan valley, became the focus of a well integrated, urban-industrial complex specialising on engineering, shipbuilding and textiles for the export trade. In contrast, the most crucial factor for manufacturing in the rest of the country was the minimal development associated with the industrial revolution (Gillmor, 1985, p.212). The dictates of the colonial system therefore relegated most of the country to providing surplus food and labour for the British industrial system. Thus, at the time of the first census in the Republic (1926), less than 10% of the workforce were engaged in industry. This contrasts with over one-third in the six northern counties which, in turn, represented some 90% of the total manufacturing employment in the island. Two distinct and specialised sub-economies therefore emerged which were fundamentally indepedent of each other yet were both dependent upon Britain.

In the two generations following partition, manufacturing activities within the two political entities experienced contrasting trends. The inter-war recession and increase in trade competition exposed the North's highly specialised and export-dependent manufacturing base, and the severity of the decline established Northern Ireland as the most problematic regional economy in Britain (Busteed, 1974). The partial revival of the traditional industries following the war, together with the failure to diversify the region's economic base, ensured that it remained the most narrowly specialised regional economy in Britain with more than one-half of the workforce involved in shipbuilding and linen in 1950 (Steed and Thomas, 1971). Recession in the late 1950s once again exposed the North to severe contraction in the economic base, and translated itself into a loss of 23,000 jobs between 1951 and 1961. The structural inadequacies of the region's economy clearly suggested a fundamental change in policy was required to revive the industrial base (Isles and Cuthbert, 1957; Black, 1977).

While the North experienced decline, a policy of protection for indigenous industry in the Republic was successful in encouraging the build up of manufacturing (Meenan, 1970). By 1951, employment had almost doubled on the 1926 level of 117,154. Although a sizeable and more diversified manufacturing base emerged in the South, fundamental weaknesses within the sector, and increased import penetration, ensured a marked slow down in the growth of employment in the 1950s (O'Farrell, 1972; Walsh, 1980). Overall, while manufacturing grew by some 61,000 between 1926 and 1961, job losses in agriculture equalled 272,000. This clearly indicated that a more vigorous policy was required if the national aim of halting emigration and providing an alternative employment base to agriculture was to be achieved (Gillmor, 1982, p.12).

Developments in manufacturing in the 1950s were a great disappointment, but the problems were compounded by the continuation of a trend for industry to polarise on the larger urban centres and more eastern regions in both economies. Thus, in the North, although the decline of traditional industries disproportionately affected the Belfast region, this area retained its dominance through the 1950s (Thomas, 1956). Not only were traditional industries entrenched in this region, but also some three-quarters of manufacturing plants attracted into the North between 1945 and 1959 located within a 30 mile radius of the capital city (Hoare, 1981, p.154). In contrast, the rural areas of the south and west, with an estimated 25% of Northern Ireland's population, could claim only 13% of the region's industrial workforce and exhibited an occupational structure more in common with the western areas of the Republic than with the rest of Northern Ireland.

Spatial inequality was also a fundamental charactistic of the Republic where polarisation tended to increase in the period up to 1960. The Dublin region doubled its manufacturing employment from 1926 to 1961 and increased its relative share of national industrial employment from 40% to 53% (Gillmor, 1982). In contrast, apart from the adjacent and urbanised North East, all other regions declined or grew more slowly than the national rate of manufacturing change. Thus by 1961, Breathnach (1985) indicates that eleven western counties, with some 30% of national population, exhibited only 13% of industrial employment as against 19% in 1926. Industrial development policies in both North and South would therefore not only have to be concerned with total job creation but also ensure a more equitable redistribution of employment in favour of the more rural and marginal areas of Ireland.

1960-1985: The National Dimension

By 1960, despite problems of decline, Northern Ireland controlled more than one-half of total Irish manufacturing employment (Table 1). As the decade unfolded, the North succeeded in attracting a significant inflow of overseas investment and proved to be a time of optimism for the problematic region (Busteed, 1974; Hoare, 1981; Rees and Miall, 1981). Problems of large-scale job losses in the traditional industries remained, however, and amounted to some 35,000 jobs during this decade; one-half of the 1960 base level (Harrison, 1982, p.273). As a result, in spite of some 3,000 jobs per year being created by new investment, total manufacturing continued to decline in the 1960s and the region lost its position as the dominant manufacturing partner within the overall Irish economy (Tables 1 and 2).

Table 1: **Manufacturing 1961-85**

	1961	1971	1981	1985
Total Ireland (000s)	372	385	355	311
Republic of Ireland (%)	47.8	55.6	67.0	67.5
Northern Ireland (%)	52.2	44.4	33.0	62.5

Source: Census Reports 1961-1981; IDA and IDB Annual Reviews, 1985

Table 2: **Irish Manufacturing Trends, 1961-85**

	Northern Ireland	Republic of Ireland
1961 Employment (000s)	194	178
1961-71 change	-23(11.9%)	+ 36(20.2%)
1971-81 change	-54(31.6%)	+ 24(11.2%)
1985-85 change	-16(13.7%)	+ 28(11.8%)
1985 Employment	101	210
1961-85 change	-93(47.9%)	+ 32(18.0%)

Source: Census Reports, 1961-81; IDA and IDB Annual Reviews, 1985

While the North had successfully attracted new investment to compensate for jobs lost in declining industries during the 1960s, the 1970s witnessed a marked curtailment of such compensatory flows. As a result, an almost frightening erosion of the region's manufacturing base occurred with total employment declining by almost one-third (Black, 1980; Rowthorn, 1981). The 1980s continued the downward spiral, suggesting the North exhibited features of significant deindustrialisation and held little prospect for future growth (Black, 1983).

In contrast to the North, the 1960s and 1970s were decades of significant expansion for industrial employment within the Republic (Gillmor, 1985; Brunt, 1987). Closure and rationalisation did affect the sector, but job losses were greatly exceeded by expansion schemes and new openings. Thus, O'Farrell (1975), in a detailed study of industrial development trends between 1960 and 1973 showed a net gain in grant aided industry of 44,882 jobs, while the same author (1984), in a components of change analysis for 1973 to 1981, indicated a net gain of 11,089 jobs. The continued growth of manufacturing and, more especially, its positive performance in the 1970s, enabled the Republic to emerge as the dominant industrial partner within the island economy (Tables 1 and 2).

This dominance was further accentuated in the 1980s, for although employment declined by some 12%, as job losses consistently exceeded job gains, the rate of decline was less than that experienced in the North. In the quarter century from 1961 to 1985, the North lost almost one-half of its manufacturing labour force and contrasts markedly with the gain of 18% in the Republic. A clear role reversal had emerged during this short period which began with the North's manufacturing sector exceeding that of the South, but terminated with the Republic possessing twice as many manufacturing workers as its neigbouring region. By 1985, over two-thirds of Ireland's manufacturing employment lay within the Republic; a position far removed from the 10% at the time of initial partition.

1960-1985: The Regional Dimension

The dominant trend of concentration and polarisation of

manufacturing on the eastern capital regions in both North and South was modified in this period (Figure 1).

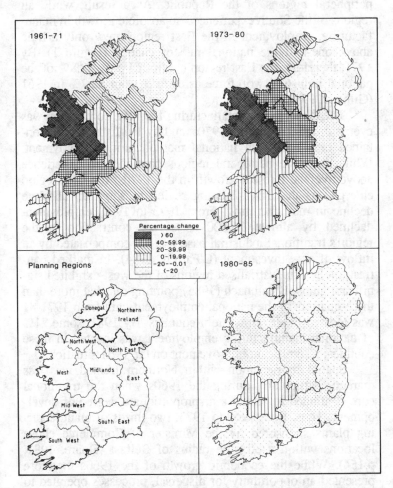

Figure 1. Changes within manufacturing employment in Ireland, 1961-1985

A preference for a more dispersed pattern of industrial location emerged and reflected both government policy and locational preference by private investment. O'Farrell (1975) highlights this pattern of dispersal during 1960 to 1973, when a disproportionate shift of manufacturing employment, industrial floorspace and capital grants gravitated to the more peripheral regions of the Republic. As a result, while all regions of the South experienced an absolute growth in manufacturing, employment in the East region grew only a little above one-half the national rate of change (Figure 1). By 1971, therefore, the East region controlled only 48.8% of the national industrial workforce as distinct from 52.7% in 1961 (Gillmor, 1982, 1985).

The dispersal of manufacturing from the core region was even more marked in the 1970s, and by 1981 this region contained only 38.8% of national manufacturing employment. While the more peripheral regions continued to experience above average rates of growth in the 1970s, the East region emerged as the only planning region to show an absolute decline in manufacturing. From 1973 to 1981, employment declined by almost 16,000 as severe contraction in the region's traditional industrial base was not compensated by an inflow of new investment (O'Farrell, 1984). In marked contrast, the less-industrialised regions of the west exhibited significant gains, Breathnach (1985) pointing out that more than three-quarters of the national employment gain from 1971-81 was located in eleven western counties. By 1981, some 21% of national manufacturing employment was located in these counties; a significant improvement on the 1961 situation.

Spatial reorientation within Northern Ireland was less marked, particularly during the 1960s when the traditional core continued to attract a disproportionate degree of development. Thus, from 1960 to 1973, two-thirds of manufacturing plants sponsored by the Minister of Commerce chose locations within a 30 mile radius of Belfast (Hoare, 1981, p.154). While the economic growth of the 1960s therefore presented an opportunity for dispersal, processes operated to effect a widening, rather than a contraction, of intra-regional differences in job accessibility. The status of the core area improved while that of the periphery deteriorated (Hoare 1982, p.215).

Paradoxically, perhaps, the more difficult conditions of the 1970s succeeded in generating a greater degree of dispersal within the North. As in the South, industrial decline was focused particularly on the core area, and Belfast emerged as the first area in Ireland to experience deindustrialisation on a significant scale. In contrast, the more rural areas of the west and south were less burdened by declining sectors and were increasingly able to benefit from a stronger tendency for private enterprise to disperse within Northern Ireland (Bull, 1984).

The 1980s have been particularly difficult for all regions in Ireland, although the newer industrial base of the more rural areas has enabled such regions to register a less than average decline in manufacturing employment. Thus, in the South, Donegal and the Mid-West emerged as the only regions to continue their growth in employment from 1980 to 1985, although the Midlands, West and North West all showed declines less than the national average (Figure 1). A loss of 21,000 jobs (21.4%) between 1980-85 for the East region indicated the continued erosion of the core's industrial base which, by the latter date, controlled only 36.9% of the national industrial workforce (Brunt, 1987).

1960-85: The Urban-Rural Dimension

Prior to 1960, the preferred location for industry appeared to be within the larger urban centres which possessed infrastructural facilities and market accessibility deemed to be advantageous by entrepreneurs. Small towns and rural areas held little attraction for industry, apart from traditional crafts and activities based on local raw materials, most notably food processing. The relative attractiveness of the larger urban centres, as opposed to the small towns and rural areas of the country, changed significantly in the 1960s, however, and is intimately associated with the changing regional pattern discussed above. In the Republic, an industrial survey by O'hUiginn (1972) pointed to an overall preference for smaller towns, while O'Farrell (1975) indicated that between 1960 and 1973, 56.7% of 418 grant-aided establishments chose to locate in urban centres of less than 5,000 population. Only thirty-six plants (8.6%) in the latter survey opted to locate in Dublin and provided 5,604 jobs. The industrialisation of the countryside and dispersal of manufacturing down through the urban hierarchy was becoming a general feature of modern economies and was clearly experienced within the Republic.

This was not replicated to the same extent in the North, apart from some decentralisation to smaller communities surrounding Belfast in Counties Antrim and Down. The urban centres of the south and west did poorly, and even Derry, the second city of the North, attracted only 6% of factory floorspace created in the region between 1945 and 1970.

The ruralisation of industry and growth of small towns gathered momentum in the 1970s (Breathnach, 1985) . While such environments were becoming increasingly attractive, the larger urban centres continued to experience difficulties. Thus, Co. Dublin lost some 25,000 (25%) jobs in manufacturing from 1973 to 1983 such that the county's share of the national total fell from 48% in 1961 to 30% in 1983 (Gillmor, 1985, p.239). In addition, Cork, the second city of the Republic, performed extremely poorly and contributed significantly to the fact that apart from the East, the South-West (centred on Cork) was the only region to perform consistently at below national trends for the entire period (Brunt, 1984).

The inability of large centres to attract new investment to compensate for the collapse of their traditional industrial base has necessitated a deliberate shift in emphasis from dispersal to some centralisation in order to meet the demands of the large numbers of unemployed located within such centres. Thus, while manufacturing remains disproportionately focused on the larger cities, the tendency to disperse has resulted in a somewhat better balance in the urban-industrial hierarchy within Ireland .

By the 1980s, the patterns of manufacturing within Ireland had altered fundamentally from the position that existed in 1960. A marked degree of convergence in the pattern of manufacturing activities emerged, especially in respect of the relative role of manufacturing in the total economy and the tendency to disperse to the more rural and peripheral areas. These changes can be better appreciated, by a consideration of five dominant processes that have operated differentially at each spatial level.

GOVERNMENT POLICY

Any assessment of manufacturing development within

Ireland demands a recognition of the central role played by government. Following partition, governments responsible for the North and South became involved in the promotion of industry, although the strategies followed were fundamentally different. This reflected the contrasting ideological stance of the newly formed and independent Republic with that of the North, which remained a peripheral region within the British economy. The Republic thus adopted a policy of protection and economic autarky to foster indigenous industry, while the North relied upon an interventionist regional policy formulated within a British context (McCrone, 1969). Neither approach proved successful in promoting a strong industrial base. In addition, the over-riding priority with national development relegated internal spatial redistribution to secondary importance and, in such circumstances, polarisation of economic activity was not effectively counteracted.

By the late 1950s the inadequacies of government policy to promote development within Ireland were recognised. In the South, this was reflected in a change in direction of policy, whereas within the North the degree of intervention to direct new investment into the region was increased. The 1960s was therefore to emerge as a critical period for government directed industrial development in Ireland.

Although some preparatory changes in the policy of the Republic had emerged in the early 1950s, the new policy crystallised in 1959 within the *First Programme for Economic Expansion*. This committed government to abandon the policy of the previous thirty years and emphasised that future industrial development would have to rely primarily upon the attraction of foreign enterprise and production for the export market. As a result, both parts of Ireland converged on a similar policy and subsequently became major competitors for mobile international investment. An extremely generous range of incentives emerged within both economies to attract investment, and included an advance factory building programme, site acquisition and a variety of grants and preferential loans to help offset the anticipated higher costs of operating in a peripheral area of Europe.

In addition to such common denominators, both governments offered special inducements to encourage development. The North benefitted from the stronger regional policy at work within Britain in the 1960s which directly encouraged job promotion for problem regions. Of special significance, however, was a series of Industrial Development (Northern Ireland) Acts which granted the Minister of Commerce for the North full control over a wide range of financial and technical assistance to industry. This degree of autonomy was not available to any other British Development Area and, together with higher grants available within the province, was successful in steering a disproportionate amount of investment and employment into the region (Law, 1980; Rees and Miall, 1981). At the same time, the autonomy of the Irish Republic enabled this state to tailor-make incentives for foreign direct investment engaged primarily in export operations. Thus, an Export Profit Tax Relief scheme (EPTR) was introduced which exempted export profits from tax for a period of 15 years. This was a major inducement for enterprises seeking a production platform from which to enter global trade and, with negotiations to join the EEC proceeding through the 1960s, became recognised as an even more potent weapon to attract foreign enterprise (NESC, 1982). In 1981, however, EPTR was replaced by a 10% Corporation Tax largely because the EC considered it discriminated too heavily in favour of Ireland as opposed to other problem regions of the Community.

The significance of government policy in creating manufacturing employment in the 1960s is well substantiated. Thus Moore et al. (1978) calculate that 35,000-40,000 jobs in the North (1960-70) and a further 75,000 in the South (1960-74) were created directly as a result of policy changes enacted by government. By 1970, government sponsored employment in the North doubled its relative importance for total manufacturing to 37.6% (Harrison, 1982, p.275). In the Republic, the role of government was even stronger with the New Industries programme accounting for 45.3% of total manufacturing employment by 1973 (O'Farrell, 1984).

While both governments adopted broadly similar policies at the national level, the internal spatial redistribution of development varied considerably and reflected a contrasting view of the need to disperse investment. In the North, spatial reorientation found formal expression in two major reports which emphasised the importance of growth centres in regional development (Matthew, 1964; Wilson, 1965). Their disproportionate presence in the east, however, signified the likely continued dominance of the area to the detriment of more peripheral locations (Figure 2).

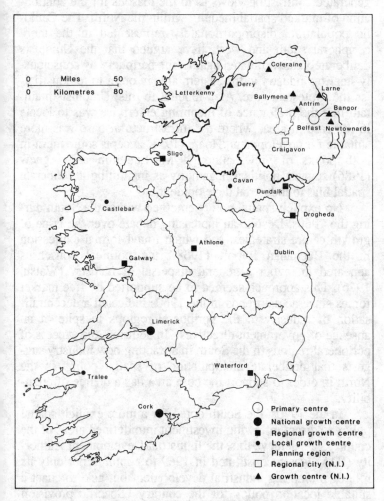

Figure 2. Planning regions and growth centres for Ireland in the 1960s

The degree of autonomy possessed by the Stormont government until direct rule in 1972, gave the dominant Unionist Party a major role in spatial policy. Certainly, the strong association between the Unionist Party and Protestant population generated contrasting views as to the reasons for the continuation of marked spatial inequity within the North. The Catholic population, disproportionately represented in the more peripheral south and west, have argued that the Unionists deliberately discriminated against the periphery by consciously steering industry to the eastern area in order to reindustrialise the Protestant core. As a rebuttal to this, Unionists maintain that the preference of incoming enterprise was to locate in the eastern areas where the infrastructural base was more suitable for development. Hoare (1981) accepts some merit in both points of view, although MacLaughlin and Agnew (1986) suggest political hegemony as instituting a deliberate spatial bias to industrial development.

No explicit spatial policy emerged within the South during the 1960s. Despite an increasing debate over the issue of growth centre strategies, and which found formal expression in the Buchanan Report (1968), the political consensus appeared to favour a general dispersal of industry (Walsh, 1976). This approach seemed to be supported by free market forces since actual trends in the 1960s indicated a decentralisation of industry to the peripheral regions in spite of an absence of government directives. In addition, the success of peripheral regions in the South in attracting new industry suggests that deliberate manipulation of spatial policy in the North in order to support the core area has a degree of credibility.

In the 1970s, the South opted for a more explicit spatial policy to direct inflowing investment into defined areas of the country. To achieve this, the Industrial Development Authority (IDA) was reconstituted in 1969 to continue not only its role of promoting industrial development but also to enact a suitable location policy for the country. Specific provision had been made for promoting industrial development within the Gaeltacht, however, as early as 1957 with the establishemt of -Gaeltarra Eireann. This was replaced by Udras na Gaeltachta in 1980. Promotional bodies were also important in the North, where the Industrial Development Body and

Northern Development Agency operated through the 1970s. These were amalgamated into the Industrial Development Board (IDB) in 1982.

The IDA published the first of two Five Year Plans (1973-77 and 1978-82) in 1972 and this appeared to end the national flirtation with growth centre policy (Breathnach, 1982). In place of nine growth centres (Figure 2), 47 clusters of towns were selected and given job targets implying a strong level of direction was available to the IDA. The plans favoured a disproportionate degree of development for the more marginal regions to add significantly to their existing low levels of manufacturing (Table 3). This period can be considered to be the most highly articulated phase of regional development in the South (Boylan and Cuddy, 1984). By 1981, 52.5% of total manufacturing employment was due to the government's New Industry Programme (O'Farrell, 1984).

Table 3: **IDA regional targets 1973-77, 1978-82**

Planning Region	1973-77			1978-82		
	Net Job Target	% 1973 base	% achieved	Gross Job Target	% 1978 base	% achieved
Donegal	2,000	31.3	40	2,800	66	77
North-West	1,300	41.9	85	3,000	68	48
West	4,200	48.8	15	7,250	62	71
Mid-West	3,800	21.2	0	4,500	47	107
South-West	7,000	19.8	14	10,450	31	83
South East	3,200	15.9	86	8,500	33	103
East	10,300	9.4	-107	19,000	20	103
North-East	3,400	20.2	-29	6,000	35	81
Midlands	2,800	30.8	113	6,500	53	86
State	38,000	16.7	5	68,000	30	89

Source: IDA, 1972; 1979

215

A commitment to growth centres was maintained in the North, although a new planning framework indicated a greater desire to disperse industrial investment (Department of the Environment, 1977). Unfortunately, the 1970s proved to be a problematic period for the North. Decline of the British economy ensured an erosion of regional policy and in the amount of mobile investment available to peripheral regions. In addition, the civil unrest, or 'the troubles', emerged as a potent force which acted to discourage inward investment upon which the region based so much of its development (Davies *et al.*, 1977; Rees and Miall, 1981; Hoare, 1981). Estimates vary as to number of industrial jobs destroyed or prevented by the conflict and range from 20,000 between 1969 and 1976 (Moore *et al.*, 1978) to 25,000 over the decade of the 1970s (Rowthorn, 1981). The deflection of employment from the North undoubtedly helped the South achieve its degree of success in job promotion and creation.

Stagnation of the global economy in the late 1970s, and which spilled over into the 1980s, made continued success of government policy based on the attraction of footloose industry difficult. Criticism of the existing policies increased and found expression in the South most notably within the Telesis Report (NESC, 1982). This questioned the degree of support and dependency given to foreign-based enterprise, and advocated a stronger commitment to indigenous industry and factor inputs in order to retain a higher proportion of value added by manufacturing within the country. Many of the suggestions have been incorporated within more recent IDA strategy, where the emphasis seems to be related more to the quality and stability of employment created rather than absolute job creation. A greater commitment to research and development and higher technology is demanded rather than simple branch plant production. Similar views are expressed in the North, where the importance of government-sponsored employment has continued to increase within the context of overall decline. Thus by 1985, two-thirds of the North's total manufacturing employment related to government assisted companies, compared with almost 60% in 1980.

Government promotion of employment remains at a high level in the Republic, but is now being more than offset by job losses. In both parts of Ireland, therefore, the central concern is with national development and less priority is attached to the spatial dimension of such development. Thus, in the Republic, the government economic plan *Building on Reality 1985-87* is devoid of a spatial dimension. Within the context of a declining commitment to spatially redistributive policies, however, there has been a noticeable shift in emphasis as to what constitutes the most pressing problem areas within contemporary economies. The increasing problems of job loss within the larger urban centres, and the failure to attract sufficient new investment to compensate for decline, has focused attention on the plight of these areas and most notably in Belfast, Dublin and Cork. Elements of centralisation and growth centre strategy have therefore reemerged to partially offset the strong decentralisation tendencies of manufacturing that highlighted the 1970s.

ORGANISATIONAL CHANGE

One of the main characteristics of post-war industrial development has been the increasing size and power of global corporations. An increasing proportion of output and employment is being focussed on a small number of organisations. An organisational and spatial hierarchy of enterprise has emerged which has consequently provided global corporations with inherent geographical and productive flexibility. This has fundamental importance for marginal areas which, in general, have attracted the lowest level of the productive and organisational hierarchy, i.e. the branch plant. Concern is focussed especially on the stability and quality of employment provided by these units.

Until the change of government policy in the Republic, the policy of protection promoted an industrial base composed primarily of small-scale, indigenous enterprises serving the domestic market. This contrasted markedly with the greater proportion of externally controlled (mainly British) and large units in the North operating to serve the export trade. Since 1960, however, government policies converged, which in turn ensured a convergence in the organisational composition of industry and the increasing dependency upon foreign enterprise. This has had important spatial consequences for manufacturing location in Ireland.

Foreign Enterprise

The most important means of creating new manufacturing employment since 1960 in Ireland has been the attraction of overseas investment. Incentive packages to encourage mobile industry to locate in Ireland have been extremely successful, although the EPTR of the Republic is recognised as being especially effective in this objective.

***Table 4*:** **The role of foreign manufacturing employment**

	Northern Ireland			
	1974	1980	1985	1974-85
Britain	37,522	30,334	19,490	-18,032 (48.1)
USA	17,096	17,914	11,435	-5,661 (33.1)
All foreign	65,230	59,647	36,942	-28,288 (43.4)
Indigenous	104,290	75,713	64,128	-40,162 (38.5)
Total	169,520	135,360	101,070	-68,450 (40.4)
% Foreign	38.5	44.1	36.6	

	Republic of Ireland			
	1973	1980	1985	1973-85
Britain	26,932	22,652	14,100	-12,832 (47.6)
USA	14,935	32,563	36,500	+21,565 (144.4)
All foreign	58,892	81,968	78,373	+19,481 (33.1)
Indigenous	158,400	166,300	134,857	-23,543 (14.9)
Total	217,292	248,268	209,841	- 7,451 (3.4)
% Foreign	27.1	33.0	37.3	

Source: IDB and IDA Reviews of Industry

Between 1960 and 1971, the South attracted 250 foreign firms; a number equivalent to that for the whole of the UK and nine times the figure that established in the North (Moore *et al.*, 1978, p.108). The larger average size of plants created in the North, however, partially offset this discrepancy, and of the 32,700 jobs created through IDB assistance in the 1960s, some 95% occurred in overseas controlled enterprise (Hoare, 1981). By 1974, 38.5% of the region's manufacturing employment was foreign controlled (Table 4). In the South, foreign controlled employment was also a conspicuous feature of the country's growth performance such that by 1973 approximately 60% of new grant-aided employment originated from this source and contributed 27.1% of total manufacturing (McAleese, 1977).

The South maintained its attractiveness for foreign investment through the 1970s, although the more difficult trade conditions of the 1980s translated itself into a slight decline (Table 4). This contrasts with the much weaker performance of the indigenous sector, especially in the 1980s. Trends in the North are far less satisfactory and the inability of the region to maintain its absolute attraction for foreign enterprise goes a long way to explaining the declining role of manufacturing in Northern Ireland (Harrison, 1982). By 1985, these differential trends with regard to attracting foreign investment resulted in a convergence in the overall importance attached to foreign employment to approximately 37% on either side of the political divide.

While the attraction of overseas investment influenced overall national industrial trends, it was also relevant to the spatial rearrangement of manufacturing at the regional level. In the South, foreign companies showed an above average preference to locate in the more disadvantaged areas rather than the core and by 1985, contributed significantly to the regional economies of the Mid-West (58.0%) and West (50.1%). Given the dynamism of foreign companies, especially in the 1970s, the relative and absolute shift of manufacturing employment to the west can be appreciated. In the East, only 33% of the regional industrial workforce was in foreign controlled enterprise, and the comparative absence of such investment contributed to the decline of the core region and capital city.

Source of Employment

The bulk of overseas industry located in Ireland in 1960 was of British origin and reflected the traditional economic ties that existed between the two islands. Since 1960, however, the United States has shown an increased involvement while British based employment has been subjected to a continuous erosion of its absolute and relative significance. The importance of US - as opposed to British - enterprise within the economies of the Republic and Northern Ireland, and the ability to attract US companies are, therefore, of major significance in the differential performance of manufacturing within Ireland in more recent times (Table 4). British enterprises have shed almost 50% of their labour force either side of the border, although in the North this source still accounted for greater than one-half of foreign employment in 1985. This dependency on a source with such a poor record of job decline is a problem for the North. The condition is less apparent in the South, where less than one-fifth of foreign employment remains in this source in contrast to 46.3% within US plants.

The origin of enterprises also has a role to play in explaining the dispersal of manufacturing employment to the peripheral areas (Blackbourn, 1972; O'Farrell, 1980). Thus, in the Republic, British industries show a disproportionate presence in the East and North East regions and, given their declining characteristics, contribute significantly to the loss of employment within this traditional area of manufacturing strength. In contrast, US companies are more dispersed, but are represented particularly strongly in the West (Galway) and Mid-West (Limerick-Shannon-Ennis) regions. This favours the above-average growth performance of these areas (Gillmor, 1982, p.43).

Closure Rates

A fundamental problem associated with a rising dependency upon foreign enterprise relates to the loss of effective control on decision making and the stability of employment created. During the 1960s and early 1970s, there was little evidence of a significantly higher incidence of closures in foreign controlled branch plants than in indigenous owned plants, and the overall closure rates were comparitively low (Townroe, 1975; O'Farrell, 1976; McAleese and Counahan, 1979). As trading conditions deteriorated, however, the flexibility possessed by multinational enterprises to rationalise operations and close down marginal units found expression in Ireland in an increasing incidence of plant closures and at a significantly higher rate among foreign owned enterprises (Rowthorn, 1981; O'Farrell and Crouchley, 1983). This clearly contributed to the deteriorating position in Irish manufacturing and posed serious questions as to the stability of employment provided under existing government-promoted industrial development.

Differences also exist with respect to closure rates by nationality and have relevance for the internal arrangement of manufacturing. British companies operating within Ireland have the highest rates of closure due in part to the weakness of the British economy but also because of the tendency to specialise in older and declining sectors (Harrison, 1982; O'Farrell and Crouchley, 1983). Given the traditional importance of attracting mobile British investment into the North, and its preferred location in the Belfast region, the high incidence of closure of such plants not only affected the entire economy but was spatially biassed in its negative impact. Similarly in the Republic, the focal area of British investment lay in the East core region and compounded the problems of its industrial decline. In contrast, the greater representation of US and other foreign companies in the more marginal regions has facilitated their industrial development, based not only on their better growth performance than British companies, but also their significantly lower rates of failure.

Linkages

Part of the rationale for attracting foreign-based industry to Ireland was the anticipated multiplier effects that would accrue to the economy through development of linkages. The indirect creation of jobs and wealth, however, did not emerge as strongly as was anticipated, largely due to the dominance of branch plant units attracted under government incentive packages. The size and complexity of multinational enterprises have allowed then to internalise many of the services traditionally supplied by the external economy such that 'organisational space becomes a much more powerful determinant of linkages than geographical space' (Hoare, 1978, p.179). In the North, Steed (1968) and Steed and Thomas (1971) point out early dissatisfaction with the degree of linkages established by foreign enterprise. Characteristic elements of a dual economy emerged with many companies operating as enclaves largely detached from their immediate environment. Similar features are present in the Republic (Stewart, 1976), and a more recent study has found that only 16.4% of raw material inputs (excluding the food sector) were purchased in Ireland (O'Farrell and O'Loughlin, 1980). The position is worse at the more localised level, where only 5% of supplies were purchased within a 20 mile radius of the plants, and some 75% of service payments gravitated to Dublin, Cork and Waterford. While the lack of linkages has to be considered a negative aspect for Irish development, the enclave nature of many plants aided their ability to disperse from well-developed urban areas and into more rural environments with their generally less-developed infrastructural facilities. The inadequate promotion of linkages is an important factor in the poor performance of indigenous industry and is being addressed under new government policy measures.

Indigenous Industry

Despite the lower closure rates of indigenous industry in both the North and South, this sector has generally shown a marked inability to create additional employment to meet the rising demands for work (Table 4). The inferior trends of indigenous industry persisted into the 1980s in the Republic, although in the North a greater concern with home-based expansion schemes ensured that the rate of decline in indigenous industry fell below that of foreign enterprise. In addition to the impact on national levels of manufacturing, the poor performance of indigenous industry had important spatial repercussions within the country. Indigenous industry has been traditionally associated with a disproportionate presence in the larger urban centres and core regions of the east. As a result, this above average dependency upon indigenous enterprise contributed to the negative employment trends experienced in such areas during the 1970s and 1980s.

In order to generate a higher profile for domestic industry and create a seedbed for future development, promotional agencies in both parts of Ireland have introduced small industries programmes. These were anticipated to act as significant generators of employment and to help further in the dispersal of industry into rural areas. In the North, the Local Enterprise Development Unit (LEDU) was established in 1971 to tackle especially the problems of job promotion in remoter areas (Busteed, 1976). Their success can be gauged by the promotion of over 11,000 jobs in the 1970s, and between 1981 and 1985 a further 15,718 jobs were promoted (LEDU, 1986). With increasing problems being faced in Belfast, however, more attention is now being given to promoting enterprise in that area.

The Small Industry programme of the IDA (1967) has also had encouraging results. From 1973 to 1981, a net gain of 11,000 jobs was achieved in small industries employing fewer than 50 people, and its share of the Republic's manufacturing base rose from 5.9% to 10.1% (O'Farrell, 1984). The important contribution of these smaller-scale operations to rural development is substantiated by O'Farrell and Crouchley (1984) who found that a strong link existed between degree of rurality and small industry establishment rates. The extension of the Small Industry programme to Dublin in 1977, however, was a recognition in the Republic of the need to bolster the declining manufacturing base of the capital city.

STRUCTURAL CHANGE AND TECHNOLOGY

Rothwell (1982) has argued that the application of new technology within the environment of increased market demand that typified European conditions prior to 1970 had a positive impact on job creation. The more difficult trading environment of the 1970s, however, encouraged companies to become increasingly concerned with process technology and rationalisation schemes in order to reduce costs and maintain

223

their market positions. Many jobs in industry were lost as a result of this process, although conversely the new technologies and market opportunities stimulated the emergence of a new group of growth industries. In general, the capital intensive nature and spatial preference of these growth sectors did not match the quantity and spatial configuration of the more labour-intensive activities that bore the brunt of the more competitive economic environment.

Multinational enterprise gained especially from the new technologies since they generally possessed the financial resources and research-and development capacity to take advantage of the changed conditions. Advances in communications effectively allowed for the spatial separation of production from decision making and this facilitated the location of branch plants within more peripheral regions where government incentives contributed in reducing the costs of operation. With its adequate infrastructural capacity and generous incentive package, Ireland became incorporated as a production platform for big business operating within a global trading system. By the mid-1970s, some 88% of multinational output was exported from the Republic (McAleese, 1977).

A great deal of Irish industry was recognised as structurally unsound in the 1960s and reflected poor levels of management efficiency, high costs and out-of-date equipment. Despite investment programmes to re-equip and modernise traditional industries in both parts of Ireland, fundamental problems remain in the efficiency levels of many Irish companies. Furthermore, much indigenous industry was involved in more traditional sectors, such as textiles and clothing, which in general exhibited characteristics of falling demand and employment. In contrast, foreign enterprise tended to concentrate disproportionately on modern growth sectors, such as light engineering (especially electrical engineering) and chemicals. The inflow of foreign investment to Ireland during the 1960s, therefore, not only provided new jobs but also encouraged a convergence of the economic structures between North and South, more especially since the gains in the Republic were not offset by such a dramatic decline of the traditional base in the North (Walsh, 1979).

Elements of convergence continued to operate through the 1970s and 1980s as the textile, clothing and footwear industries experienced major declines in employment (Table 5). The problems facing these sectors are further highlighted in the North, where even the more recently introduced and capital-intensive man-made fibre industry almost totally collapsed in 1980-81.

Table 5: **Relative importance and changes in key manufacturing sectors**

	1970	1985	1970-1985	
Northern Ireland				
Engineering and metals	30.1	30.2	-22,918	(42.9)
Textiles	17.2	11.0	-19,462	(63.6)
Clothing and footwear	18.0	16.6	-15,254	(47.7)
Chemicals	1.2	2.4	+ 312	(14.8)
Republic of Ireland				
Engineering and metals	14.9	29.8	-30,686	(96.6)
Textiles	11.8	5.4	-13,856	(55.1)
Clothing and footwear	13.7	8.1	-12,227	(41.8)
Chemicals	7.1	6.1	- 2,439	(16.0)

Source: IDB and IDA

Major contrasts, however, emerge with respect to the engineering sector either side of the border. In the North, the need to compensate for continuing job losses in the more traditional and heavier engineering and metal-working industries, the comparative failure to maintain a high rate of new firm formation and the instability of much of the employment created, resulted in a major decline of 43% in the sector's employment between 1970 and 1985. In contrast, the Republic almost doubled its employment base in this sector as the continued influx of foreign enterprise created a substantial number of jobs. Electrical engineering became of particular significance, and it has been estimated that some 90% of employment is foreign controlled. By 1985, the relative importance of this sector to total manufacturing in the South matched the position in the North; a region more traditionally associated with engineering.

Contrasting sectoral performance also contributes to an understanding of the dispersal of manufacturing that occurred during this period. The footloose nature of the modern industries, largely controlled by foreign enterprise, showed little reluctance to move into more peripheral regions and rural communities. In contrast, the declining sectors exhibited a strong locational base in the larger urban centres and core regions and thus contributed markedly to the overall decline of employment in these areas.

Although the introduction of modern growth sectors is vital for development, a serious question exists as to the levels of technology used and promoted by plants located in Ireland. The dominance of branch plants utilising simple production line technology, possessing low decision-making powers and generating little internal research and development capacity, point to units that typify the third stage of the product cycle. Policy changes in both the North and South reflect a desire to encourage the attraction of research and development into Ireland and, via enterprise development programmes, to create better seed-bed conditions for supporting Irish initiative.

THE CHANGING ROLE OF LABOUR

The consistently high natural rate of population increase in Ireland, occurring in a context of low employment opportunity, traditionally translated itself into high rates of unemployment and emigration. Such circumstances strongly influenced government in the North and South to develop industrial policies aimed initially at maximising job creation. Not only did the population base influence government strategy, however, but it also emerged as as a major attractive influence for incoming investment. The high levels of unemployment, underemployed farm workers, low activity rates and emigrants who would remain at home, should work present itself, suggest a sizeable pool of labour existed for potential industrial development. In addition, the cost of labour was low, and prior to the accession of the Mediterranean countries to the EC, the Republic exhibited the lowest national wage rates in the Community. The North, despite some convergent tendencies, remains the lowest labour cost region in Britain (Black, 1985). These factors, together with the low levels of skills demanded by standardised production units, which could be quickly assimilated by an adaptable and eager workforce, were extremely attractive for multinational investment. As a result, Ireland became a component in the new international division of labour favoured by the increasing dominance of multinational enterprises within the global economy (Perrons, 1981).

Differential labour advantages within Ireland also influenced the spatial realignment of manufacturing. While government incentives and the organisational structure of big business influenced the dispersal of industries, Massey (1978) suggests that many industries actively sought locations beyond more traditional sites within large urban centres. New push-pull forces had clearly emerged, some of which relate to the contrasting labour attributes of urban versus rural workforces. While the absolute supply and cost advantage of labour in rural communities is now, at best, only marginal over larger urban centres, the stronger tradition of trade unionism and greater militancy of workers in large urban centres have contributed to deflect industry out of such areas as agglomeration economies are seen to give way to diseconomies. In addition, within more rural environments, the almost monopolistic properties that can be achieved by large companies over small communities, together with the increasing involvement of part-time farmers and the feminization of the workforce, all contribute to a low level of militancy and a high level of labour adaptability. The lack of traditional industrial skills does not present a major problem since the new technologies of branch plant production have effectively deskilled many functions. The net result has encouraged the decline of Dublin and Belfast core areas and helped convert the traditional regional problem of industrial decline and high rates of unemployment into a more urban-related issue.

FREE TRADE AND THE EUROPEAN COMMUNITY

The open trade policy favoured in the post-1960 period and formalised in the Republic by the Anglo-Irish Free Trade Agreement (1965) and, for both economies, by the accession to the European Community in 1973 had major implications for manufacturing in Ireland. Access to a large and prosperous trading zone added significantly to the attraction of locating industrial units within Ireland, and there seems little doubt that the degree of inflowing investment would have been substantially less than experienced in the 1960s if accession had not been contemplated (NESC, 1982; Brunt, 1987). The more competitive trading environment, however, brought problems for indigenous industry, which, in the Republic, had benefitted from a strong policy of protection. Some of the pessimism regarding the ability of industry in the Republic to compete effectively within a European framework can be gleaned from the interim report of 1957 prepared by the secretaries of the main government departments in response to OEEC proposals for a European Free Trade Area. They concluded: '...as regards a large sector of existing industry, the Department of Industry and Commerce can see no prospect of their survival, even as suppliers of the home market, except with permanent protection' and that furthermore, the Department '...can see no prospect of a significant expansion of industrial exports from Ireland to the continental part of the free trade area' (quoted in Maher, 1986, p.63).

The degree of pessimism proved inaccurate. While indigenous industry did experience problems, this was more than compensated by an influx of foreign industry and the improved performance of manufacturing in export trade. Thus in the Republic, the proportion of manufacturing in total exports increased from 19% to 36% in the decade of the 1960s, and dependency on the British market for exports declined from 74% to 66% (Gillmor, 1985, p.19-23). In discussing accession to the Community, therefore, the government was able to express an optimism that membership would not 'add significantly to the competition which Irish industry will face in the home market' (Stationery Office, 1972, p.35). Furthermore, the government stated that 'membership provides much more favourable conditions for industrial output and employment than any alternative option open to us' (Stationery Office, 1972, p.40).

The Republic has benefitted particularly from membership of the Community and has competed effectively in the export sphere. Manufacturing output increased to account for 65% of total exports by 1985, when continental members of the Community took over one-third of the South's exports. Dependency on Britain declined to only 33%. The resulting diversification of market outlets was important for manufacturing development since the depressed state of a single market did not have such severe knock-on effects as previously was the case. Thus, in spite of two recessions, the degree of development within the Republic was such that it emerged as the only member state to exhibit an absolute increase in manufacturing employment during the 1970s.

While Ireland has benefitted from the European Community accession, the degree of success in manufacturing has been less than anticipated. Accession corresponded with the start of a major recession which has persisted for the island economies to the present. This telescoped the need for rationalising existing industry while, at the same time, slowing down the rate of inflowing investment. As a result, Ireland remains one of the least-developed areas of the Community. The difficulties facing the Republic were recognised in a special protocol attached to the Treaty of Accession and some concessions were granted to Irish industry to make transition into the free market economy easier to manage. This has been followed by the European Regional Policy and Fund which recognises the entire country as a disadvantaged area within Europe and eligible for the full range of financial incentives emanating from Brussels. The economic potential of the island economies, however, remains at a low level in a European context (Clark et al., 1969; Keeble et al., 1982). If the future of Irish industry is to remain within Europe, then the emergence of a much stronger regional policy at the European level would seem essential.

CONCLUSION

Governments in small, open economies are severely limited

in what they can achieve in terms of industrial development, more especially when they function essentially as client economies. The Republic has become increasingly a client state of international business. While this is also true of the North, this spatial unit has the added problem of being a client region of a state experiencing significant industrial decline. Political independence in the South and its more stable political circumstances have been important in allowing for a more successful promotion of manufacturing in the Republic and a role reversal of manufacturing strength between the two states.

Future manufacturing patterns will continue to depend on the interlocking nature of the five processes outlined in this chapter. Given the scale of change that has already occurred at all spatial levels, however, it is unlikely that the degree of momentum will continue, at least in the shorter term. Internal policies appear to be focusing attention on the quality and stability of employment associated with indigenous industry, and promotion of research and development units, rather than the maximisation of jobs. While this policy should ultimately create an increased level of employment through enhanced value and multiplier effects within Ireland, the overriding demand is for immediate and maximum job creation. Ireland has failed to emerge from the 1980-82 recession and unemployment, rising emigration rates and unfulfilled job expectations are critical issues. Even in the Republic, where some success in attracting foreign companies has been achieved, unemployment has consistently risen from approximately 75,000 at the start of 1973 to some 250,000 at the end of 1986. It therefore seems likely that potitical expediency and economic reality will demand that foreign investment remains fundamental to the manufacturing profile of Ireland, even though the degree of dependency will fall from the excessive levels that characterised the 1970s.

With fewer jobs likely from mobile international investment, and the overriding concern with national economic performance, the forces of dispersal may be diluted. This is further supported by the apparent government desire to ensure that any future spatial policy will not react so negatively on the larger urban centres where problems have focused to an increasing degree. In opposition to this, however, counterurbanisation has become an established trend, and the extent of rural industrialisation in Ireland suggests that it will be a difficult force to deflect. Furthermore, if foreign enterprise is to remain a major element of government strategy for national development, it will be difficult to resist the locational preferences exhibited by such activities unless a much stronger spatial policy emerges within both economies. Given the national problems in Britain and the Republic of Ireland, this seems, at present, to be unlikely. The ultimate expression of the changing pattern of manufacturing in Ireland will depend on a consideration of all these forces.

REFERENCES

Black, B. (1985) Regional earnings convergence: the case of Northern Ireland, *Regional Studies*, 19, 1-7.

Black, W. (1977) Industrial development and regional policy, in Gibson, N.J. and Spencer, J.E. (eds) *Economic Activity in Ireland, a Study of Two Open Economies*, Gill and Macmillan, Dublin.

Black, W. (1980) The economy of Northern Ireland: performance and prospects, *Irish Banking Review*, June 1980, 25-30.

Black, W. (1983) Northern Ireland after the recession, *Irish Banking Review*, December 1983, 3-12.

Blackbourn, A. (1972) The location of foreign-owned manufacturing plants in the Republic of Ireland, *Tijdschrift voor Economische en Sociale Geographie*, 63, 438-43.

Boylan, T.A. and Cuddy, P. (1984) Regional industrial policy: performance and challenge, *Administration*, 32, 255-70.

Breathnach, P. (1982) The demise of growth centre policy: the case of the Republic of Ireland, in Hudson, R. and Lewis, J.R. (eds) *Regional Planning in Europe*, Pion, London, 35-56.

Breathnach, P. (1985) Rural industrialisation in the West of

Ireland, in Healey, M.J. and Ilbery, B.W. (eds) *The Industrialisation of the Countryside*, Geo Books, Norwich, 173-95.

Brunt, B.M. (1984) Manufacturing change in the Greater Cork Area, *Irish Geography*, 17, 101-8.

Brunt, B.M. (1987) *Ireland*, Paul Chapman, London.

Buchanan, C. and Partners (1968) *Regional Studies in Ireland*, An Foras Forbartha, Dublin.

Bull, P. (1984) Economic planning for rural areas in Northern Ireland, in Jess, P.M. *et al.,* (eds) *Planning and Development in Rural Areas*, Queens University, Belfast, 41-62.

Busteed, M.A. (1974) *Northern Ireland*, Oxford University Press, Oxford.

Busteed, M.A. (1976) Small-scale economic development in Northern Ireland, *Scottish Geographical Magazine*, 92, 172-81,

Clark, C., Wilson, F. and Bradley, J. (1969) Industrial location and economic potential in Western Europe, *Regional Studies*, 3, 197-212.

Davies, R., McGurnaghan, M.A. and Sams, K.I. (1977) The Northern Ireland Economy: progress (1968-75) and prospects, *Regional Studies*, 11, 297-307.

Department of the Environment (1977) *Northern Ireland Regional Physical Development Strategy 1975-95*, HMSO, Belfast.

Gillmor, D.A. (1982) *Manufacturing industry in the Republic of Ireland: its development and distribution*, Bank of Ireland, Dublin.

Gillmor, D.A. (1985) *Economic Activities in the Republic of Ireland: A Geographical Perspective*, Gill and Macmillan, Dublin.

Goddard, J. (1980) Industrial innovation and regional economic development, *Regional Studies*, 14, 159-266.

Harrison, R.T. (1982) Assisted industry. employment stability and industrial decline, *Regional Studies*, 16, 267-85.

Hoare, A.G. (1978) Industrial linkages and the dual economy: the case of Northern Ireland, *Regional Studies*, 12, 167-80.

Hoare, A.G. (1981) Why they go, where they go: the political imagery of industrial location, *Transactions of the Institute of British Geographers*, New Series, 6, 152-75.

Hoare, A.G. (1982) Problem region and regional problem, in Boal, F.W. and Douglas, J.N.H. (eds) *Integration and Division: Geographical Perspectives on the Northern Ireland Problem*, Academic Press, London.

Industrial Development Authority (1972) *Regional Industrial Plans, 1973-77*, Dublin.

Industrial Development Authority (1979) *Regional Industrial Plans, 1978-82*, Dublin.

Isles, K.S. and Cuthbert, N. (1957) *An Economic Survey of Northern Ireland*, HMSO, Belfast.

Keeble, D., Owens, P.L. and Thompson, C. (1982) Regional accessibility and economic potential in the European Community, *Regional Studies*, 16, 419-32.

Law, C.M. (1980) *British Regional Development Since World War 1*, David and Charles, Newton Abbot.

Local Enterprise Development Unit (1986) *The Annual Report, 1986*, Belfast.

Lyons, F.S.L. (1971) *Ireland Since the Famine*, Weidenfeld and Nicholson, London.

McAleese, D. (1977) *A Profile of Grant Aided Industry in*

Ireland, IDA, Dublin.

McAleese, D. and Counahan, M. (1979) 'Stickers' or 'Snatchers'? employment in multinational corporations during the Recession, *Oxford Bulletin of Economic Statistics*, 41, 345-58.

McCrone, G. (1969) *Regional Policy in Britain*, George Allen and Unwin, Hemel Hempstead.

MacLaughlin J.G. and Agnew, J.A. (1986) Hegemony and the regional question: the political geography of regional industrial policy in Northern Ireland, 1945-1972, *Annals of the Association of American Geographers*, 76, 247-61.

Maher, D.J. (1986) *The Tortuous Path: The course of Ireland's entry into the EEC 1948-73*, Institute of Public Administration, Dublin.

Massey, D. (1978) Capital and location change: the UK electrical engineering and electronics industries, *Review of Radical Political Economics*, 10, 39-54.

Matthew, R. (1964) *Belfast Regional Survey and Plan*, HMSO, Belfast.

Meenan, J. (1970) *The Irish Economy since 1922*, Liverpool University Press, Liverpool.

Moore, B., Rhodes, J. and Tarling, R. (1978) Industrial policy and economic development: the experience of Northern Ireland and the Republic of Ireland, *Cambridge Journal of Economics*, 2, 99-114.

NESC (1982) *A Review of Industrial Policy*, Report 64, Stationary Office, Dublin.

O'Farrell, P.N. (1972) A shift and share analysis of regional employment change in Ireland, 1951-1966, *Economic and Social Review*, 4, 59-86.

O'Farrell, P.N. (1975), *Regional Industrial Development Trends in Ireland, 1960-1973*, IDA, Dublin.

O'Farrell, P.N. (1976) An analysis of industrial closures: Irish experience, 1960-73, *Regional Studies*, 10, 433-48.

O'Farrell, P.N. (1980) Multinational enterprises and regional development : Irish evidence, *Regional Studies*, 14, 141-50.

O'Farrell, P.N. (1984) Components of manufacturing change in Ireland 1973-1981, *Urban Studies*, 21, 155-76.

O'Farrell, P.N. and Crouchley, R. (1983) Industrial closures in Ireland 1973-1981: analysis and implications, *Regional Studies*, 17, 411-27.

O'Farrell, P.N. and Crouchley, R. (1984) An industrial and spatial analysis of new firm formation in Ireland, *Regional Studies*, 18, 221-36.

O'Farrell, P.N. and O'Loughlin, B. (1980) *An Analysis of New Industrial Linkages in Ireland*, IDA, Dublin.

O'hUiginn, P. (1972) *Regional Development and Industrial Location in Ireland*, Foras Forbartha, Dublin.

O'Malley, E. (1985) Industrial development in the North and South of Ireland, *Administration*, 33, 61-85.

Perrons, D.C. (1981) The role of Ireland in the new international division of labour: a proposed framework for regional analysis, *Regional Studies*, 15, 81-100.

Rees, P.D. and Miall, R.H.C. (1981) The effect of regional policy on manufacturing investment and capital stock within the UK between 1959 and 1978, *Regional Studies*, 15, 413-24.

Rothwell, R. (1982) The role of technology in industrial change: implications for regional policy, *Regional Studies*, 16, 361-69.

Rowthorn, R.E. (1981) Northern Ireland: an economy in crisis, *Cambridge Journal of Economics*, 5, 1-31.

Stationery Office (1972) *The Accession of Ireland to the European Communities*, Stationery Office, Dublin.

Steed, G.P.F. (1968) Commodity flows and inter-industrial linkage of Northern Ireland's manufacturing industries, *Tijdschrift voor Economische en Sociale Geographie*, 59, 245-59.

Steed, G.P.F. and Thomas, M.D. (1971) Regional industrial change: Northern Ireland, *Annals of the Association of American Geographers*, 61, 344-60.

Stewart, J.C. (1976) Foreign direct investment and the emergence of a dual economy, *Economic and Social Review*, 7, 173-97.

Thomas, M.D. (1956) Manufacturing industry in Belfast, *Annals of the Association of American Geographers*, 46, 175-96.

Townroe, P.M. (1975) Branch plants and regional development, *Town Planning Review*, 46, 47-62.

Walsh, F. (1976) The growth centre concept in Irish Regional Policy, *Maynooth Review*, 2, 22-41.

Walsh, F. (1979) The changing industrial structures of Northern and Southern Ireland, *Maynooth Review*, 5, 3-14.

Walsh, F. (1980) The structure of neo-colonialism: the case of the Republic of Ireland, *Antipode*, 12, 66-72.

Wilson, T. (1965) *Economic Development in Northern Ireland*, HMSO, Belfast.

10 THE CHANGING NATURE OF IRISH RETAILING

Tony Parker

INTRODUCTION

Change is the keynote feature of both Irish retailing and consumers. Retail organisation, technique and environment are all in an interdependent state of flux and are responding to, and creating, changes among consumers. Retailing in Ireland is experiencing the changes that have been occurring in western Europe (see Davies (1979)), North America and elsewhere. Retailing is one of the most dynamic activities in the economic and social landscape, as new ideas, formats, techniques, companies and locations are developed. It is a highly competitive activity and one in which the importance of location cannot be overestimated. However, the relative importance of different locations alters through time in response to changes in the consumer and retail contexts.

The past quarter century has seen massive changes in the retail system both in the Republic and in Northern Ireland. This has been in direct response to both the changing nature and redistribution of consumer demand and to international trends in retailing. Society has changed perhaps more rapidly in the last quarter century than at any other time in history and this has been reflected in the demand for and consumption of goods.

Twenty five years ago retailing throughout Ireland could be characterised as being 'traditional'. There were no planned shopping centres and retailing was largely confined to the

237

centres of towns and cities throughout the island. There were very few multinational enterprises in Irish retailing. Most outlets sold goods over the counter: the self-selection process in both foodstores and comparison goods outlets was not yet established. Outside the urban areas, there was a preponderance of family-operated businesses typified by the bar-cum-grocery or bar-cum-general store. Today, increasingly retailing is dominated by largescale multiples, often with international links.

Many of the trends that have taken place and are continuing to develop in Irish retailing are a reflection of what has happened in other countries. Nevertheless, although retailing is very much an international activity, with ideas and skills being transferred between different countries, it is rare for developments to be adopted without some modification to local conditions, and this is certainly true of the Irish situation. Variations in local consumer demand, different retail responses in terms of organisation, selling technique and style, and the relative importance of external factors such as the legislative framework affecting retailing, all result in different countries having different retail environments. Even within the island, with two different political states and differing traditions, retailing has not developed in an identical manner. For example, although planned shopping centres have developed in both the north and the south, their locational pattern has varied. While multiples have developed throughout the island, in the Republic there has been a substantial conflict with independents in the grocery sector. The troubles have had a specific impact in Northern Ireland that has not been replicated in the Republic.

The border between the Republic and Northern Ireland also allows consumers access to different markets, as is the case in Europe. The cheaper prices obtaining in Northern Ireland in recent years have led to cross-border shopping trips from the Republic, differential economic circumstances for retailers on either side of the border and eventually government legislation in the south to halt the outflow of shoppers' money.

The focus of this chapter therefore is on the changes that have been affecting retailing within both the Republic of Ireland and Northern Ireland in recent years. The consumer context is an important framework to these changes and this is considered in the following section. Thereafter the changes in the retail context are discussed. Inevitably change brings conflict and problems, and the major issues and responses in Irish retailing are considered in the third section of the chapter. Finally, an overview provides some comments upon the way that retailing may develop in the future.

THE CONSUMER CONTEXT

Retailing responds to the overall demographic situation as well as the specific attributes of consumers, which include the changing nature of consumer organisation and lifestyles, changing consumer technology and consumer environments. The term 'demographics' has been employed to indicate the increasingly segmented consumer of today, with quite different lifestyles and sets of demands to a generation ago. Different groups of consumers occupy different locations, and therefore 'geodemographics' has become vital to retail success. Technological changes, such as car ownership, household appliances and credit facilities have aided the development of different lifestyles. Furthermore consumers are exposed not only to their physical environment, but also to their economic, social and psychological environments, each of which contributes to the consumer context for retailing.

The Demographic Background

Perhaps the most notable geodemographic feature in Ireland is the distribution of demand. In comparison to the rest of western European, including Britain, the island has a low density of population. As a result there are still large rural areas served by small towns and villages with a consequently widespread, and low, distribution of demand. This inevitably leads to problems in terms of both the range of retail provision and the supply and distribution of goods. Naturally the extent of these problems varies throughout the island depending upon local circumstances. Both the Republic and Northern Ireland are each dominated by one urban centre. Dublin, the capital of the Republic, contains almost one-third of the country's 3.5 million population, while the greater Belfast area contains about half the 1.5 million population of Northern Ireland. Within these relatively small areas there is a great concentration of spending power, which has proved attractive to retail innovators, and from which new ideas and techniques have diffused throughout the respective states.

The age structure of the population is also important for the retail market. Ireland has a very youthful population with 47% of the Republic's population and 44% of Northern Ireland's population being under 25 years of age. By comparison the European Community average is 36%. Clearly this provides a large consumer segment within the marketplace, but not a uniform segment, for it ranges from the young professional to the unemployed school leaver, as well as including children. However, it does provide the potential for future population growth and sustained demand for retail goods in the future, provided emigration does not become excessive.

Unemployment is a major feature of many western societies and the unemployment rate in both parts of the island is high by European Community standards, being 18% in the Republic and about 21% in Northern Ireland. Unemployment patterns vary spatially though with consequent implications for retail demand. The high level of youth unemployment is also important for retailing. In common with Britain, 18 to 25 year olds are no longer the employed, free-spending market segment that they were fifteen years ago, rather it is the 25 to 40 year olds with secure employment and homes and families that are the major spenders.

Consumer Organisation

The consumer market today is very different from that of a generation ago. Changes in society have influenced consumer shopping behaviour and the retail environment. Changes in work practices have led to the development of flexitime, which involves different working and leisure hours, and have increased unemployment. Developments in social attitudes and the working environment have resulted in an increased female participation rate in the workforce and an increase in part-time or short-term contract employees. Working wives are now a significant part of the labour force but restrictions upon their time have meant demands for convenience - particularly for late night opening, which both supermarkets and city centre retailers have responded to.

In the home the nuclear family unit is breaking down. Household members increasingly conduct independent lifestyles and are concerned about time-management. One person and single-parent households are growing in Ireland and all of these aspects demand responses from the retail system. 'Lifestyles capture many external influences - cultural, demographic, social and family influences' (Wilkie, 1986, p.107) and changing lifestyles result in a highly segmented consumer market. The DIY-consumer, the sports-orientated consumer and the health-conscious consumer are readily recognised and catered for by retailing. The 'dinki' ('dual income, no kids') phenomenon may be more apparent elsewhere in the western world than in Ireland, but it is increasing within certain districts of the major cities and represents one of the highest spending consumer segments. These consumers are 'money rich, time poor' and part of their lifestyle is the efficient management of time. Demands for style and design in both the goods they purchase and their shopping environments are matched by demands for convenience, both in the home where technology becomes important and in the market place, particularly where foodstuffs are concerned. Price is far less important than guarding precious leisure-time. Yet such consumers contrast radically with the 'time-rich, money-poor' unemployed where price is of utmost importance and time management is replaced by money management.

Lifestyles have increased consumers' consciousness regarding health which has led to greater leisure activities, catered for by specialist sports shops, gymnasiums, sports and health clubs, and increased concern about healthy eating. The retail trade has responded by producing healthier products, emphasising natural and wholefoods at the expense of processed foodstuffs, and providing greater consumer information regarding foodstuffs. However, this concern is more evident among middle class consumers, for lower status households are almost exclusively concerned about cheap prices and will often undertake long journeys on foot to shop at the cheapest supermarket.

Consumers are also becoming increasingly aware of their rights, which have been enshrined in legislation as goods and services, and consumer information acts. Furthermore the establishment of the Northern Ireland Consumer Council and the Consumer's Association of Ireland (in the Republic) have done much to heighten consumer awareness.

Collectively these different aspects of consumer organisation interact with each other and consumer technology and consumer environments to produce distinctive market segments. When related to the spatial dimension, where such groups live and work, then the importance of local, distinctive retail markets becomes clearly evident.

CONSUMER TECHNOLOGY

Technology has developed ways of making life easier and providing greater free time. Mobility, household appliances, financial transactions and information technology all contribute to the changing consumer context.

The increase in personal mobility experienced throughout the western world with mass car ownership has affected consumers in both the Republic and Northern Ireland. Three-fifths of households in Northern Ireland and two-thirds of households in the Republic have cars, although the figures are lower than many west European countries. The advantages of car ownership for shopping are threefold. It frees the consumer from the 'tyranny of distance'; it permits the purchase of a large volume of items on a single trip; and it enables family-based shopping trips to be undertaken. Car ownership has therefore radically altered the pattern of shopping, with one major grocery shopping trip per week being the norm for most people.

Access to a car enables consumers to 'shop around' particularly within urban areas and it has therefore facilitated the development of large-scale shopping opportunities, many of which draw custom from well beyond their immediate area. Furthermore it has turned shopping into a family outing and this has had important implications for the design, character and style of newly planned shopping facilities, as well as the retail mix, with demands for more comparison goods outlets, such as fashionwear, more service facilities such as cafes and restaurants, and less convenience outlets, particularly in the largest-scale developments. However, car ownership is not uniform and increased mobility is apparent particularly among the middle classes and in suburban areas.

Conversely, car ownership has reduced the dependence of many people on public transport. But for those without cars, in cities and rural areas alike, accessibility to shops has become that much more difficult, particularly in view of the structural changes that have occurred in retailing. Furthermore the use of the car on the shopping trip has increased pressure for parking facilities adjacent to shopping opportunities, which has created problems in town and city centres; problems which are compounded by the increased use of cars by commuters working in these areas.

Technology in the home has complemented the car in terms of 'one-stop' shopping. Widespread ownership of fridges and freezers or fridge-freezers enables people to store food for much longer periods than in the past. In conjunction with appliances like food processors and microwave ovens technology has aided the development of a 'convenience-orientated' lifestyle for many consumers. The retail responses to these technological changes have included freezer food shops - subsequently overtaken by large areas of frozen food cabinets in supermarkets - and the development of 'convenience foods' including upmarket gourmet products, a niche exploited notably by Marks and Spencer.

Technology has also affected the way that people pay for their goods and services when shopping. Worthington (1987a) has noted that attitudes towards debt have changed substantially compared to previous generations. The advent of credit and charge cards has permitted a 'live now, pay later' lifestyle to develop, and increasingly throughout not only Ireland but also western Europe, it is possible to pay for many types of goods and services with these cards, including, in recent years, groceries. Many retail organisations are increasingly introducing their own charge cards, for the information yielded by the transaction data is vital in identifying consumer segmented purchasing among their own customers.

Other technological developments have aided shopping. While banks continue to keep relatively short opening hours, the development of ATMs (automatic teller machines), which will dispense cash and handle other transactions on a 24 hour basis and are located in many major shopping centres, on main shopping streets and even in-store, has meant that cash is more readily available. The logical extension of financial transaction technology is information technology. While television and advertising have exposed consumers to other lifestyles and new goods and services, satellite television channels have introduced on-air remote selling techniques into Irish, as well as many European homes, whereby goods can be purchased by simply telephoning in one's credit card number.

Consumer Environments

Where a consumer lives, his or her economic circumstances, and their social and psychological environment all influence retailing. Differences in consumer organisation and access to consumer technologies vary in intensity with different groups in society. Given the nature of the housing market and planning policies, not only throughout Ireland but also in other countries, these groups occupy different locations with consequently varying patterns of consumer demand, shopping behaviour and retail provision. The decentralisation of population from Dublin and Belfast in particular during the last quarter century has produced new patterns of demand in those city regions.

The strategy in Northern Ireland of expanding towns such as Antrim, Ballymena and the development of Craigavon new town have increased demand for modern shopping facilities which have been met in each of those localities, as well as in other towns in Northern Ireland. By contrast, the decentralisation of Dublin's inner city population has been primarily to the outskirts of the built-up area, most notably to the western satellite towns of Tallaght, Clondalkin-Lucan and Blanchardstown during the 1970s and 1980s. Even so, these localities are still severely undershopped. By and large the general rule of suburban residential growth and inner city decline noted by Dawson (1979) holds in both Northern Ireland and the Republic. The broad resultant contrast in purchasing power is therefore between buoyant, often middle class and car owning, suburbs and depressed, usually lower status inner city areas, which has implications for the range and quality of retail provision.

Elsewhere, the countryside has seen migratory changes as high status commuters have moved into village communities with a consequent change in the nature and patterns of demand. In Northern Ireland, underlying the distribution of demand are the cleavages between Catholic and Protestant communities which create the distinctive retail issues considered later.

Due to changing lifestyles, shopping journeys now originate not just from the home but increasingly from the workplace. Up to a quarter of weekday shoppers in Grafton Street in Dublin city centre might be in town primarily for work purposes. As offices decentralise, their suburban locations can benefit from lunchtime and evening shopping activities of office workers as Fleming (1988) has documented for Blackrock and Dun Laoghaire.

Shopping is increasingly a leisure-time activity. In North America it is reputedly the favourite leisure time pursuit and Irish shopping developments, in common with their European counterparts, are endeavouring to provide an attractive ambience for this type of activity. The development of cafes, restaurants and bars in shopping areas are a response to this demand and also support an increasing trend towards eating out.

The economic environment in which consumers operate has suffered considerably during the 1980s with a fall in real disposable income. Yet this trend has affected various groups differentially, strengthening the segmentation of the population into money-rich and money-poor sections. There has been an increased awareness of value for money which has taken a number of forms ranging from support for price competitive strategies of both multiple supermarkets and discount comparison goods outlets (as well as cross-border shopping trips) to a demand for increased quality and range among goods, even at the expense of 'rock bottom' prices as consumerism becomes more evident.

The social and psychological environment of consumers has also changed in recent years. Consumer tastes have altered and expanded, so that the 'average consumer' no longer exists. Mass communication and overseas holidays have also expanded consumer horizons. This has resulted in changes in tastes, attitudes and consumer demand, not only for goods, but also in terms of presentation and design, shopping environment and services. For example, hamburgers, pizzas and kebabs would have been virtually unheard of fifteen years ago but they are now readily available from the rapidly growing fast food outlets or from the frozen food unit in the local supermarket. Demand for style and design has encouraged the development of quality fashion boutiques and has led to a radical improvement in shopping centre developments in recent years.

Inevitably, the changes in demography, consumer organisation, technology and environments, interact with each other producing a multi-faceted mosaic of consumer demand throughout the island. In many respects the higher status consumers of the Dublin or Belfast suburbs have more in common with each other, and with their counterparts in Britain, Europe or North America, than they do with residents of lower status areas in their own cities. Consumers have become internationalised and so too has retailing, with trends evident in Ireland that are occurring elsewhere.

THE RETAIL CONTEXT

Major changes have occurred in retailing within the last quarter century, and change is taking place with increasing rapidity. Many of the changes have been in response to the changing consumer context, but equally many have developed from within the industry itself. The changes have resulted in a very different retail system to that of the 1950s and 1960s. Change in the retail context can be broadly considered under the headings of organisation, technique and location, although inevitably they are interrelated and also to changes in the consumer context.

Retailing is of increasing importance in terms of employment-generation in many western countries. As countries move into the post-industrial society, employment in manufacturing industry is declining while employment in the tertiary and quaternary sectors of the economy increases.

Ireland is typical of this trend. In 1984, retailing provided employment for 121,800 people in the Republic and some 41,000 in Northern Ireland. The distributive sector in total, including the associated activity of wholesaling, employed 15.7% of the Republic's workforce and 18.4% of Northern Ireland's workforce. By comparison, manufacturing employed just 19% of the workforce in both parts of the island. Furthermore the proportions employed in the distributive sector have been growing in both the Republic and Northern Ireland, while manufacturing employment has continued to decline. Retail employment increasingly involves females and also part-time workers, with consequent changes in the consumer context.

Retail Organisation

During the past two decades there has been a shift in the balance of market power, and the development of multinational organisations. The market power has shifted in both the Republic and Northern Ireland from being dominated by independent retailers to being dominated by multiple organisations, each of which has many outlets. This is particularly evident in the grocery trade, but is also true of other retail trades. Benefits of economies of scale operate through larger organisations in terms of purchasing power, marketing, advertising, selling techniques and funding the development of additional outlets.

British-based multiples, such as Boots and Burtons, have been long-established in Northern Ireland and in the 1960s they were joined by others including Marks and Spencer and British Home Stores (now BhS). Although the troubles of the 1970s virtually halted the influx, the recent decline in the commercial bombing campaign in town and city centres together with the high turnover figures being achieved by multiples' branches in the north has led to increased interest in Northern Ireland with organisations like Boots extending their branch network and new arrivals including Chelsea Girl and Olympus Sport (Brown, 1985a). In the Republic, Marks and Spencer and BhS are both represented in Dublin city centre. Marks and Spencer have opened a second store in central Dublin and one in Cork and companies like Next, Principles,

Top Shop, Laura Ashley, Virgin Records and the British Shoe Corporation have all moved into Dublin and other urban areas in the Republic. Local multiples have also developed in such trades as electrical goods. In the face of multiple competition, independents ranging from small-scale grocers to department stores have experienced considerable problems and have either had to adapt or even close down.

Table 1: **The grocery market in Ireland and Britain**

	Northern Ireland (1985)	Republic of Ireland (1987)	Britain (1984)
Multiples	67%	64%	67%
Cooperative Society	3%	-	17%
Symbol groups	18%	17%	12%
Independents	12%	19%	4%

The most dramatic changes in the balance of market power are in the grocery sector. During the last two decades, the growth of multiple supermarket organisations has continued steadily with expansion taking the form both of newly opened outlets and the take-over of other grocery organisations. The result has been a steady decline in the number of small independent grocery outlets, particularly in the Republic. The grocery market in Northern Ireland and the Republic is estimated to be some £900 million and at least £2,500 million respectively and Table 1 indicates market share among different types of organisation in Northern Ireland, the Republic, and for comparative purposes, Britain. The multiples command virtually three-quarters of the grocery market on both sides of the border and this figure has increased from around 50% at the turn of the decade. In Northern Ireland a small percentage goes to British-based multiples such as Marks and Spencer, Littlewoods, BhS and Woolco (who have since left Northern Ireland). Compared to Britain the cooperative sector is very weak in Northern Ireland and non-existent in the Republic. The symbol groups include Spar, VG, Mace, Centra and Super Valu and command almost one-fifth of the grocery market north and south of the border. Both they and the independents have a larger market share in Ireland than is the case in Britain.

Market share differs throughout the island. In the Republic, for example, the multiples command 86% of the grocery trade in Dublin but less than one-third in the sparsely populated Connacht and Ulster counties. The impact of the growth of the multiples on small grocery outlets can be judged from information in the Census of Distribution, the latest of which were published in the mid-1970s. In the Republic there was a 35% decline in grocery outlets between 1966 and 1977, while in Northern Ireland there was a 27% decrease between 1965 and 1975.

Increasing concentration of the grocery market is being vested in fewer and fewer companies, for the number of multiples has decreased. Operating on much lower margins than the 5% to 6% that generally holds in Britain, the companies have developed a highly professional approach to retailing. In the Republic, there are now just three multiple supermarket companies - Dunnes Stores, Quinnsworth/Crazy Prices and Superquinn, while in Northern Ireland there is a similar number: Stewarts/Crazy Prices, Dunnes Stores and Wellworths. As in recent years in Britain, there has been considerable volatility in the grocery trade. In the Republic in the early 1970s, the British Liptons organisation withdrew, their larger supermarkets being sold to the locally based Five Star company. In 1977, Albert Gubay, who had developed the Kwik Save discount organisation in Britain and 3 Guys in New Zealand (Lord et al., 1988), arrived in Ireland and rapidly developed a number of 3 Guys 'no frills' discount supermarkets primarily in the Dublin region (Parker, 1978). In 1978/79 he sold out to Tesco, the British supermarket multiple (Parker, 1979), who expanded rapidly throughout the Republic by developing new supermarkets and shopping centres. 1979 saw the demise of Five Star, whose purchase by Quinnsworth almost doubled their number of outlets to 70 (Parker, 1979). In 1986 a variety of factors led to Tesco withdrawing from Ireland (Parker, 1986) and selling their operation to H.Williams, the smallest - in terms of market share - and probably most traditional of the other four multiples. H.Williams in turn collapsed in 1987 and their outlets are being sold off in a piecemeal fashion.

Quinnsworth have been a major beneficiary of these changes, gaining outlets from the Five Star takeover, prime Tesco stores from H.Williams' takeover of that organisation, and most recently gaining more outlets at the time of H.Williams' collapse. Their locational policy appears at times to be almost one of saturation in the Dublin market and it is certainly aimed at minimising Dunnes Stores' presence. At the same time, a rationalisation process has taken place over the years as Quinnsworth have closed their smaller outlets. Quinnsworth are part of the Weston family retail empire, which includes Stewarts and Crazy Prices supermarkets in Northern Ireland, Loblaws supermarkets in Canada and, until recently, Fine Fare supermarkets in Britain, as well as the Brown Thomas department store in central Dublin, the A-Wear/Gaywear chain of fashion boutiques in the Republic and Penneys non-fashion clothing chain, which trades in Northern Ireland as Primark.

Dunnes Stores, by contrast, have expanded largely through new outlet development, often acting as developer and anchor unit of shopping centres throughout the country. An Irish family-based company, Dunnes are a major non-fashion drapery business, but since moving into foodstuffs in the 1960s the company has taken a major share of the grocery market. Although they have fewer stores than Quinnsworth, both companies are now estimated to each have about a quarter of the Irish grocery market. Dunnes Stores have built their grocery reputation on cheap prices, supporting their operation through the drapery and household goods business that the company operates - often in adjoining units or within the same outlet as their grocery business.

In Northern Ireland, the largest and longest-established multiple, Stewarts, took over Crazy Prices in 1983, a multiple operating cut price supermarkets in the Belfast area and as Stewarts have moved upmarket, Crazy Prices have represented the group's competitive wing (Brown and Bell, 1986). Like Quinnsworth, Stewarts have rationalised over the years by closing down their smaller outlets. The Crazy Prices name and low price structure has also been incorporated into the Quinnsworth organisation with some of their Dublin stores reopening as Crazy Prices outlets. In a reverse direction, Dunnes Stores opened their first outlet in Northern Ireland in 1971 and have grown until they now have about 19% of the grocery market. As in the Republic, Dunnes Stores in the north are important for non-fashion draperies and household goods. The third Northern Ireland multiple, F.A. Wellworths, was originally a family-run business selling a range of goods very similar to F.W. Woolworths. The design, layout and selling formats as well as the names were also very similar. In 1983, Wellworths were taken over by the Dee Corporation, who have also taken over a number of multiples in Britain, and the chain is in the process of expansion.

As part of the internationalisation of Irish retailing, Dunnes Stores has two outlets in Spain, one in northern England at Billingham and has just purchased a supermarket and adjoining unit in Barnsley, a former Woolco store in the centre of Sheffield and is taking space in a shopping centre in Southampton. Furthermore there is the possibility of a management buy-out of Quinnsworth, which would leave that company in a position to possibly expand into Britain.

The expansion of the multiples in the grocery trade has led to an increase in the number of independent retailers joining symbol groups, as has happened elsewhere in western Europe. Contractual chains enable independents to gain the benefits of corporate purchasing power when dealing with manufacturers and to share the advantages of group advertising and marketing strategies. Contractual chains such as Spar and VG are part of multinational organisations. In the Republic one of the major wholesalers has developed a chain of independent supermarkets under the Super Valu name and, having expanded successfully outside the capital, the organisation has recently gained a foothold in Dublin with the acquisition of some of the former H.Williams outlets.

Furthermore, franchising, particularly of American companies, such as McDonalds and Wendy's hamburger restaurants, as well as 7-Eleven convenience stores (C-stores), is an increasing trend in Ireland, with Kentucky Fried Chicken outlets expanding rapidly in recent years throughout the north.

Retail Technique

In parallel with the shifts in the balance of market power and the development of multinational organisations, changes in retail technique have been taking place, most notably in terms of the number of shops, operating technique, the size of shops and consequent polarity of the operating scale, the development of scrambled merchandising and changes in technology.

Table 2: Retail outlets in the Republic of Ireland, 1977 and 1983

	Number of Outlets 1977	Outlets 1983	% Change 1977-83
Grocers	9,042	7,736	-14.4
Grocers with Off-licences	377	541	43.5
Grocers with Public Houses	1,933	936	-51.6
Tobacco, Sweets and Newsagents	2,428	2,357	-2.9
Public Houses	7,231	7,653	5.8
Off-licences	55	84	52.7
Chemists	1,199	1,132	-5.6
Butchers	1,759	1,947	10.7
Fishmongers/Poulterers	133	143	7.5
Greengrocers/Fruiterers	346	430	24.3
Bread and Flour Confectioners	350	454	29.7
Footwear	585	691	18.1
Drapery	3,169	3,900	23.1
Hardware	1,242	1,780	43.3
Electrical Goods	913	1,228	34.5
Garages/ Filling Stations	1,785	2,062	15.5
Other Retail Outlets	4,765	6,946	45.8
Total Retail Outlets	37,312	40,022	7.3

Source: Nielsen (1984)

The total number of retail outlets have generally been growing in recent years in both the Republic and Northern Ireland. During the period 1977-83, the total number of retail outlets in the Republic increased by 7% (Table 2), with the largest numerical increase in the 'other retail outlets' category which includes florists, jewellers, furniture and furnishing shops, and other non-food outlets. Increases also occurred in drapery, hardware and electrical goods, as well as among specialist foodstuffs. However, grocery outlets declined by 14%, in common with trends wherever a move from independent, family run, counter service grocers to multiple, self-service supermarkets has taken place.

In Northern Ireland there was a decline of one-fifth between 1965 and 1975, a figure which contrasted with the 25% growth from 1966 to 1977 in the Republic. In overview, rates of change vary between different business categories and different locations. For example, in the inner Dublin suburb of Rathmines between 1960 and 1983 the proportion of outlets selling convenience goods declined from almost half to only one-fifth while major increases occurred in clothing and footwear, household goods and service activities, including cafes, public houses and fast food outlets (Parker, 1984a).

A radical change has taken place in retailing in terms of selling technique. Self-selection has taken over almost completely in many retail trades and this has resulted in shops operating with lower labour, but larger selling space, requirements. A polarity in operating scale has developed, particularly with the growth of larger and larger supermarkets. Stores of 1,000m^2 were considered large twenty years ago but Albert Gubay's 3 Guys outlets standardised on about 2,000m^2 in the late 1970s, while increasingly superstores in excess of 2,500m^2 are being constructed in both parts of the island; changes which parallel the development of hypermarkets elsewhere in Europe.

At the other end of the spectrum, there is a demand for small retail outlets, often located in proximity to the major anchor outlets of supermarkets in planned shopping centres or major department or variety stores in the city centre. Growth has taken place particularly in fashion boutiques, belonging both to multiple organisations and independent retailers. Furthermore, as the operating scale increases, so too does the hinterland necessary to support large-scale operations. This allows the development of small-scale retail enterprises which fill the spatial gaps, particularly in the grocery trade. In Northern Ireland, Spar have developed '8-till-Late' outlets and VG have opened 'Late Stop' outlets: more than a third of both organisations' outlets trade as C-stores (Worthington, 1987b). In the Republic, the development of C-stores has been somewhat slower, but since the beginning of 1987, two 7-Eleven franchised outlets have opened in Dublin and the Spar group have developed over thirty outlets throughout the country.

C-stores fulfill the role of the old corner grocery shops of yesteryear, but their emphasis is very much on convenience, related to changes in consumer lifestyles: convenience in terms of location, opening hours (hence their names) and product range, with a much wider range of products than just grocery items. These outlets provide not only for the immediate population, but also for a passing trade and this is increasingly being recognised by oil companies who are developing C-stores in filling stations.

C-stores are one of many examples of scrambled merchandising, the mixing of product categories in a single outlet. In association with consumer demands for convenience and 'one stop shopping', which have facilitated and been facilitated by the development of large-scale multiple organisations operating large supermarkets and superstores with plentiful parking facilities, shops are increasingly selling a wide variety of products. The multiples' large superstores can sell non-fashion clothing, light hardware, electrical goods, gardening items, toys and car accessories from within the one unit. Conversely, Marks and Spencer, reknowned for clothing, also sell foodstuffs and in Belfast, they have recently extended and refurbished their foodhall. Further examples of company expansions include the proliferation of multiple off-licences in the North with Stewarts and Marks and Spencers both opening separate wine shops. The BhS stores in both Belfast and Dublin have recently been reorganised to include Mothercare outlets, while Habitat, also part of the major Storehouse group, is accommodated in the Dublin BhS store and will be soon opening a 'flat pack' goods warehouse facility in the Boucher Road retail park in Belfast.

Changes in technology include the palletisation of goods which substantially aided supermarkets, most notably Albert Gubay's 3 Guys organisation (see Lord *et al.*, 1988), to reduce their prices. But the major changes are taking place in terms of information technology. The introduction of EPoS (Electronic Point of Sale) data from cash points as well as changes in methods of payment makes information on stock flow and customers' requirements more accessible and significantly improves companies' decision-making (Jones, 1987). Laser scanning has recently been introduced in some supermarkets in both Northern Ireland and the Republic which provides a detailed printout of goods bought and prices paid for the customer and valuable sales information for the company. Similar information can be gained from stores's own credit cards such as those issued by Marks and Spencer (Worthington, 1987a).

Retail Location

The most notable change in terms of retail location has been the development of planned shopping centres throughout the island (Brown, 1987a; Parker, 1987a). They offer the comfort of a vehicle-free shopping environment, but usually have plenty of free car parking space surrounding them. To facilitate customers, most stay open late at least two evenings a week. The pattern in Ireland has been very much to emulate their North American origins catering for mobile, suburban residents, and the decentralisation of population to the suburbs of towns and cities in both the Republic and Northern Ireland as well as the planning policy of developing overspill or satellite towns has aided these trends. Originally seen as primarily bringing convenience goods shopping closer to people's homes, increasingly larger-scale developments have been constructed which provide comparison goods as well. This has created problems for town centre retailers most notably in Dublin and Belfast, but also in smaller urban areas, particularly in the Republic.

The development of shopping centres in Ireland contrasts with Britain where up to the mid-1970s planning was concerned with maintaining the commercial core of city centres, with the result that planned shopping centre development was largely in town centres and only recently has a more liberal attitude allowed the development of suburban or even 'greenfield' sites, as well as the development of retail parks. In both parts of Ireland, initial shopping centre development was in suburban locations in Dublin and Belfast and only subsequently did developments take place in town and city centres. In Dublin, the suburban expansion occurred largely because of the tremendous pressure for local shopping facilities in the rapidly growing suburbs. In Belfast the commercial bombing campaign of the 1970s meant that developers were reluctant to invest in high risk and high cost town centres (Brown, 1985a).

A number of trends are identifiable in the development of planned shopping centres in both parts of the island (Parker, 1982a). Centres first opened in the suburbs of the largest urban areas; subsequent developments, notably in the Republic, diffused down the urban hierarchy until now quite small towns have shopping centres. A third trend was the move towards town centre schemes, while a fourth trend was a later infill process between existing centres in the suburban areas of Dublin and Belfast. Rehabilitation of old buildings as shopping centres has taken place in central Dublin and Cork, and organisationally, supermarket multiples are increasingly acting as developers as well as anchor tenants in shopping centres.

The first planned shopping development in Northern Ireland was the Supermac supermarket which opened in the Belfast suburb of Newtownbreda in 1964, two years prior to the development of the Republic's first planned centre at Stillorgan. Both of these were located in affluent, expanding suburban areas and primarily provided convenience goods. Through to the mid-1970s, much of Northern Ireland's planned shopping development was in suburban or out-of-town locations (Parker, 1984b, 1985a). Derriaghy, Springhill, Clandeboye and the massive 20,500m^2 Ards shopping centre were all a part of this trend (Figure 1). Each centre was anchored by a major supermarket multiple, with additional unit shops selling a variety of convenience and comparison goods; the Ards centre having both a Woolco superstore selling a variety of merchandise and a Stewarts supermarket/Primark clothing outlet.

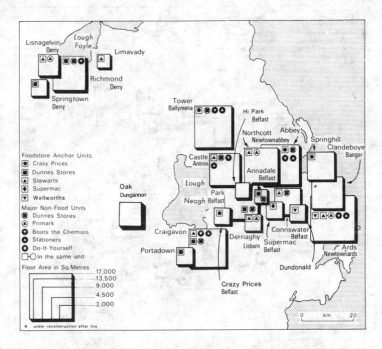

Figure 1. Shopping developments in Northern Ireland

In the Republic, initial development took place rapidly in the Dublin suburbs with fourteen centres providing an additional 100,000m² of retail space by 1974 (Parker, 1980a, 1987b). The centres ranged from relatively small developments serving an essentially local market - for example, Finglas Main, Kilbarrack and Killiney - to large-scale developments in excess of 10,000m² - for example, Northside and Ballymun. Supermarket multiples anchored each of the new developments and in a number of instances non-fashion clothing outlets provided a second anchor unit as shown in Figure 2. Many of these earliest developments in Dublin, in common with a number in the Belfast area, were open plan centres. This contrasts with later, enclosed developments throughout the island, and in some cases some of these earlier centres have been subsequently enclosed.

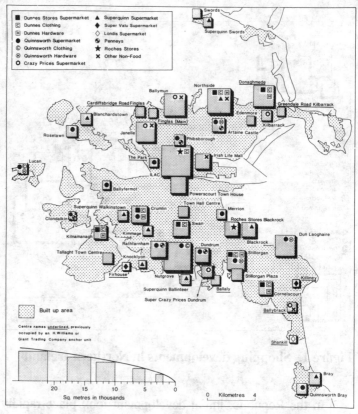

Figure 2. Planned shopping developments in Dublin

By the mid-1970s a number of other trends in shopping centre development were emerging. In the north, the trend towards town centre development commenced with the opening of the Craigavon shopping centre in 1976; a trend which continued with in-town, enclosed developments in Antrim, Ballymena and Derry during the early to mid-1980s. In the Republic, shopping centres spread outside Dublin with the opening of three developments in Cork during 1970-71 (Parker, 1981a). By the mid-70s, planned shopping centres had opened in other urban areas including Limerick, Galway, Waterford, Dundalk, Drogheda, Athlone, Letterkenny and Portlaoise (Parker, 1982a). None of these developments were located in the town centre though, with consequent distortion of the pre-existing shopping patterns.

Between the mid-1970s and 1980, eighteen centres and 70,000m^2 of planned space were added in Dublin (Parker, 1987b, 1987c). Virtually all of these are located in suburban or edge-of-town locations including the satellite towns of Tallaght, Clondalkin-Lucan and Blanchardstown to the west of Dublin. Albert Gubay was a leading developer of shopping centres during 1977-78, building standardised small-scale, 3 Guys shopping centres emphasising 'no frills' retailing in both appearance and prices (Parker, 1978; Lord *et al.*, 1988). Multiples were starting to act as developers as well as anchor unit occupants during this period. Outside Dublin, Gubay commenced, and Tesco continued, the development of their own edge-of-town centres in many towns, while Dunnes Stores and increasingly Quinnsworth acted as developer and major anchor in a number of centres. Inevitably the arrival of multiples in small towns created major problems for local independent retailers.

In-town shopping schemes commenced in the latter part of the 1970s in the Republic. Arguably, the first such development was in the centre of Dun Laoghaire. The town operates as a regional scale retail node within the built-up area of Dublin, but the shopping centre is a town centre scheme rather than a true suburban-style development. During 1979-80, Irish Life Assurance Company, operating as financier, developer and centre management, opened two in-town centres in Dublin: the Irish Life Mall, a part of an office and residential complex, and the 20,000m^2 ILAC Centre, located on Henry Street, one of the city's two major shopping streets (Parker, 1985b).

In-town centres were also developed in Limerick and Cork; the Savoy Centre and Queens Old Castle Centre in Cork city being rehabilitation schemes (Parker, 1981a, 1982a). In common with the two Irish Life schemes in Dublin, the in-town centres do not have a supermarket multiple as an anchor unit, rather they are devoted primarily to non-convenience goods. This is also true of the Powerscourt Town House development in central Dublin, which is a rehabilitation scheme of the 1774 townhouse and courtyard of Lord Powerscourt (Parker, 1985b). Developed in the style of

Covent Garden in London and Faneuil Hall in Boston, there is a concentration upon cafes, restaurants, up-market boutiques and craft shops.

Since the turn of the decade the trends of suburbanisation and shopping centre expansion to other towns have continued. Virtually all towns of 5,000 population in the Republic have at least one planned shopping centre and sometimes two. In Dublin, in-fill within the urban area is continuing as centres open in inner city residential areas (Park centre) and older suburban areas (Swan centre), as well as in established suburbs (Blackrock and Nutgrove). These latter two centres are what has been termed 'second generation' centres, with increasing emphasis upon a pleasant shopping ambience and up-market retail outlets, and many of the earlier centres in Dublin are in the process of environmental improvement, as suburban shopping facilities become increasingly competitive.

In Northern Ireland, outlying suburban centres have opened in Derry (Parker, 1985a), and a number of inner suburban centres have opened in Belfast, most notably the Connswater and Park centres (Brown, 1986a). Free-standing superstores have also developed in Northern Ireland in both foodstuffs and other retail trades. These include Dunnes' Annadale development and Wellworth's Dundonald and Coleraine superstores (Brown, 1986b), and Texas DIY and MFI furniture superstores. Although there are relatively few examples of grocery superstores in the Republic - most multiples being located in shopping centres - the development of non-food superstores, often located in industrial estates or along arterial roads is common, with both electrical goods (Power City) and DIY (Chadwicks and Atlantic) having opened in recent years. Just as the grocery trade has responded with suburbanisation to changes in the consumer context, particularly in terms of mobility, so too are bulky, non-food goods retailers.

The north is also seeing the development of in-town centres as an integral part of city centre retail structure. In Derry, the Richmond centre opened two years ago and in Belfast, there have been a number of recent in-town developments, the most notable of which is Castle Court which will bring an additional 40,000m^2 of modern shopping facilities to the city centre (Brown, 1985b). Perhaps the most radical development in Northern Ireland though has been the decision of Marks and Spencer to develop a 'greenfield' 24,000m^2 centre at the junction of the M1 motorway and the main Belfast-Dublin road near Lisburn (Brown, 1987b) extending the trend in retail decentralisation firmly into the comparison goods sector and paralleling earlier decisions in the Republic by Roches Stores, a department store multiple, to locate in the Wilton centre in suburban Cork (Parker, 1981a) and open a free standing store in Blackrock, a Dublin suburb (Parker, 1985b).

ISSUES IN CONSUMERISM AND RETAILING

Changes in the consumer and retail contexts have brought a variety of issues and conflicts to the fore. Consumers have become increasingly price conscious, although they also demand service and quality. This has led to cross-border shopping trips to Northern Ireland and has been used by multiples to engage in price wars in order to increase their market share. The growth of the multiples and their power to negotiate special terms with suppliers has also produced a strong reaction, particularly in the Republic, from the independent sector. Increasingly the Republic's planning system has been found unable to cope with the changing nature of retailing, and the legislative framework has not only been questioned but has also been amended. In the north, the troubles and the sectarian conflict have had considerable impact on shopping facilities, particularly in Belfast city centre, while in Dublin casual trading and security have created problems. Competition between the suburbs and town centres has resulted in the latter emphasising their role for comparison goods, but recent developments raise questions as to their continued success.

Inexpensive prices have been a major factor influencing consumer choice of grocery store after convenience (Parker, 1976; Coopers and Lybrand, 1985). However, increasingly many shoppers are prepared to pay a little more in order to shop in pleasant surroundings; and quality and range of goods is becoming increasingly important. As a result some multiples have installed in-store bakeries, pizza kitchens, salad bars, and most have improved their range of fresh produce.

Prices differ between the Republic and Northern Ireland and there has been much recent debate over the magnitude which has been detailed by the Restrictive Practices Commission (1987a). In recent years large numbers of people travelled to northern towns to buy groceries, electrical items and other goods, and it has been estimated that some £300 million per annum was being spent by southern shoppers in Northern Ireland. Retailers in most trades on either side of the border experienced either a boom or bust situation, depending on their location, during the past decade. Special bus trips for shoppers were organised to Newry and the Belfast area from as far afield as Limerick, Cork and Tralee. In 1987, the government of the Republic made it illegal for any goods to be imported unless the consumer had been out of the state for at least 48 hours. Denmark, with cross-border shopping problems in relation to Germany, also has similar legislation. Both countries are deemed to be in violation of the principles of the European Community, but it will take a European Court case to resolve the situation and in the interim, retailers in the Republic, particularly in border counties, are seeing an improvement in their trade, while retailers in Newry, Enniskillen, Derry and the Belfast area have experienced a decline.

Price has also been a major weapon in the expansion of the multiples. Less expensive than the independent sector (Parker, 1974, 1979b), the multiples have used their corporate power to obtain discounted prices and long-term credit from suppliers. Periodic price wars have been used as a weapon to increase market share particularly at the expense of other multiples. Dunnes Stores have consistently had a reputation for cheap prices, and in Northern Ireland initiated a price war with Stewarts in 1983 (Bell and Brown, 1986).

In the Republic there has been evidence that multiples have subsidized outlets in competitive locations in Dublin by charging higher prices for the same goods in stores outside the capital (Parker, 1980b, 1984c). Furthermore the multiples in the south have been accused of both advertising and selling below cost, and creating problems for the food processing industry by importing goods readily available in the Republic.

Consumer pressure for cheaper prices has led the multiples to introduce generic brands. Initially these were lower quality products sold for low prices and many were imported.

Of recent time the quality of the goods has improved and, in the Republic, many products are manufactured locally. Generics offer a further example of the internationalisation of retailing, with the same yellow pack products being available in Crazy Prices, Stewarts and Quinnsworth.

Multiple expansion outside the Dublin region has had a catastrophic effect upon independent grocers in the Republic. The independent sector including the symbol groups and their suppliers, responded by organising. The wholesalers formed the Irish Association of Distributive Trades (IADT) and later this merged with the independent grocers organisation, RGDATA. Planning applications for shopping centre developments and multiple expansion were opposed and forced to oral hearings. Issues of equity and efficiency in retail legislation and the need for legislation along European lines were discussed (Parker, 1981b). Questions were raised about individual multiples' share of the market, the local impact of multiples (which led to a Ministerial Directive to planning authorities) and the issues of below-cost advertising (which was banned some years ago) and below-cost selling. Under the Ministerial Directive (Local Government (Planning and Development) General Policy Directive of 1982) the planning appeals board (An Bord Pleanala) and the planning authorities must take into account a number of considerations when faced with a large-scale addition to the existing retail shopping capacity. These include the adequacy, size and location, and quality and convenience of existing retail shopping outlets; the effect on existing communities and established retail outlets and employment; the needs of the elderly, infirm or disabled and other persons who may be dependent upon local shops; and the need to counter urban decline and promote urban renewal. The government's Joint Committee on Small Businesses noted the importance of maintaining the small-scale grocery retailer and addressed the problems of food imports and below-cost selling (Oireachtas Eireann, 1984; Parker, 1985c). A Restrictive Practices Commission report on the grocery trade recommended the banning of below-cost selling (Restrictive Practices Commission, 1987b) and recently such a ban was imposed. In Northern Ireland, the conflict has been nothing like as apparent, but as the grocery market intensifies with the expansion of Wellworths and Dunnes

Stores and as Stewarts/Crazy Prices attempt to hold their market share, then it may well create problems for the independent sector.

The retail planning system differs between the Republic (Parker, 1979c) and Northern Ireland. In the Republic, planning has often been unable to cope with the changing nature of retailing and consumer behaviour (Parker, 1982b). It has been reactive rather than prescriptive, planning for trends long gone rather than in tune with current social trends, often employing outdated notions of a hierarchy of shopping nodes, and with planning applications frequently being judged on engineering and traffic criteria rather than social and commercial needs. In Northern Ireland, the Department of Environment (DoE) acts as the planning authority and has also acted as the developer of a number of planned shopping centres. The 14,000m^2 Richmond Centre in Derry was developed by DoE primarily because no one else would take the risk (Brown, 1985c). As such there has been a much closer relationship between planning and the development of shopping opportunities in the north.

Security problems have beset retailing in Northern Ireland in recent years. In the early to mid-1970s there was a sustained commercial bombing campaign in the north, and it has been estimated that a quarter of the total retail floorspace in central Belfast was destroyed between 1970 and 1975 (Brown, 1985d). As a result, security barriers were erected in many towns throughout Northern Ireland (Brown, 1985e) and pedestrianisation schemes were developed in a number of towns including Belfast, Coleraine and Portadown. Nevertheless a shift in shopping habits took place with consumers avoiding high-risk town centres and also avoiding shopping centres which were located across the perceptual, and physical, sectarian divide. Together with the decentralisation of population to satellite towns and the urban periphery, these factors positively encouraged the development of the shopping centre phenomenon (Brown, 1985c). Sectarian differences have also impacted in other ways on the retail system: in the presence of two branches of the same symbol group outlet in one village, each owned by shopkeepers of different religious affiliations; in attitudes towards Sunday trading; in attitudes towards products that are manufactured in the

Republic, with the consequence of some retailers in Protestant districts being 'asked' to remove these goods from their shelves; and in patronage patterns of the Park Shopping Centre, situated just beyond the perceptual sectarian divide in Catholic West Belfast (Brown, 1986b).

The suburbanisation of retailing throughout the island has meant that city centres have been facing a challenging period in recent years. Both local authorities and the business community are increasingly recognising the importance of providing a pleasant town centre shopping environment. Late night shopping has been instituted in both Belfast and Dublin to counteract the competition of the suburbs. As elsewhere in Europe, in-town shopping centres have been developed as has car parking in close proximity to the retail core, while the development of the DART rapid rail system in Dublin has also generated more city centre shoppers. Environmental improvements particularly in the form of the pedestrianisation of streets have been introduced in Belfast and Dublin and have been suggested for other towns, while the environmental improvement of buildings and the streetscape, of street furniture, and the influx of restaurants and cafes, have all aided the revitalisation of the city centre retail areas, as too has increased investment and confidence in town and city centres. In Northern Ireland this confidence has been in response to the decline of the commercial bombing campaign and a reorganisation of security arrangements (Brown, 1985d). In the Republic the introduction of designated areas with financial incentives for commercial and residential development has done much to increase investment in many town and city centres throughout the country. At the same time the security presence is a necessary feature of the north, while the south has problems with crime against property and the person in Dublin city centre - some £19 million per annum being spent by businesses on crime/security problems - and illegal casual trading also creates problems of congestion in Dublin's major shopping streets.

THE FUTURE

Change will be the major feature of the future for retailing in Ireland, as elsewhere. New forms of organisation and technology will affect both consumers and retailers as will potential locational shifts. Increased leisure will almost certainly result in the development of combined retail and leisure activity centres, as has happened on the continent. However, whether these occur in town and city centres or on greenfield sites remains to be seen. The Sprucefield development by Marks and Spencer in the north and the proposed Blanchardstown town centre, in the Dublin satellite town, with 100,000m^2 of space of which almost a third will be leisure-related activities, may well herald a future trend for Irish retailing. Already the latest shopping centre to open in Dublin (Janelle) has a significant leisure component and retail parks on the British model may also evolve as significant retail features.

Furthermore, increasing consumer polarisation may occur with the technological developments of 'teleshopping' ('viewdata'), whereby goods and services can be ordered directly by in-home computer systems (Guy, 1985). This could lead to a substantial shift in retail property particularly with large grocery outlets becoming unnecessary in middle class areas. At the same time, the needs of consumer convenience and time-management will lead to an increase in C-stores in the future and also the provision of attractive convenience foods. The prospects for the future development of retailing are as exciting as they are uncertain, but to quote Reagan's election rallying cry 'you ain't seen nothing yet'.

REFERENCES

Bell, J. and Brown, S. (1986) Anatomy of a supermarket price war, *Irish Marketing Review*, 1, 109-17.

Brown, S. (1985a) The 'Retail Revolution' in Northern Ireland, The Northern Ireland Economy : *Quarterly Review and Outlook*, 2(2), 40-9.

Brown, S. (1985b) Smithfield Shopping Centre, Belfast, *Irish Geography*, 18, 67-9.

Brown, S. (1985c) Conflict and its influence upon the retail environment of Northern Ireland, in Shaw, S., Sparks, L. and Kaynak, E. (eds), *Marketing in the 1990s and Beyond*, Proceedings of the Second World Marketing Congress, University of Stirling.

Brown, S. (1985d) City centre commercial revitalisation: the Belfast experience, *The Planner*, 71(6), 9-12.

Brown, S. (1985e) Central Belfast's security segment - an urban phenomenon, *Area*, 17(1), 1-9.

Brown, S. (1986a) The Park Centre, *Estates Gazette*, 280, 156-9.

Brown, S. (1986b) The impact of religion on Northern Ireland retailing, *Retail and Distribution Management*, 14(6), 7-11.

Brown, S. (1987a) Shopping centre development in Belfast, *Land Development Studies*, 4, 193-207.

Brown, S. (1987b) Sparks fly in Belfast, *Estates Gazette*, 282, 1332-3.

Brown, S. and Bell, J. (1986) Multiple grocery retailing in Northern Ireland, *Retail and Distribution Management*, 14(1), 57-60.

Coopers and Lybrand (1985) *Grocery Shopping in Northern Ireland*, Coopers and Lybrand Associates (NI) Ltd, Belfast.

Davies, R.L. (ed.) (1979) *Retail Planning in the European Community*, Saxon House, Farnborough.

Dawson, J. (1979) Retail trends in the EEC, in Davies, R.L. (ed.) *Retail Planning in the European Community*, Saxon House, Farnborough, 21-49.

Fleming, M. (1988) *Shopping Behaviour of Office Workers in Blackrock and Dun Laoghaire*, Centre for Retail Studies, University College Dublin.

Guy, C. (1985) Some speculations on the retailing and planning implications of 'push-button shopping' in Britain, *Environment and Planning B*, 12, 193-208.

Jones, G. (1987) EPoS and the retailer's information needs, in McFadyen, E. (ed.) *The Changing Face of British Retailing*, Newman Books, Dublin, 22-32.

Lord, D., Moran, W., Parker, A.J. and Sparks, L. (1988) Albert Gubay: an international entrepreneur, *International Journal of Retailing*, 3(3), 2-54.

Nielsen (1984) *Nielsen Retail Census 1983*, A.C. Nielsen of Ireland Ltd, Dublin.

Oireachtas Eireann (1984) *Second Report of the Joint Committee on Small Businesses - Retail and Distribution*, Pl. 2705, Dublin.

Parker, A.J. (1974) Intra-urban variations in retail grocery prices, *Economic and Social Review*, 5(3), 393-403.

Parker, A.J. (1976) *Consumer behaviour, motivation and perception : a study of Dublin*, Department of Geography, University College Dublin Research Report.

Parker, A.J. (1978) Discount grocery comes to Dublin, *Retail and Distribution Management*, 6(2), 36-9.

Parker, A.J. (1979a) A volatile year for Irish grocers, *Retail and Distribution Management*, 7(6), 25-8.

Parker, A.J. (1979b) A review and comparative analysis of retail grocery price variations *Environment and Planning A*, 11, 1267-88.

Parker, A.J. (1979c) Retail Planning in Ireland, in Davies,

R.L. (ed.) *Retail Planning in the European Community*, Saxon House, Farnborough, 177-202.

Parker, A.J. (1980a) Planned retail developments in Dublin, *Irish Geography*, 13, 83-8.

Parker, A.J. (1980b) *Spatial Variations in Intra-Organisational Retail Prices*, Department of Geography, University College Dublin Research Report.

Parker, A.J. (1981a) Planned retail developments in Cork city, *Irish Geography*, 14, 111-16.

Parker, A.J. (1981b) The retail legislation debate, *Industry and Commerce*, May, 30-1.

Parker, A.J. (1982a) The development of planned shopping centres in the Republic of Ireland, *Retail and Distribution Management*, 10(2), 25-9.

Parker, A.J. (1982b) Shopping Centres: Changing Trends in Retailing, *Pleanail*, 1(1), 6-19.

Parker, A.J. (1984a) Stability and change in the Rathmines retail system, *Irish Geography*, 17, 108-12.

Parker, A.J. (1984b) Planned shopping developments in Northern Ireland, *Estates Gazette*, 272, 131-6.

Parker, A.J. (1984c) Spatial variations in retail grocery prices within organisations: Supermarkets in the Republic of Ireland, in Bender, R.J. (ed.) *New Research on the Social Geography of Ireland*, Mannheimer Geographische Arbeiten 17, 223-59.

Parker, A.J. (1985a) Shopping Centres in Northern Ireland, *Irish Geography*, 18, 63-6.

Parker, A.J. (1985b) Planned shopping developments in Dublin, *Estates Gazette*, 275, 524-7.

Parker, A.J. (1985c) Small shops in Ireland: the government takes a hand, *Retail and Distribution Management*, 13(4), 22-6.

Parker, A.J. (1986) Tesco leaves Ireland, *Retail and Distribution Management*, 14(3), 16-20.

Parker, A.J. (1987a) Retail developments and planning in Ireland, in Metton, A. (ed.) *Geographical Research on Commercial Activities*, Proceedings of the IGU Study Group: Geography and Commercial Activities International Symposium Paris 1985, Collection Universities d'Orleans, 69-86.

Parker, A.J. (1987b) The Changing Nature of Retailing, in Horner, A.A. and Parker, A.J. (eds) *Geographical Perspectives on the Dublin Region*, Geographical Society of Ireland Special Publication No.2, Dublin, 27-40.

Parker, A.J. (1987c) *Dublin Shopping Centres : A Statistical Digest*, Centre for Retail Studies, University College Dublin.

Restrictive Practices Commission (1987a) *Report of Review of Restrictive Practices (Groceries) Order, 1981*, Pl.4678, The Stationery Office, Dublin.

Restrictive Practices Commission (1987b) *Report on Alleged Differences in Retail Grocery Prices between the Republic of Ireland and the United Kingdom (including particularly Northern Ireland) 1987*, Pl. 5236, The Stationery Office, Dublin.

Wilkie, W. (1986) *Consumer Behaviour*, John Wiley & Sons.

Worthington, S. (1987a) The credit explosion, in McFadyen, E. (ed.), *The Changing Face of British Retailing*, Newman Books, Dublin, 33-41.

Worthington, S. (1987b) Convenience Stores in Northern Ireland - Vive La Différence !, Irish Retailing and Distribution Symposium, Northern Ireland Retailing and Distributive Trades Research Centre, The Queen's University of Belfast.

11 TRANSPORTATION

James Killen and Austin Smyth

INTRODUCTION

The position of transport within the overall context of geography, resource management and planning is fundamental. From the viewpoint of the individual, transport provides the means by which movement can be achieved in order to satisfy personal needs through such journeys as to work, to educational institutions, to purchase goods, to avail of services and to engage in social activities. Even when one is at rest one is likely to be affected by the noise and pollution transport causes. From an economic viewpoint, transport provides the means by which economic activity is facilitated by such movements as raw materials to processing plants and products to markets. From a geographical/environmental viewpoint, transport facilities comprise some of the most highly visible man-made features in the landscape. Furthermore, the location of such facilities as motorways and railway stations can affect fundamentally such other aspects of landscape design as rural settlement patterns and industrial location. Finally, from a political viewpoint, transport provides on the one hand an entity where positive planning can materially benefit both individuals and the economy as a whole and on the other, an entity where, in certain cases, careful Government and/or international control is required. To some extent, the transport issues that arise in an Irish context are the same as those which arise in any country which has developed

271

from an essentially agricultural base towards a more mixed economy with relatively high living standards. Two important factors must be imposed on this general picture in the case of Ireland: first, Ireland comprises a relatively small island; second, the island is divided by an international boundary.

The island location of Ireland means that internal journey distances are low by European standards. Journeys outside the island require a sea or air transfer which underlines the importance of these modes and raises such questions as to how and by whom they are controlled.

An immediate effect of the political boundary between Northern Ireland and the Irish Republic has been and continues to be that the control of transport policy on either side of the boundary has rested with different parties. Given that transport planning comprises facilitating communication between places, it could be argued that the adoption in many instances of different transport policies on either side of the political boundary has not always been to the benefit of the island as a whole. An important influence in this respect in recent times has been the European Community (EC) which aspires to certain overall policy objectives *vis-à-vis* transportation. This chapter considers Irish transport within the foregoing context. A brief review of transport infrastructure leads to a definition of the transport issues which are currently of importance. The main body of the chapter discusses these issues in detail.

TRANSPORT INFRASTRUCTURE

Roads

The Irish Republic is served by 92.3 thousand kilometres of roads. There were 26.09 kilometres of road per thousand persons in 1986. The most important routes, designated 'National Primary Routes' (Figure 1) comprise 2.8% of this length. These were estimated to be carrying 23.8% of all traffic in 1980. Although 94.7% of all roads have an improved surface there are less than 15 kilometres of motorway in the Irish Republic.

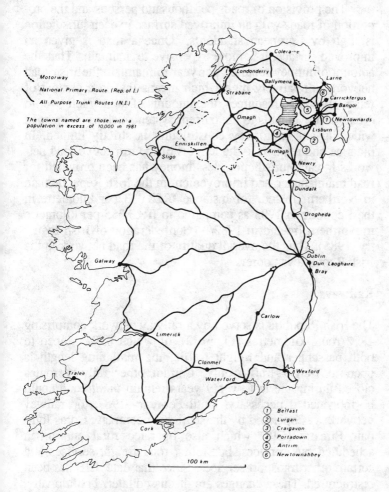

Figure 1. The Road Network of Ireland

Northern Ireland is served by 23,500 kilometres of roads including 111 kilometres of motorway. There were 15.04 kilometres of road per thousand persons in 1981. The major routes (other than motorways) are designated 'All Purpose Trunk Routes' (Figure 1). The proportion of roads with an improved surface is similar to that in the Irish Republic.

The provision of roads per thousand persons and the proportion of roads with an improved surface in each jurisdiction is high by European standards. Once a road is given an improved surface, it is more expensive to maintain. Thus the amount of money required per year to maintain the Irish road network is relatively high which means that in a context of relatively scarce resources, the amounts of money left over to effect improvements to the current system such as major road widening and motorway construction is strictly limited. In practice and relative to the lengths of the respective road networks in each jurisdiction, less money has been provided for road maintenance and improvement in the Irish Republic than in Northern Ireland, for instance IR£2,033 per kilometre in the Republic in 1983 as compared to IR£3,355 per kilometre in Northern Ireland in 1982/3 (Confederation of Irish Industry, 1985). Thus the overall quality of the road network in the Irish Republic is poorer.

Railways

The Irish Republic is served by a railway network comprising 1872 route kilometres (Figure 2), of which 85% is open to both passenger and freight traffic; the remaining length is open to freight traffic only. Certain other railway routes closed during the last twenty years remain *in situ* and could be reopened if necessary. Until February 1987, the railway network was operated by the semi-state company Coras Iompair Éireann (CIE) which also operated rural and urban scheduled bus services, bus tours, road freight services and certain other transport services. Since that date, CIE has been restructured. These changes are discussed later. Traditionally, and in common with most European countries, the railway in the Irish Republic incurs a financial loss and is revenue-supported in accordance with EC regulations. In 1985, the loss after payment of revenue support of IR£90.63 million was IR£2.62 million.

Northern Ireland is served by a railway network comprising 328 route kilometres (Figure 2). With the exception of 24 route kilometres which are used for empty rolling stock transfers, the network is open to passenger services. There are no internal railway freight services in Northern Ireland although freight trains conveying goods from the Republic to Belfast and to Londonderry for ultimate transfer to County Donegal are handled. The railway in Northern Ireland has consistently shown a small surplus after payment of revenue support in accordance with EC regulations. The level of this support peaked in 1983/4 at Stg£4.6 million (1982 prices) and has declined by more than 15% in real terms since.

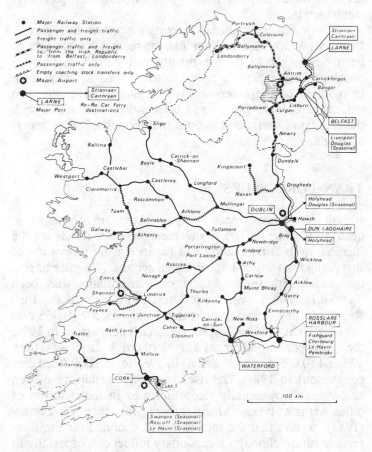

Figure 2. Railways and airports in Ireland

275

Airports and Seaports

Ireland possesses four major airports (Figure 2). The largest of these, Dublin, handled 2.28 million passengers in 1985 and offered direct scheduled flights to various destinations in the United Kingdom, continental Europe and to New York and Boston. In addition, there are a number of regional airports five of which, Londonderry, Belfast (Harbour Airport), Waterford, Galway and Connaught (Knock), offer direct scheduled flights to Great Britain. Air travel within the island accounts for an extremely small proportion of all trips undertaken therein.

The major international seaports are shown in Figure 2. In recent years, Larne, Dun Laoghaire and Rosslare have been developed to handle mainly roll on/roll off (ro-ro) passenger/freight ferry services to the United Kingdom and, in the case of the latter, to France. Such services are also offered at Belfast, Dublin and Cork which also, along with the Shannon Estuary and various smaller ports, handle general freight.

TRANSPORT ISSUES

Recent population trends within Ireland are discussed in detail in Chapter 4. The key elements of these trends are the increase in total population and the increasing proportion of that population which lives in urban centres. A further important trend within the context of transportation has been increasing vehicle ownership. Thus, in the case of the Irish Republic, car ownership increased from 186,302 (0.066 cars per person) in 1961 to 774,594 (0.23 cars per person) in 1981; the corresponding increase for Northern Ireland was from 135,264 (0.095 cars per person) in 1961 to 356,000 (0.24 cars per person) in 1981. The rises in car ownership have been accompanied by equally steep increases in the numbers of other vehicle types. Most authors, for example Feeney (1982), predict that the increases will continue for the foreseeable future although recessionary influences, especially in the Irish Republic, where car ownership has actually declined most recently, may act as a limiting factor.

The main result of the foregoing trends has been a dramatic increase in traffic flow levels especially within the major urban areas and on the main roads between them and this in its turn has raised important issues in a planning sense. Many Irish towns and cities now suffer from serious traffic congestion. This raises the issue as to the extent to which transport facilities such as roads and car parks can and should be improved even if this involves destroying part of the urban fabric itself. Another issue concerns what the role of urban public transport - buses, suburban railways and taxis - should be and whether, for example, improvements in public transport performance brought about by such steps as the electrification of railway lines and the provision of bus lanes will cause a significant shift away from the private car and ultimately improve traffic conditions as a whole.

The increase in traffic on the major interurban roads raises the issue of whether the major thrust of roads policy should be to invest heavily in these roads or whether such finance as is available should be used to bring the complete road network up to some minimum standard thereby ensuring that users in the remoter areas are not unduly discriminated against. The matter of inter-urban traffic also raises the highly important and difficult issue of what should be the role, if any, of the railway as a mover of passengers and freight. The view adopted concerning the railway has implications in its turn for the role of rural and long distance bus services and for freight movements by road.

The foregoing represent the major issues which are discussed further in this chapter. They are inter-related both with each other and with the matters discussed in other chapters, for example, population distribution and the level and distribution of agricultural and economic activity. The importance within a transportation context of the international dimension which raises such issues as ports policy and the role of air and sea transport has been noted already. In this respect, the comparison between Northern Ireland and the Irish Republic which is undertaken throughout the chapter is of particular relevance.

URBAN TRANSPORT

As has been the case almost worldwide, the twin trends of increasing urbanisation and increasing car ownership have combined to create problems of some magnitude for urban transport in Ireland. The extent to which the symptoms of these problems which include traffic congestion, parking problems, inadequacies in public transport, difficulties for pedestrians, traffic accidents and environmental impacts appear tends to vary by city size and structure. It also tends to reflect the urban transport planning strategies which have been followed.

It is worth emphasising at the outset that by European and North American standards, much of Ireland remains an essentially rural society. Apart from Dublin, Belfast and Cork, no urban centre has a population in excess of 100,000. The transportation problems of the smaller centres generally comprise localised traffic congestion and car parking difficulties which can usually be ameliorated through traffic management schemes and road improvements. A further feature of these towns is that their size is such that they cannot support a comprehensive public transport system. Thus, while many journeys can be made on foot, those who reside in these towns and who do not have access to a car are often severely disadvantaged.

In contrast to the smaller urban centres, the transportation problems of Dublin, Belfast and, to a lesser extent, Cork are more severe and have been subject to intense study during the past twenty years. (For a review, see for example: Heanue *et al.* (1971/2), Killen (1979), Transport Consultative Commission (1980) in the case of Dublin; Travers Morgan and Partners (1976) and Smyth (1986) in the case of Belfast; Skidmore *et al.* (1978) in the case of Cork.) Some characteristics of Dublin and Belfast in relation to specific transport indicators are given in Table 1. With regard to traffic movement, Belfast enjoys higher average traffic speeds due to a better road system which gives rise to less traffic congestion. With regard to safety, comparative statistics are not readily available on a city by city basis : figures at the national level - Department of Transport (1986), Royal Ulster Constabulary (Various) - suggest that the Republic and, to a lesser extent,

Northern Ireland have accident rates which exceed considerably those of the rest of the United Kingdom and much of Western Europe.

Both cities suffer severe transport related environmental problems. In the case of Dublin, these problems generally stem from noise, vibration and air pollution on congested streets clogged with traffic; in Belfast, the construction of a number of new purpose-designed routes and major reconstruction on the principal radial routes has ameliorated traffic congestion but it has also led to different problems including severance of access for pedestrians, high profile visual impacts and, if the overall figure for Northern Ireland can be applied to Belfast, higher accident rates than elsewhere in the United Kingdom.

Car parking provision in Dublin and Belfast is relatively generous. Thus there is little restraint in using the private car for city centre based trips. Although the available statistics do not relate to exactly comparable definitions of the central business districts, it is evident that in comparison to Belfast, Dublin has a higher proportion of private car parking facilities and on-street space and a lower proportion of publicly available off-street car parking. The high level of on-street parking coupled with low levels of enforcement of parking regulations adds significantly to traffic congestion in Dublin.

With regard to public transport, both cities are served by conventional bus services, limited suburban rail networks, taxis and, in the case of Belfast, para transit (taxis operating a frequent service along fixed routes and charging a fixed fare per passenger). Comparative statistics (Table 1) show that the level of service offered by conventional bus services in Belfast is low; Dublin enjoys a higher level of service but one which is inferior in terms of vehicle speeds. Para-transit which operates in competition with conventional bus services on certain routes in Belfast makes a substantial contribution to the supply of transport in the areas it serves and is one factor explaining why, in contrast to Dublin, bus patronage in Belfast has declined by some 80% in the last fifteen years. *Per capita* bus use in Belfast is now approximately one-half that in Dublin.

Table 1: Comparative transport statistics:Dublin and Belfast

Dublin

Private transport

Average traffic speed km/h	20	(1976)
Private CBD* parking spaces/km^2	952	(1977)
Public CBD off-street parking spaces/km^2	278	(1977)
Public CBD on-street parking spaces/km^2	635	(1977)

Bus

Vehicle km operated (million)	45.3	(1985)
Route length (km)	806	(1985)
Length of bus priority lanes (km)	14.26	(1985)
Fleet size (vehicles)	865	(1985)
Maximum frequency on any route in peak hour (mins)	6	(1985)
Average speed (km/h)	11	(1977)
Total passenger journeys (millions)	162	(1985)

Suburban railway

Average no. of passengers carried (weekdays)	45,000	(1985)

Taxis and Para-transit

Number of taxis	1,691	(1984)
Number of para-transits	0	(1986)

Modal split for journeys entering CBD

Private vehicles (%)	56	(1979)
Bus (%)	35	(1979)
Suburban railway (%)	9	(1979)

Total trips to CBD per hour at peak	66,400	(1979)

Transportation

Belfast

Private transport

Average traffic speed km/h	Peak:26-27.5
	(Est. 1986)
	Non-peak:35-40
	(Est. 1986)
Private CBD* parking spaces/km²	432 (1982/3)
Public CBD off-street parking spaces/km²	295 (1982/3)
Public CBD on-street parking spaces/km²	166 (1982/3)

Bus

Vehicle km operated (million)	11.9 (1983)
Route length (km)	197 (1985)
Length of bus priority lanes (km)	1 (1985)
Fleet size (vehicles)	360 (1985)
Maximum frequency on any route in peak hour (mins)	7.5 (1985)
Average speed (km/h)	12 (1985)
Total passenger journeys (millions)	32 (1982)

Suburban railway

Average no. of passengers carried (weekdays)	15,000 (1985)

Taxis and Para-transits

Number of taxis	1290 (1984)
Number of para-transits	343 (1986)

Modal split for journeys entering CBD

Private vehicles %	68 (1976)
Bus %	24 (1976)
Suburban railway %	8 (1976)
Total trips to CBD per hour at peak	23,130 (1976)

* Dublin CBD comprises 21 km centred on O'Connell Bridge.
Belfast CBD comprises 17 km centred on Belfast City Hall.
Sources: Coras Iompar Eireann; Lavery (1986); Transport
Consultative Commission (1980); Travers, Morgan and
Partners (1976); Official records and timetables.

The policies which have been followed with regard to suburban railway services in both cities contrast markedly: large-scale investment in Dublin's suburban railway network including the electrification of the coastal railway line, the introduction of feeder bus services and the improvement of park-and-ride facilities has raised the quality of rail services significantly; in the case of Belfast, investment levels have been lower and the railway has been further handicapped by its poor penetration of the city centre area.

The foregoing contrasts reflect to a large extent the different transportation planning strategies and, more generally, the different urban planning strategies which have been followed in the two largest cities on the island. The vast majority of new investment in Belfast has been in roads, car parking and traffic management and the advantages that this has brought, for instance in terms of ease of access for car users, have been offset by such disadvantages as negative environmental impacts and difficulties for those without access to a car. In Dublin, the emphasis on the ground (although not necessarily in planning proposals) has been on low-cost piecemeal solutions such as the provision of bus lanes. These solutions while providing local temporary relief have been unable to avert severely deteriorating conditions for all road users and, more generally, for all within the city. The development of Dublin's suburban railway system represents a significant exception to this emphasis.

INTERURBAN AND RURAL TRANSPORT

The essentially rural nature of much of Ireland's population together with the dominance of just two major urban centres, Dublin and Belfast, which are separated by an international boundary, results in inter-urban and non-urban flows of passengers and freight which, by European standards, are small. Five key issues have dominated planning policy discussions:

1. Defining the role of the railway

2. The maintenance and upgrading of roads

3. The provision of rural bus services

4. The particular problem of providing public transport in remote rural areas

5. The impact of the political boundary between the Republic and Northern Ireland.

These are discussed in the following sub-sections.

The Role of the Railways

Of all of the elements comprising transport infrastructure, the railway systems in both Northern Ireland and the Republic have probably received more attention than any other. The reason for this is not hard to find: railway operations have traditionally incurred large financial losses.

The reaction to the so-called 'railway problem' has differed markedly on either side of the political boundary. In Northern Ireland, most of the original railway network has been closed. In the Republic the attitude has been more benign: while many minor routes have closed (Killen, 1973), most of the main routes remain open. In 1981, there were 0.54 route-kilometres of railway per thousand persons in the Republic as compared to 0.21 route-kilometres per thousand persons in Northern Ireland. The former figure represents the most generous provision in the entire European Community. The figures relative to the areas of the two jurisdictions are 0.027 route-kilometres per square kilometre for the Republic and 0.024 route-kilometres per square kilometre for Northern Ireland. In contrast to the previous figures, these represent the poorest provisions relative to area in the European Community with the exception of Greece.

The foregoing statistics encapsulate very neatly the key to the so-called railway problem in Ireland. In order for any railway network to survive without incurring significant financial losses, what is required is a large total population distributed between sizeable population centres and with substantial distances between them. These conditions simply do not exist in either Northern Ireland or in the Republic. In the former, distances are relatively short and approximately one-third of the total population is living within the Belfast urban area. In the latter, while distances are longer, the population is still relatively small and with approximately thirty per cent concentrated in one centre, Dublin.

The foregoing conditions mean that Irish railways cannot be expected to generate large profits in future years; indeed, in the case of the Republic in particular and assuming a continuance of the current level and method of operation, the present financial performance is likely to be sustained into the foreseeable future. Accepting this to be the case, the continuing existence of the railway can presumably be justified only if it can be shown that the social and economic benefits it confers exceed the financial loss it incurs. A recent Government publication in the Republic (Department of Communications, 1985) identifies two such benefits: the railway is a significant employer, an important consideration at a time of high unemployment and second, the railway could assume strategic importance at a time of national emergency. Other such benefits might include the diversion by the railway of traffic from the road system thereby easing traffic congestion and safety considerations. The difficulty with all of these benefits is that while they surely exist, they generally cannot be quantified easily in financial terms and thus offset accurately against the financial losses being incurred. The outcome of this has been that decisions concerning the railway have been and will continue to be highly politicised. Indeed, the varying fortunes of the railway systems in Northern Ireland and the Republic reflect very much alternative political views which have been adopted as to their potential role and value.

The short to medium term future of the railway systems in Northern Ireland and in the Republic seems relatively secure. In the case of the former, there has been limited investment recently, for example in new rolling stock. The linking of the railway route from Larne to the rest of the railway system in Belfast via a cross-city link initially approved in principle by Government in 1978 (Department of the Environment, 1978) has now been reaffirmed. In the case of the Republic, the Government has effectively decided to retain the railway in the medium term but without substantial new investment. Given that the most efficient use of such new investment would be in schemes which would ultimately reduce financial losses through the introduction of technology which reduces employment levels such as centralised signalling and automated level crossings, the number employed on the railway in the Republic is unlikely to fall significantly. This may fulfill a short-term social objective but may ultimately threaten the railway's long-term future.

The Maintenance and Upgrading of Roads

The down turn in the fortunes of the Irish railway system since the Second World War has been paralleled by the rise of road transport. The dramatic increase in the number of private cars, together with comparable increases in freight vehicle numbers, has placed severe pressures on the road system and, as in the case of railway policy, these pressures have met with contrasting reactions in each jurisdiction.

In Northern Ireland, the government's response has been to invest very heavily in road improvements including the provision of a metalled surface on virtually all roads and the provision of new limited access highways. In some years, the level of investment in road improvements in Northern Ireland has been unparalleled in any region of the United Kingdom and, on a mileage basis, has been as much as four or five times that allocated to improvements in the Republic. The outcome today is that Northern Ireland possesses a road system which is unsurpassed for quality in the United Kingdom or, indeed, Western Europe; indeed, in the case of Northern Ireland, there is evidence that there has been considerable over-provision in this area of government expenditure (Northern Ireland Economic Council, 1981).

In the Irish Republic, the initial policy thrust after the Second World War was to provide a metalled surface on all major and minor roads. This reflects the fact that at that time, responsibility for road maintenance and improvement effectively rested solely with individual County Councils whose interests were essentially local rather than national. The outcome was that by the mid 1960s, the Republic possessed a road network which was virtually completely metalled and therefore expensive to maintain but which had scarcely been improved, for instance in terms of width and geometry. The major roads carrying heavier traffic were thus perceived as being particularly deficient at this time.

In order to rectify this situation it was decided in 1969 that the national government would assume responsibility for the maintenance and improvement of the main road system now taken as comprising roads designated as 'National Primary' (Figure 2) and 'National Secondary' routes. A road development plan published in 1979 (Department of the Environment, 1979) envisaged an expenditure of IR£1,100 million (1984 prices) on road improvements over the period 1980-9 with a further IR£950 million (1984 prices) set aside for maintenance. Expenditure on road improvements did not reach the levels anticipated during the period 1980-5 but the National Plan published in 1984 (Stationery Office, 1984) and elaborated upon in 1985 (Department of the Environment, 1985) projects an expenditure on road improvement in excess of that anticipated in the earlier plan during the period 1985-7. This expenditure is justified by the argument that a high quality road system is necessary to facilitate economic development.

Whatever may transpire *vis-à-vis* future expenditure on roads in the Irish Republic, it is still the case that these expenditures are low by Northern Irish and European standards. This lower level of road investment in the Irish Republic reflects to some extent the poor state of the Irish economy in recent years and underlines the importance of obtaining non-Government funding for road improvement where possible. One potential source of such funding is the European Regional Development Fund which has already provided some finance. Another possibility is the introduction of toll roads financed wholly or partly by the private sector and operated under the terms of some agreement sanctioned by the State.

The Provision of Rural Bus Services

Rural bus services were introduced in Ireland in the 1920s. By 1950, the vast majority of services were being operated by two semi-state organisations which were protected by monopoly and which also controlled the railways and road freight services in their respective areas of operation: the Ulster Transport Authority (UTA) in Northern Ireland and Coras Iompair Éireann (CIE) in the Republic. This legislative situation whereby railway and road services were effectively controlled by one semi-state company meant that government policy *vis-à-vis* the railways could be implemented expeditiously: if that policy was to effect railway closures, alternative bus services could be provided immediately by the same operator; on the other hand, if that policy was to protect the railway, bus services and road freight services could be operated strictly as an adjunct to it.

These are, in effect the policies which were pursued in Northern Ireland and the Republic respectively. In the former case, the programme of rail closures was accompanied by a vigorous expansion of bus services including the introduction of limited stop and express services from the early 1960s which, of course, benefitted from the improvements to the road system. In contrast, a network of express bus services only emerged in the Republic in the 1970s and, to this day, this network is concentrated on non-railway routes. It is worth noting that express bus services in the Republic return a profit.

The bus traditionally has many advantages over the railway and in the particular case of Ireland where population is low and there are few large urban centres, the economics of bus operation are difficult to overlook. The closure of much of the railway system in Northern Ireland means in effect that the financial advantages of bus operation have already accrued there : Ulsterbus, which is now responsible in succession to the UTA for virtually all inter-urban and rural bus services is relatively efficient in financial terms and profitable (Northern Ireland Transport Holding Company, Various). In the case of the Republic, it is interesting to note that attempts are being made by a number of private bus operators who are debarred by law from running scheduled services to exploit the financial advantage of bus over railway: what are effectively scheduled bus services are being offered under the auspices of 'private travel clubs' which patrons must join prior to travelling. In many cases, these services parallel CIE's scheduled train and bus services but charge lower prices. Their existence obviously indicates that current official policy in the Republic is deficient; yet, to date, no action has been taken to regularise the situation.

Problems of Public Transport Provision in Remote Rural Areas

A key problem which faces bus passengers, bus operators and, ultimately, the government in both parts of Ireland concerns how best to provide a reasonable public transport service in remote rural areas. This problem has been exacerbated by rural depopulation and increasing car ownership: in many rural parts of Northern Ireland for example, over 80% of all households have access to a car. Yet, there are always persons such as the very old and the young who must rely on public transport.

As is the case in much of Western Europe and North America, it is becoming increasingly clear that the needs of those in remote rural areas requiring public transport cannot be satisfactorily served by conventional bus services. In the Irish Republic, for example, there have been significant service cutbacks in recent times and the fares charged are relatively high. One possibility of alleviating this situation in the case of the Republic would be to utilise the school bus fleet (currently used for that purpose only) for local service and to allow ordinary passengers to travel on school bus runs. In Northern Ireland, many of Ulsterbus's rural services are operated primarily for schoolchildren but also convey other passengers. Another possibility might be to introduce post buses similar to those operated in various European countries : an experiment introduced in County Clare in 1982 has been declared a success (Department of Communications, 1985) and could presumably be extended. There might also be a cost saving to be achieved in subsidising private bus operators, especially those with small vehicles, to operate certain scheduled services currently operated by CIE or Ulsterbus. In summary, exciting possibilities would appear to exist for improving the level of rural bus service provision, the financial performance of these services and vehicle utilisation. It is perhaps surprising that these have not as yet been pursued vigorously.

The Impact of the Border

Numerous differences in the transport policies which have been pursued in Northern Ireland and in the Republic have been mentioned already. It is worth noting here that the existence of the border in itself creates particular difficulties for the areas on either side of it and indirectly, for the provision of transport on the island as a whole. These difficulties have been examined, for example, by the recent New Ireland Forum. The relevant report (New Ireland Forum, 1984) also summarises various earlier studies carried out under European Community auspices. In 1977, for example, a study examining all aspects of transport and communications in the Londonderry and Donegal area identified such inadequacies as a poor road network, a low quality/high cost public transport system and poor connections to ports and airports. Improvements were recommended, including a substantial investment in the road system and the provision of a second bridge over the River Foyle in Londonderry. In 1980, a similar European Community sponsored study made recommendations concerning the best route through the border area for the Dundalk to Newry main road. A 1984 report suggests that major improvements to the road network in the areas on either side of the political boundary including the reopening of certain roads which cross the boundary but which are currently closed for security reasons, is required (Economic and Social Committee of the European Communities, 1984).

FREIGHT TRANSPORT

Two related aspects of freight transport merit discussion - first, the size and nature of the freight transport market and second, the extent to which that market is being serviced by the various types of freight transport operator. With regard to the former, the point has been made already that the geography of Ireland is such as to give rise to passenger traffic flows which are small by European standards. The same is true in the case of freight. No major natural resources exist which cause large demands for freight transport. The peripheral island location means that through freight, something which is of considerable importance to many European railway administrations, is absent. Irish industry, much of it export based, tends to be located in the larger urban centres many of which are ports or are adjacent to ports.

While Irish freight flows are small by European standards, they are still important. Agricultural activity produces significant demands for transport such as for the movement of milk on a daily basis to creameries while increasing economic activity generally has caused a rise in the size of the freight transport market. In the case of the Irish Republic, this market is estimated to have increased from 1,714.6 million ton-kilometres in 1964 to 4,817.7 million ton-kilometres in 1981.

The Irish freight transport market is served almost entirely by four types of operator - the railway; road services provided by railway companies; road services provided by commercial hauliers licensed by Government to engage in the business; and own account haulage, where individual firms transport their own goods in their own vehicles. Internal air transport and coastal shipping play virtually no role in freight transport within the island and internal waterways including the canals are no longer used for this purpose.

With the exception of own account hauliers, each of the foregoing methods of freight transport is controlled either directly, in the case of the railway and their road services, or indirectly, in the case of licensed commercial hauliers by government. Thus the market shares which have been held and are currently held by each type of operator reflects to a large extent government policy. In the case of the Irish Republic, the thrust of the policies initially formulated to cope with the emergence of road freight transport in the 1920s and early 1930s was to preserve the dominant position of the railway by requiring that road freight services operated by the railway companies ran as an adjunct to the railway and by restraining the growth of the commercial haulage sector by limiting the number of vehicle licences issued. In the case of Northern Ireland, the Northern Ireland Road Transport Board (NIRTB) was formed in 1935 and was empowered to take over various commercial hauliers operating at that time. It was intended that the NIRTB would then coordinate its operations with the railways but this did not happen. On July 1, 1966, the Northern Ireland road haulage industry was liberalised permitting commercial hauliers to enter the market and to expand provided they could meet criteria relating to such matters as safety.

Following upon the closure of much of the railway system, the 1963 Benson report (Benson, 1963) recommended the withdrawal of all internal rail freight services in Northern Ireland and this proposal took effect in 1965. Its implementation reflected a recognition of the unsuitability of the railway mode for transporting freight over the very short distances involved in internal movements in Northern Ireland coupled with a view that the newly improved road system could cater adequately for the level of traffic being generated and likely to be generated. The outcome of these policies has been that today, internal freight transport in Northern Ireland is performed entirely by road. The only freight moving by railway is through freight to/from the Irish Republic destined for Belfast or Londonderry for ultimate transfer to County Donegal. There are indications that the current railway administration in Northern Ireland is keen to re-enter the freight market; any specific proposal will be judged solely on commercial criteria.

In the case of the Irish Republic and in contrast to Northern Ireland, the legislative structures referred to earlier were still essentially *in situ* in the late 1970s. An estimate of the share of the total freight market held by each type of operator at that time is given in Table 2. The figures indicate that the legislation of the 1930s did not succeed in transferring the majority of freight movements to the railway and its associated road services; rather, in the absence of a well developed commercial haulage sector, individual shippers purchased their own vehicles and engaged in own account haulage to a much greater extent than in other European countries. The figures highlight the two main policy issues facing freight transport policy formulation in the Irish Republic in the late 1970s namely defining the long-term role of the railway with its associated road freight services and defining the extent to which the commercial haulage sector might be liberalised along Northern Irish and European lines.

The increase in size of the freight market in the Irish Republic has been referred to already. When viewed against this expanding market, the performance of the railway has been disappointing : the number of tons of freight carried in 1985 was virtually the same as in 1971. A further feature of current railway freight operations is the relatively high dependence on a relatively small number of traffic types: in 1985, over half the tonnage transported by the railway comprised cement, fertiliser and mineral ores. While these traffics are particularly suited to transport by railway, the dependence on them implies that a decline in the demand for cement and/or fertilisers and/or the exhaustion of certain mineral ore deposits will seriously weaken the railway's overall position in the freight market.

Table 2: **Share of the freight market in the Republic of Ireland, 1978**

Operator type	Percentage of total ton-miles performed	Trend during 1970s
Railway (CIE)	12.5	Decreasing
Road Freight (CIE)	2.4	Decreasing
Licensed haulier	14.0	Increasing
Own account	71.1	Decreasing

Source: Barrett, S. (1982) Transport Policy in Ireland, Dublin, (Chapter 7, Tables 2 and 3)
Note: The market share for the railway in 1978 has been assumed to equal that for 1979.

With regard to the medium term future, it is unlikely that the railway will withdraw from the freight business as has been the case within Northern Ireland. It is however possible that there will be a withdrawal from those parts of the market where large losses are currently being incurred, most notably the transport of small loads which are referred to as sundries.

The future legislative position *vis-à-vis* commercial hauliers in the Republic has become clear with the enactment of the 1986 Road Transport Act which in effect and in accordance with EC policy completely liberalises this sector of the industry from October 1988. Such limited liberalisation as has already taken place has caused an increase in the share of the road freight market held by the private haulage sector to 41% of ton - kilometres in 1983 and full liberalisation is likely to increase this further. Increased competition between individual hauliers is likely to reduce rates to the customer in real terms. It may also further erode the position of the railway as a freight carrier.

EXTERNAL LINKS

The relatively open nature of the Irish economy has been referred to in previous chapters. Within the context of transport, this underlines the importance of external links. As far as these links are concerned, Ireland is at a disadvantage by virtue of being located on the periphery rather than near the centre of Europe and on an island which forces a transhipment from/to the air or sea modes for incoming/outgoing shipments. Accordingly, it is not surprising to find that the share of the Irish Republic's wealth production devoted to transport is approximately two-and-half times the average for the European Community (European Community, 1981; Barrett, 1982).

The foregoing geographical facts continue to act as a constraint to movement despite improvements in both shipping and airline technologies. In the case of the former, the development of ro-ro ferry services for cars and freight vehicles together with the widespread use of containers have reduced transport costs in real terms. They have also encouraged the development of a strong indigenous, long distance commercial haulage industry first in Northern Ireland and, more recently, in the Irish Republic.

The increase in vehicular traffic both cars and road freight vehicles, offering for transport across the Irish Sea has reinforced the attraction of the shortest routes. Thus the ports associated with these routes, Larne (for Cairnryan/Stranraer), Dun Laoghaire/Dublin (for Holyhead) and Rosslare (for Fishguard) have captured an increasing share of the traffic while the role of other ports, associated with routes which traditionally operated mainly overnight and for passengers without cars has declined. In the mid 1960s, Belfast was linked by passenger ferry services to Liverpool, Douglas (seasonal), Heysham, Ardrossan and Glasgow. Today, only the first two remain with the former reduced to operate with one vessel only. In contrast, traffic through the nearby port of Larne has increased dramatically with ten or more departures per day to Cairnryan /Stranraer.

The decline in the longer Irish Sea passenger shipping routes reflects not only the increasing importance of ro-ro services but also the increasing importance of air travel for passengers without cars. Frequent jet services link both Dublin and Belfast with London while smaller numbers of flights are offered on various other routes between the Irish Republic, Northern Ireland and the rest of the United Kingdom. Air fares have declined dramatically in real terms in recent times, initially to/from Northern Ireland but, more recently and with the arrival of a new aggressive low fare carrier, Ryanair, from the Irish Republic as well. No less than five companies now operate on the Dublin/London route. The financially troubled semi-state Dublin based B+I shipping line has cited increased airline competition as the main reason as to why it has not met its financial objectives.

Turning to links with countries other than the United Kingdom, the Republic's politically independent status has encouraged the development of such links by Irish-based companies. Irish Shipping Limited was set up in 1941 specifically to provide the Republic with access to self-controlled deep sea shipping during the Second World War. In 1969 and in order to avail of the advantage of ro-ro technology, what was ultimately to become Irish Ferries commenced operating ro-ro passenger ferry services directly between the Irish Republic and France. These services have been a success: while Irish Shipping Limited has ceased trading, Irish Ferries currently operates year-round passenger ferry services between Rosslare and Le Havre/Cherbourg and a seasonal service between Cork and Le Havre.

Recent changes in marine techology, most notably the increased use of larger, more specialised vessels on the deep sea routes have had important implications for ports policy and ports design. What is now required are deeper water berths with specialised loading/unloading facilities. Such berths and facilities are expensive to provide. Thus a small number of ports, most notably Dublin, Cork, Waterford and the Shannon Estuary in the Republic and Belfast in Northern Ireland have developed to handle such traffic while other smaller ports have fallen into disuse. Unlike most other aspects of transport discussed in this chapter, the control of ports in both Northern Ireland and in the Irish Republic has remained vested mainly in individual locally based competing port authorities. Accordingly, while the ports which have developed in recent times as major freight handlers have been those located near the main industrial concentrations, the extent of that development has depended greatly upon the quality and vigour of the local managements which have had to overcome considerable difficulties, most notably the achievement of substantial labour force reductions associated with technological changes. An interesting recent development in the case of Cork has been the creation of a duty free industrial zone in the port area.

Considering the development of airline links to non-United Kingdom destinations, the maintenance of such links by Aer Lingus, the national airline of the Irish Republic, has always been considered at government level to be an important objective. Accordingly, in addition to its United Kingdom services, Aer Lingus currently operates scheduled services to a number of European destinations and to the United States. In contrast to this, Northern Ireland, as part of the United Kingdom relies on feeder services from Belfast to London which offer onward international connections. In this respect, Belfast is no different to Glasgow, Edinburgh and certain other British cities all of which represent the ends of 'spokes' oriented towards a 'hub' in London. The only non-stop scheduled air service from Belfast to continental Europe operates to Amsterdam.

Of the major international airports (Figure 2), the case of Shannon is particularly interesting. The airport replaced a transatlantic flying boat base at Foynes on the south bank of the River Shannon in 1945 and immediately became the refuelling point for aircraft making the transatlantic crossing. When aircraft technology developed to a point where refuelling was no longer necessary, steps were taken to maintain the airport's position most notably through the opening of the world's first duty free shopping facility which was later copied at many other locations, and then, through the opening of a duty free industrial estate which has been copied recently at Belfast, and the construction of a new town based on the airport. Transatlantic flights from Dublin are required to land at Shannon and at present the three US airlines operating to Ireland land at Shannon only. Recently, the Russian airline Aeroflot has commenced using Shannon as a refuelling stop for flights en route from the Soviet Union to Cuba and South America. In the longer term, with the advent of very large freight carrying aircraft, Shannon could become a bridgehead for the European distribution and collection of North American freight.

CONCLUSION

A recurrent theme throughout this chapter has been the fact that the geography of Ireland is such as to create relatively unfavourable conditions for the management and development of transport both internally and externally. One consequence of this has been that many decisions relating to transport such as the role to be played by railways, the role of the private and public sectors in providing various types of transport service and the amount of money to be invested in transport have been taken on the basis of political and strategic rather than economic considerations.

The point was made at the outset that transport comprises an entity where positive planning can materially benefit both individuals and the economy as a whole and therefore, that it is something which can be used to generate political benefits for the relevant decision takers. Against this, certain political disadvantages which are often inherent in transport-related decisions must be noted. First, many transport-related decisions like the undertaking of major improvements to a road network are expensive to implement. Second, short-term negative effects such as increased traffic congestion during roadworks and long-term negative effects such as environmental destruction may accrue. Finally, the positive returns from the decision such as shorter journey times for road users may not be enjoyed for a number of years by which time a different political administration might have assumed power and be in a position to take the credit. One outcome of the foregoing characteristics of transport-related decisions is that in a political sense, there is often perceived to be an advantage in commissioning numerous studies concerning transport which demonstrates concern whilst at the same time making few decisions which, in the short term, is relatively inexpensive and alienates no particular interest group. A second outcome is that such decisions as are taken often reflect little more than the power of the various interest groups within the political arena. Thus, in the case of the Republic where a key role of national elected representatives is perceived to be to support the interests of the local constituencies which elected them, much of the railway system remains intact because railway closures generate opposition at the local level; in contrast, in the case of Northern Ireland where there is a strong roads lobby and where in any case other issues, most notably of a sectarian nature, have traditionally assumed prime importance at local level, much of the railway network has been closed.

When viewed at the level of the political arena as a whole, transport must compete with a plethora of other demands for scarce economic resources. The fact that transport often assumes a low priority is indicated by the fact that the major role of government in transport matters has often been to act as a regulator rather than as an innovator, for example, in respect of the growth of the commercial road haulage sector and the role of private bus companies. On a day-to-day basis, various aspects of transport matters have been overseen traditionally by different government departments which again suggests that it is not perceived to constitute an issue of central importance. Thus for example, in the case of the Republic, the Department of the Environment oversees roads policy, the Department of Communications oversees public transport, while the Department of Justice oversees the Garda Siochána who play a key role in respect of traffic control. It is significant that no in-depth review of the administrative structures appropriate to transport has ever been undertaken either in Northern Ireland or in the Irish Republic; rather, such changes as have occurred have tended to be of an *ad hoc* nature and have related to specific sectors of the industry only.

Commenting on future trends and changes within the transport sector is obviously difficult as these will depend on such imponderables as future levels of economic activity, population trends and fuel prices. In the case of Northern Ireland, an essentially irrevocable decision was taken in the mid 1960s to adopt a roads based policy and one in which public transport plays a relatively minor role. In the case of the Irish Republic, a lesser emphasis on road investment together with the retention of much of the railway system together with an extensive system of rural and school bus services has meant that a greater element of choice currently exists as to the direction of future transport policy. A recent step which suggests the ultimate direction of this policy has been the division of CIE into three semi-independent companies which are now responsible for the railway, Dublin city bus services and all other bus services formerly operated by CIE respectively. One medium-term outcome of this administrative development is likely to be increased competition between bus and passenger railway services which, when taken with increasing liberalisation within the road freight sector and the current policy of investing in roads but not in the railway is likely to raise ultimately the question as to what the long-term role of the railway is to be. In the case of external links, the current international trend of deregulating air and shipping operations is likely in the case of the Irish Republic to raise fundamental questions concerning the future role of the national airline and merchant fleet. The manner in which these and other future transport issues are handled is likely to be prompted as much by political and strategic as by economic considerations.

REFERENCES

Barrett, S.D. (1982) *Transport Policy in Ireland*, Irish Management Institute, Dublin.

Benson, H. (1963) *Northern Ireland Railways*, HMSO, Belfast.

Confederation of Irish Industry (1985) *Irish Road Statistics*

1985, Dublin.

Coras Iompair Éireann (Various) *Annual Reports*, Dublin.

Department of Communications (1985) *Transport Policy : A Green Paper*, Stationery Office, Dublin.

Department of the Environment (1978) *Belfast Urban Area Plan, Review of Transportation Strategy. Public Inquiry; Statement by the Department*, Belfast.

Department of the Environment (1979) *Road Development Plan for the 1980s*, Stationery Office, Dublin.

Department of the Environment (1985) *Policy and Planning Framework for Roads*, Stationery Office, Dublin.

Department of Transport (1986) *Transport Statistics Great Britain 1975-85*, HMSO, London.

Economic and Social Committee of the European Communities (1984) *Irish Border Areas*, Information Report, Brussels.

European Community (1981) *The European Community: Transport Policy*, Office for Official Publications of the European Community, Luxembourg.

Feeney, B. (1982) *Car Ownership Forecasts 1995-2005*, An Foras Forbartha, Report RT.262, Dublin.

Heanue, K. *et al.* (1971/2) *The Dublin Transportation Study*, Two Volumes, An Foras Forbartha, Dublin.

Jane (Various) *Jane's Urban Transportation Systems*, Jane's Publishing Company Limited, London.

Killen, J.E. (1973) The passenger rail system in Ireland : present trends and future prospects, *Irish Geography*, 5, 632-8.

Killen, J.E. (1979) Urban transportation problems and issues in Dublin, *Administration*, 27(2), 151-66.

Lavery, I. (1986) Public transport by shared taxi : the Belfast experience, *Proceedings : Seventeenth Annual Public Transport Conference*, University of Newcastle Upon Tyne.

New Ireland Forum (1984) *Integrated Policy and Planning for Transportation in a New Ireland*, Stationery Office, Dublin.

Northern Ireland Economic Council (1981) *Public Expenditure Priorities: Roads*, Report No. 21, Belfast.

Northern Ireland Transport Holding Company (Various) *Annual Report*, Belfast.

Royal Ulster Constabulary (Various) *Road Traffic Accident Report*, Belfast.

Skidmore *et al.* (1978) *Cork Land Use / Transportation Plan*, Cork.

Smyth, A.W. (1986) Transport Planning and Conflict: the Evolution of a Transportation Strategy for the Belfast Urban Area, Occasional Paper, 12, Department of Town and Country Planning, The Queen's University of Belfast.

Stationery Office (1984) *Building on Reality*, Stationery Office, Dublin, 57-63.

Transport Consultative Commission (1980) *Passenger Transport Services in the Dublin Area*, Stationery Office, Dublin.

Travers Morgan, R. and Partners (1976), *Transport for Belfast*, Belfast.

12 PATTERNS IN IRISH TOURISM

John Pollard

'The two prime factors in the choice of holiday destination are the friction of distance and the lure of the sun' (Patmore, 1983, p.156).

Viewed in the context of such decision-making priorities of holidaymakers in the 1970s and 1980s, tourism development in Ireland would appear to be at an immediate and considerable disadvantage. After all, here is an island that lies on the extreme periphery of Europe, cast out into a North Atlantic that not only bestows on it the most fickle of European climates, but also denies it the land connection that might otherwise guarantee some foundation of cross-border traffic.

While there is no doubt that these constraints represent formidable obstacles to the exploitation of a mass tourism market according to the Mediterranean model, visitor numbers totalling almost 2.5 million in 1985 equally attest to Ireland's drawing power. Whereas the island may lack the tourist product of the most favoured European resorts, other more specialised resources provide some compensation. Tourism is, after all, a complex phenomenon, incorporating and exploiting a multiplicity of motivations, markets and resources so that no countries lack any attractions upon which to capitalise. Ireland has, in its own way, taken advantage of its unique combination of resources and market links to carve itself a special niche in world tourism, and has thereby developed an important limb to its economy.

301

It is against that background that the contemporary tourist industry of Ireland will be examined, paying particular attention to its recent development, the influences that have shaped it throughout the island as a whole and have differentiated it regionally, and its impact in economic terms. The study is divided accordingly into four sections, covering (i) the visitor market and its development over the past quarter century, (ii) the basis of Irish tourism, namely the 'tourist product' that attracts Ireland's clientele, (iii) the geographical spread of tourism activity within Ireland and, (iv) its economic consequences.

It will be apparent from what has already been said that the discussion will emphasise the international aspect of tourism, that is those temporary visitors staying at least 24 hours in the country visited and whose purpose of journey can be classified under one of the following headings:

1. Leisure (recreation, holiday, health, study, religion and sport);

2. Business, family, mission, meeting (UN Conference on International Travel and Tourism, 1963).

Unfortunately, this generally accepted definition raises problems of application in an Irish context resulting from the complicating presence of both the land border and the physical separation of Northern Ireland from the rest of the United Kingdom. Strictly speaking, cross-border movements involving a stay of more than 24 hours should be included in the tourist statistics of both political units, whereas visitors from Britain to Northern Ireland should be ignored as no more than inter-regional travellers. To do the latter when at the same time including the equivalent movement of persons, many of whom are of Irish origin and nationality, from Britain to the Republic, would smack of inconsistency, while to try to take accurate account of all persons crossing an indifferently controlled and partially recognized land border is, in practice, impossible. Thus the procedure adopted here is to examine travel to the island of Ireland from beyond its shores. Although this introduces some anomalies, such a policy will simplify the drawing of comparisons between the tourism industries both sides of the Irish border, while simultaneously minimising reliance on dubious data.

Clearly, concentration on this modified form of international tourism will exclude an important domestic tourism element, that is persons staying away from home for periods in excess of 24 hours, but remaining entirely within the Republic or Northern Ireland. However, although such travellers may be equally important for the maintenance of tourism services, incomes and employment in parts of the island, the concern of this study is to examine the individuality of tourism in Ireland in an international context. This has meant that references to domestic tourists occur principally when enforced by data sources, such as some hotel statistics, that do not distinguish the indigenous and external traveller.

Table 1: **Visitors to Ireland, 1960 and 1985 (thousands)***

Destination	Republic of Ireland		Northern Ireland	
Origin	1960	1985	1960	1985
Great Britain	827	1,119	430	419
Continental Europe	25	334		28
North America	69	422	43	63
Other Overseas	20	69		22
TOTAL	941	1,944	473	532

*Unless stated otherwise, figures quoted throughout this paper are extracted from Bord Failte's and/or the Northern Ireland Tourist Board's annual publications Tourism Facts (Bord Failte and NITB), Trends in Irish Tourism (Bord Failte), and Tourism in Northern Ireland (NITB). If no reference is made to either the Republic of Ireland or Northern Ireland, then figures apply to the combined totals from both sources.

VISITORS AND VISITOR TRENDS

A broad indication of the level of tourism activity in Ireland in the mid-1980s and its development since 1960 is given in Table 1. The stagnation of the Northern Irish market immediately stands out in marked contrast to the doubling of the Republic's visitor total. In that respect the impact of violence upon tourist activity was spelt out bluntly in successive Annual Reports of the Northern Irish Tourist Board in the 1970s employing such sweeping statements as 'Northern Ireland holidays could not be sold through any tour operators or travel agents anywhere in the world at large' (NITB, 1978). Although the tone is more optimistic in 1986, attention is still drawn to the universally negative media image of Northern Ireland (NITB, 1986).

In comparison, tourism south of the border has managed to push ahead despite any spill-over effect of civil unrest in the North and the growing competition offered by cheap alternative Mediterranean packages. Admittedly the expansion over a somewhat foreshortened period (1960-1982) is a modest one in comparison with the European average and the experience of all Western European nations except those of Scandinavia which, in any case, exclude travel between themselves from their calculations (Table 2).

The contrast between Ireland's lethargic performance and the majority of other countries is probably more stark than the data indicate for, in many cases, tourist numbers are severely understated through the omission of visitors arriving by land. Even taking those figures at their face value, Ireland's share of Western European tourism has declined significantly from 1.9% to 1.1%. In many respects, Ireland possesses little ability to influence these trends, suffering, as have the Nordic countries, from changing fashions as well as physical peripherality and poverty of access. Set against such largely immutable external controls, domestic matters such as the impact of inflation on holiday costs, and the effect of violence in the North or crime in Dublin on overseas perceptions of Ireland, are but minor obstacles in the fight to retain market share. In these circumstances, it is perhaps to the Irish tourism industry's credit that careful marketing, particularly of new growth areas of activity and special interest holidays, has ensured some progress in the Republic, while similar attractions have been exploited in the North, not without some success, to avoid the complete collapse of tourism in the face of adversity.

Table 2: **Tourist arrivals in Western European countries, 1960 and 1982**

	1960	1982	% growth
	(thousands)		p.a.
Austria	4,595	14,253	5.3
Belgium	3,892	6,785	2.6
France	5,613	33,156	8.4
Germany	5,476	11,075	3.3
Greece	344	5,033	13.0
Ireland (Rep.)	941	1,730	2.8
Italy	9,100	22,223	4.1
Netherlands	1,477	3,080	3.4
Portugal	353	3,164	10.5
Scandinavia*	5,558	8,453	1.9
Spain	6,113	25,291	6.7
Switzerland	4,949	9,186	2.9
UK	1,669	11,646	9.2
Total	50,080	155,075	5.3

Source: United Nations (1961 and 1983) Statistical Yearbooks, 1960 and 1982 (UN: New York).
* Includes Denmark, Finland, Iceland, Norway and Sweden

The reaction of the two political units' tourism industries to events can be seen in the more detailed presentation of tourism trends in Figures 1(a) and 1(b). In both Northern Ireland and the Irish Republic, the trends exhibit marked irregularity with underlying growth until the late 1960s being followed by the sharp falls of 1971 and 1972 and later recovery, particularly in the South, impeded somewhat by the onset of reduced economic activity on a world scale at the beginning of the 1980s.

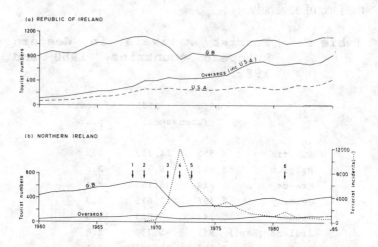

Figure 1 Tourist Traffic to Ireland 1960-85 (Key: 1 - First Civil Rights Marches; 2 - Civil unrest, Derry and Belfast; 3 - Special Powers Act; Internment; 4 - Direct rule from Westminster; 5 - (1972-74) High level of sectarian murders; 6 - Maze Prison (H-Block) protests)

In many respects the curves parallel one another, although the downturns have been more severe and the subsequent recoveries less pronounced north of the border. While the factors that underlie those fluctuations are many and complex, it is inevitable that the susceptibility of tourism to political and personal insecurity will be reflected in marked downturns in the annual patterns recorded in tourist arrival graphs wherever risk is perceived. In this respect 1968 represented a turning point in Northern Ireland from which it has never really recovered. In October of that year the disaffection felt by some members of the population manifested itself in the first civil rights marches in Londonderry. Continuous growth in tourist traffic up to that time was followed by a marginal decline over the next two years and thence total collapse as civil disturbances deteriorated into violence, murder and indiscriminate bombings which might affect resident and visitor alike. The five years 1972-1977 saw tourism hovering around a base level of travellers composed of those on essential business or family affairs, supplemented perhaps by a small band of visitors aware of the localised nature of trouble spots and thus willing to discount the minimal risk involved in visiting other parts of the Province (Table 3).

Table 3: Purpose of visit of Northern Ireland visitors

	1967	1973	1985
	(percentage of arrivals)		
Holiday	36	7	10
Visiting friends/relations	38	65	60
Business	25	6	27
Other	1	2	3
All purposes	100	100	100

Sources: NITB Annual Reports

Figure 1(b) includes reference to the numbers of terrorist incidents (shootings, bombs and incendiaries) recorded by the Royal Ulster Constabulary, as well as some of the most news-striking events. Visual inspection provides strong circumstantial evidence for the close link between the political and security situation and tourism trends, although with occasional and understandable exceptions. In particular, it will be noted that the tourist response to the onset of disturbances was immediate and dramatic, whereas a lag effect delaying tourism recovery is evident once the level of violence fell in the mid-1970s, for it is not until 1978/9 that some pick-up in arrivals is discernible.

In the Northern Ireland context the apparent link between violent activity and tourism's response provokes no surprises. Unfortunately, the geographical impact of such events is rarely narrowly confined. Thus, the succession of events in Northern Ireland appears to have had a profound impact on the Republic, if not to quite the same degree (Figure 1(a)). The human tendency to telescope distance when viewed from afar, combined with any blurring of distinctions between North and South in the minds of politically naive potential visitors, would all operate to the Republic's disadvantage, particularly in repelling the longer distance traveller. However, it is the visitors from Great Britain that overwhelmingly dominated the Irish market in the 1960s (Table 1), and it is their reaction that has provoked a disproportionate impact on overall trends. In just two years, 1970-1972, the British visitor total suffered a 30% decline, accounting for 300,000 persons. Visitors to Ireland are representatives of a public that quickly became all too aware of the blacker side of Irish affairs through a media coverage that reached saturation point. Many might have had genuine fears that violence would spread south into the border counties and Dublin itself. While additional local knowledge and pull of family ties might overcome those fears for many ethnic Irish, those Britons without an Irish connection would be specially vulnerable to adverse publicity relating to what is just one of many competing potential destinations.

Lest the impression be conveyed that the performance of Ireland's tourism industry is solely a reflection of politically related events in the North, it bears repeating that pressures of a less nationally selective kind have acted to the detriment of tourism. Undoubtedly the most important of these has been the rapidly widening scope for international travel prompted by improved income levels, holiday arrangements and transport facilities, and encouraged by persuasive promotional campaigns. Southern European countries in summer and Alpine countries in winter have been the principal beneficiaries of these trends, and it is no coincidence that their expansion parallels this period of slow tourism growth in a country that can offer neither of the vogue holiday attractions. Superimposed on this general trend may be seen the adverse impact of cyclical economic factors, and particularly the downturn in the international economy in the early 1980s that has brought with it an inevitable pruning of expenditure on this most income and price elastic of activities in all countries. Clarke and O'Cinnheide (1981) also point out that an earlier period of price inflation was felt particularly strongly in the Republic in the mid-1970s, and had a serious impact on tourism activity at that time. Even the inherited market composition of the 1960s would have acted to Ireland's disadvantage, for the dominant market segment of social and family visitors from Great Britain would contain little capacity for rapid expansion. However, ironically, the disruption that affected the British market in the early 1970s has had a positive effect in that a far more diversified market has been steadily developing (Figure 1(a)), an occurrence that is welcome not only for the greater inherent stability, but also for its economic benefits of lessening dependence on low-spending visitors from Britain who rely largely on accommodation provided by friends and families.

Table 4: **Purpose of visit of Republic of Ireland visitors, 1985**

	All visitors	Great Britain	North America	Cont. Europe	Other
(percentage of arrivals)					
Pure holiday*	37	16	67	55	58
Visiting friends and relations	45	60	32	19	32
Business	18	21	10	22	10
Other	14	20	7	13	6

N.B. Totals exceed 100% due to multiple purpose visits.
* Holidaymakers not staying with friends or relatives.

Despite the changes that have occurred over the past quarter century, the tourist markets of both parts of Ireland still show exceptional concentrations in the Visiting Friends and Relatives (VFR) and Great Britain categories (Tables 3 and 4; Figure 2). Geographical proximity is a powerful influence on the composition of any country's tourist market, but the special relationship between Britain and Ireland inflates the proportion far more than might otherwise be expected for countries lacking a land border. The second major geographical source of visitors, North America, also displays the importance of cultural links in Irish tourism. While the proportion of North American visitors with Irish ancestry is not known, it is significant that 40 million Americans claim Irish ancestry, and almost one-third of visitors from that continent to the Republic of Ireland listed visiting friends or relatives as the motivation for their trip (Table 4).

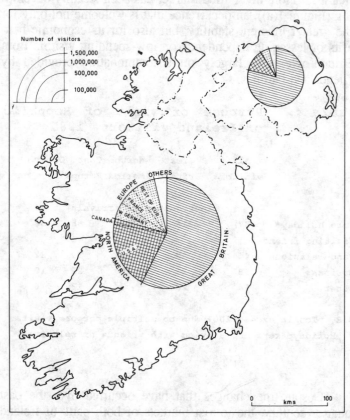

Figure 2 Visitors to Ireland and their origins

A similar motivation is apparent among the Other Origins category which is dominated by visitors from Australia and New Zealand, both having migrant connections with the mother country. It is only in the Continental European category that the holiday visitor without an accompanying social motivation becomes really dominant. The continental visitors are principally French, for whom a direct sea connection is possible through Le Havre, Cherbourg or Roscoff, and Germans. Other notable contributions originate in the Low Countries, Scandinavia, Switzerland, Spain and Italy (Table 5).

Table 5: **Origins of Ireland's tourists, 1985**

	Republic of Ireland		Northern Ireland	
	Tourists ('000s)	%	Tourists ('000s)	%
Great Britain	1,119	57.6	419	78.8
North America	422	22.0	63	11.8
Canada	30	1.5	-	-
United States	392	20.5	-	-
Continental Europe	334	16.0	28	5.3
Belgium, Luxembourg	22	1.1	-	-
Denmark	17	0.9	-	-
France	95	4.9	-	-
Italy	16	0.8	-	-
Netherlands	33	1.7	-	-
Norway, Sweden	10	0.5	-	-
Spain	15	0.8	-	-
Switzerland	16	0.8	-	-
West Germany	98	5.0	-	-
Other Europe	12	0.6	-	-
Other areas	69	3.5	22	4.1
Australia, New Zealand	37	1.9	-	-
Rest of World	32	1.6	-	-
Total	1,944	-	532	-

THE TOURIST PRODUCT

Although a minority of Ireland's visitors both North and
South are drawn independently of any social or business
incentive, such persons still account for approximately
700,000 arrivals, and it is thus worthwhile considering the
attractions that bring these 'pure' holidaymakers to the island,
or, as it is known, the 'tourist product' that Ireland has to
offer. Conceptually that tourist product encompasses the
complete experience from the time the tourist sets foot in Ireland to the time he or she leaves its shores, and will comprise
a potentially wide range of natural and man-made components. However, these components will vary between individuals according to their demands, while the relative value
of different elements of the tourist product is best measured
through expressed or effective demand.

Surveys are conducted regularly by both Bord Fáilte and
the Northern Irish Tourist Board eliciting, *inter alia,* information on the attractions to holidaymakers and their usage of
various products. Unfortunately, the categorisation of
responses is sometimes a broad one in which specific resources, particularly natural resources, are difficult to ascertain.
Thus a 1982 survey determined that the principal preoccupa-
tion of visitors was with sightseeing and exploring the countryside (Table 6), and indeed touring is a recurring theme
among visitor responses year by year as visitors incorporate a
variety of natural and cultural attractions into their holiday
programmes.

Table 6: **General attractions to holidaymakers in the Republic of
Ireland***

Sightseeing/exploring the countryside	59%
Warm and friendly Irish people	42%
Relaxed way of life	20%
Traditional music	8%
Sporting facilities	7%

*Excluding those visiting friends and relatives above
all else.

The breadth of attractions indicated in visitor surveys is such that consideration of both physical and human resources is warranted, although in less detail than otherwise might be justified as many of these resources are discussed in other contributions to this volume, albeit in rather different contexts.

Ireland's exceedingly varied geological and geomorphological history has ensured a generous distribution of scenic attractions, although those aesthetic qualities do provide a disproportionate advantage to the ancient Palaeozoic mountain rim over the drift-covered central lowlands. This rugged western edge can be traced almost continuously from the Inishowen peninsula in the north of Co. Donegal through to Co. Cork in the south, reappearing on the eastern margins in the Mountains of Wicklow and Mourne, and supplemented in the north-east by the Antrim Plateau whose Tertiary origin provides the only geologically youthful element. This impressive upland rim produces some of its most dramatic landscapes in its coastal margins which incorporate the variety of the intriguing columnar basalts, laterites and chalk of Co. Antrim, the grandeur of the precipitous cliffs of Moher (Co. Clare), and the picturesque inter-penetration of land and sea in the peninsulas of Kerry and Cork.

The ice and water that have acted as agents of erosion and deposition in the moulding of the Irish landscape have also presented a series of more specific resources for tourist exploitation through activity-based holidays. The lake and river systems, fed by that high precipitation that is otherwise a detraction to holiday-making, come high on the list of such attractions whether it be for cabin-cruising on Lough Erne or the Shannon system, or fishing in the loughs and streams that abound throughout the length and breadth of Ireland (Table 7).

Complementing fishing are the land-based sports of which horse-riding and golf stand out, taking advantage of the unspoiled or open countryside that contrasts with the premium on space and pressure upon golfing facilities in more urban-industrial communities elsewhere in north-west Europe. In coastal areas a high proportion of superior quality fine sand beaches washed by clean Atlantic waters provide a superior resource in comparison to the littoral on the other side of the Irish Sea. Moreover, the low population densities characteristic of all Ireland except in the vicinities of Dublin and Belfast maximise the availability of beach space for visitor and resident alike.

Table 7:   **Holidaymakers' use of tourist resources in Ireland[a]**

Activity	Northern Ireland[b] Total Nos. ('000)	Republic of Ireland Total Nos. ('000)	%
Cabin cruising	4	14	2
45shing specialist		35	5
Fishing	{ 12	107	10
Golf	22	36	5
Horse riding	n.a.	14	2
Festivals	7(1983)	43	6
Historic visits	60(1983)	533	75
Stately homes	28(1983)	220	31
American heritage/ properties	11(1984)	(c)	(c)

a. N. Ireland figures include persons staying with friends and relatives; Republic of Ireland figures do not include this group.
b. 1985 unless stated otherwise.
c. Included in Historic visits.
n.a. Not available.

Despite the undoubted natural attractions, it is unlikely that they alone could sustain anything like the present level of tourism. It is rather the cultural attractions which, either on their own, or in conjunction with the physical resources, are responsible for tourist activity in Ireland as in most of the climatically less-appealing European locations. The definition of those cultural attractions is normally sufficiently wide to include anything of a human character whether relating to the past or present. Keane (1972) recognises under this heading the local character and heritage including ethnic and religious attractions, houses and gardens, national monuments and architecture. An integral part of that character is the temperament and disposition of the Irish population, who are frequently reported as warm, friendly and relaxed (Table 6), and who respond to tourists in a more accommodating and appreciative manner than in some other parts of the world where their appearance may sometimes be regarded as something of an intrusion.

While cultural resources is an all-embracing term, Table 7 gives a clear idea of the more tangible aspects, or tourist 'products', that it incorporates for Ireland's visitors. Traditional music and visits to historic sites and stately homes receive frequent mention in tourism survey responses. Of the core components of a country's culture, namely its language, literature and folklore, its religion and its music, the latter is the most accessible, easily absorbed and understood by the average tourist. It may be the case that traditional music is not necessarily actively sought out but, even so, through visits to festivals, or more frequently to bars, there is passive exposure to it often leaving a lasting impression of a uniquely Irish experience. Against this, language, literature and religion play a far less overt role in forming part of the Irish tourist product, although they clearly do exercise a fascination and even a prime motivation for pilgrims to Ireland's holy shrines and sources of Yeats' or Joyce's inspirations.

The more distant past, too, provides its contribution to Ireland's tourist product through monuments that span four millennia to incorporate both megalithic remains and unique examples of Christian art and architecture exemplified in Celtic crosses and Mediaeval round towers. Though widespread throughout Ireland many of the best examples are located in Co. Louth, so that an excursion through the Boyne Valley makes for a singularly fascinating journey through time.

With such a background it is hardly surprising that the Historic Visits and Stately Homes categories exercise such strong drawing power for overseas visitors, although rather more so among North Americans and continental Europeans than Britons, whose exposure to similar remains and architectural styles would be that much greater.

Finally, it is worth repeating that it is the cultural links that form the mainstay of Irish tourism, and it is in direct response to them that certain resources have been deliberately highlighted by Bord Fáilte and the NITB to encourage visitors and especially those from North America. It is here that Northern Ireland has been particularly active in making the most of the Ulster links of so many American presidents or fathers of the Declaration of Independence, opening the former family homes and sometimes developing them, as at the Ulster-American Folk Park (Co. Tyrone), into a fully fledged theme park based on life-styles on both sides of the Atlantic a century or more ago. This is undoubtedly a rich vein, but one whose potential still lies largely untapped thanks to the counter-productive image of Ulster that has prevailed on American television screens and among more extreme Irish-American opinion for so many years.

THE PATTERN OF TOURISM IN IRELAND

For the pure holidaymaker the distribution of natural and cultural resources described in the previous section must play the leading role in the control of tourism patterns within the country, although the friction of distance from points of entry, the violence within Northern Ireland, and the presence of the political border might restrict the otherwise free circulation of visitors.

Unfortunately, the pattern of holidaymaking is not immediately apparent from published statistics as the data relate to tourism in all its aspects, including that for social and business reasons. Thus the pattern displayed in Figure 3 which allocates tourists on a county and, in Northern Ireland, a district council basis, will inevitably display a resident population effect (the larger the population, the more friends and relatives likely to visit), as well as the attractive power of the main commercial, industrial and administrative centres of the island. Such effects should underlie the dominance in the South of Ireland of Co. Dublin, which receives twice the number of overseas visitors as Co. Cork, itself not without a considerable pull on non-holidaymakers.

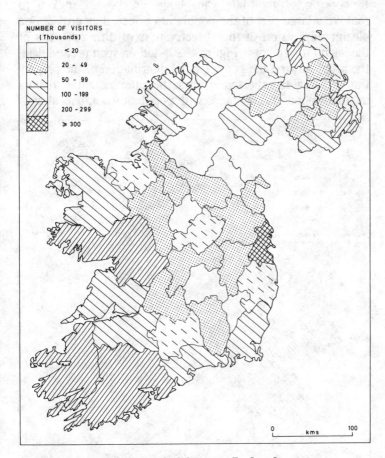

Figure 3 Distribution of visitors to Ireland

In the North, Belfast was responsible for 305,000 tourist trips, a figure not much in excess of that for the very much smaller population centres of Coleraine and Down DCAs. Nevertheless, large cities, and particularly capital cities, cannot be dismissed as holiday centres in their own right, and Dublin at least might be expected to exert considerable drawing power, whether it be for reasons of cultural and historical resources, or for playing host to a biennial invasion of

317

French, Welsh, Scottish or English rugby supporters. Reworking of the data can, however, be undertaken to remove a large part of the population effect, thus providing a clearer impression of the attractiveness of different parts of the island for visitors. Figure 4 is based on such revised data, the map showing the geographical distribution of the proportion of actual to expected visitors, the latter assuming visitors are distributed in accordance with population numbers alone.

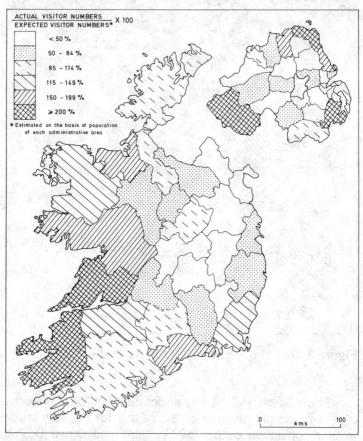

Figure 4 Distribution of visitors to Ireland, adjusted for population

With respect to the Republic of Ireland, the comparison of Figure 4 with Figure 3 is immediately striking for the downgrading of Dublin and to a lesser extent Cork, while Sligo, Clare and Waterford have their positions enhanced. While tourists in the broader sense show a marked peripheral distribution, the pattern of pure holidaymaking emphasises the attractions of the west.

Only Donegal fails to attract its 'expected' total, probably a reflection of distance from points of entry and limited approaches that do not involve a Northern Ireland border crossing. Excluding the western counties, only Waterford and Wexford exceed expectations and both these have the advantage of sea links to Wales and France through Rosslare as well as having rural scenic attractions of their own. Nonetheless, it is the pull of picture-postcard Ireland - 'the white cottages set in their bright green fields, the tawny colours of the unimproved land, the black slashes of peat diggings', (Whittow, 1975) - that, with their mountain or marine backdrop, is epitomised in the Gaeltacht of the west, and so dominates the true holiday scene. An apparent peak is reached in Kerry which receives almost three times its expected visitor total, followed by Clare and Galway, but Mayo with its much larger expanses of blanket bog and glacial drift fails to match up to the inspiring Dalradian summits of Connemara's Twelve Pins and Maumturk Mountains (Co. Galway), the distinctive ecology and physical formations of the Burren of Co. Clare, or the romantic and picturesque image of the Killarney lakelands of Kerry, and accordingly falls short of the adjacent coastal counties in its visitor total. Co. Cork, too, appears surprisingly unattractive given the similarity of landscape of its Slieve Miskish and Bantry peninsulas with the bays and peninsulas of Dingle and Valencia immediately to the north in Kerry. However, this is probably little more than a scale effect in operation with Co. Cork, Ireland's largest county, incorporating substantial areas that are relatively less appealing, a factor that may well also dilute Galway's holidaymaking figures. Eastern Galway penetrates into Ireland's hollow centre as far as tourism is concerned, only Westmeath closely approaching the expected visitor level. There, Athlone, a major centre on the Shannon Navigation that also lies astride the routeway from Dublin to the west, plus the hills and lakes in the vicinity of Mullingar, all help to provide some relief from the relatively uninspiring Lower Carboniferous and glacial lowland features of Ireland's Midlands.

North of the border, however, such lowland glacial landscapes of tills, drumlins and moraines, often infertile and so waterlogged that lakes abound, provide one of the chief holiday regions of the Province (Figures 3 and 4). In the Lough Erne system, Fermanagh possesses boating facilities to rival those of the Shannon and fishing opportunities that are unsurpassed in Europe, and the great majority of the 4,000 cruising holidaymakers and 9,000 fishermen visiting Northern Ireland in 1984 were concentrated in this county.

Unfortunately the Northern Ireland area data do not distinguish home-based holidaymakers from visitors to Ireland, so that the outnumbering of the latter in the ratio of six to one effectively swamps the overseas holidaying pattern. On the other hand, both sets would be expected to take advantage of many of the same resources and facilities, although the beach attractions and associated caravan parks and guest houses of the Derry, Antrim and Down coasts would be relatively heavily used by the domestic holidaymaker. The attractions show through particularly clearly in the District Council-based statistics for Northern Ireland (Figure 4). With the one exception of Fermanagh the pattern is very much a coastal one with Coleraine, Moyle and Down recording influxes well in excess of expectations based on population figures. Coleraine DCA contains within it the twin resorts of Portrush and Portstewart, while Newcastle plays a similar role within Down DCA. The Moyle DCA, which extends from the east of Portrush through Ballycastle and Fair Head and thence south to Cushendun, possesses Ireland's most renowned natural attraction in the Causeway Coast, with the Giant's Causeway itself providing a remarkable geological centrepiece to a whole series of rock formations and geomorphological features, backed up by legends and an historical legacy of medieval castles and tales of the Spanish Armada, that elevates the area to Northern Ireland's premier leisure location. The importance of the Giant's Causeway complex in international terms was recognised in 1986 with its designation as a World Heritage Site by UNESCO, one of only two in the British Isles, the other being the uninhabited island of St Kilda. Elsewhere the pulling power reduces substantially, although Larne DCA containing many of the Glens of Antrim as well as the continuation of the Antrim Coast Route, and Newry and Mourne DCA stand out. However, it is inevitable that Estyn Evans' beautiful Mourne Country has suffered from the high level of violent incidents occurring in south Down and south Armagh since 1970, although Fermanagh's numbers are also certain to have reacted similarly to events in that county, both areas so losing potential props to otherwise weak economies. It is to this aspect of tourism in the economy that attention may finally be turned.

TOURISM AND THE ECONOMY IN IRELAND

There are no countries that do not participate in the international tourism trade, although attitudes towards its encouragement may vary between the extremes of Albania's glowering acceptance and its open espousal by many tropical islands. However, most national governments view international tourism in a positive light and play an active role in its development, specifically encouraging it for a variety of economic, social and political reasons. Of these the economic are normally paramount for the advantages it brings for economic diversification, income and employment generation, the Balance of Payments, tax revenue growth and regional development that accrue to those countries that are successful in attracting foreign visitors, but not forgetting the social spin-offs that derive from local inhabitants' use of recreation facilities and the possibilities offered for arresting population decline in what are often peripheral areas of the national territory. All these factors take on greater importance when other forms of wealth creation have proved elusive or of limited ability to satisfy the aspirations of the population.

The Republic of Ireland, although by no means as economically deprived as some developing countries that have wholeheartedly welcomed tourism as a potential panacea for their ills, clearly accepts tourism in that light and provides every encouragement through the operations of Bord Fáilte. In the North, the NITB performs many similar functions to those of Bord Fáilte, although with less of a central administrative and controlling role (Mowat, 1984). Furthermore, perhaps in keeping with its role as an adjunct to a regional rather than a national economy, tourism has played a less prominent part in a Province that traditionally has emphasised the agricultural and manufacturing sectors of its economy. However, the decimation of old-established engineering and textile trades, coupled with the failure of many replacement industries, and the modernisation of agriculture, have resulted in labour-shedding of such a high order that alternative tertiary employment opportunities including tourism are desperately needed.

Studies of the economic aspects of tourism in an Irish context are limited, and published data are largely confined to travel receipts rather than employment and tax generation, the latter being difficult to obtain accurately given the wide variety of facilities used, the problem of divorcing tourists' use of these facilities from that of the country's own residents, and the tracing of tourists' expenditure flows through the economy. Thus, in the main, the following remarks are confined to the immediate boost to the economy based on tourism revenues estimated through Bord Fáilte questionnaires directed to tourists at ports of exit.

As far as the Republic of Ireland is concerned, a total of 1,944,000 overseas tourists in 1985 spent an average of 10.8 nights. To put this in perspective, that number of tourist nights expands the population of the country, by the equivalent of 1.7%. However, the stimulus to the economy would be expected to considerably exceed the demographic effect, for visitor expenditures always outstrip those of the country's own residents both by virtue of the additional costs involved in the frequent requirement for supplementary accommodation, and the less-restrained expenditure pattern of the holidaymaker. In Ireland's case, expenditure amounted to IR£238 per person in 1985 (or IR£22 per day), both substantial figures, if reduced from many other tourist venues' receipts by the high proportion (49%) of bed-nights spent with friends or relatives.

Table 8: Distribution of tourist expenditure by region, 1985 (IR£M)

Bord Failte Region	Overseas	NI	Home	All	TEQ*
Cork/Kerry	93.0	3.7	70.1	166.8	1.44
Donegal/ Leitrim/Sligo	34.3	14.6	20.9	69.8	2.14
Dublin	103.0	7.1	33.4	143.5	0.48
East (1)	43.3	3.5	29.2	76.0	1.11
Midlands (2)	32.9	0.8	14.1	47.8	0.72
Shannonside (3)	55.1	0.9	22.7	78.7	1.19
South-East (4)	39.9	1.2	40.5	81.6	1.09
West (Galway/ Mayo)	60.7	4.1	38.3	103.1	2.18
All regions	462.2	35.9	269.2	767.3	1.00

Expenditure by tourist group

*TEQ - Tourist Expenditure Quotient = Expenditure by All Tourist Groups in the Region as % of All Tourism Expenditure divided by total Personal Income of Region as % of All Personal Incomes

(1) Kildare, Louth, Meath, Wicklow;
(2) Cavan, Laois, Longford, Monaghan, Offaly, Roscommon;
(3) Clare, Limerick, N. Tipperary;
(4) Carlow, Kilkenny, S. Tipperary, Wexford.

Visitor expenditure could theoretically have an important regional impact, and one which should benefit those relatively disadvantaged areas of the rural west of Ireland that have historically suffered from both economic and demographic decline and yet which, according to the previous analysis, possess a high proportion of the island's favoured tourist resources. Table 8 summarises tourism revenues by Bord Fáilte region, and provides a crude indication of the relative benefits accruing from those revenues through quotients that relate the proportion of tourist expenditure received to the proportion of personal incomes accruing to each region. While there is some evidence of distribution of overseas tourism revenues in favour of more disadvantaged regions the effect is far from overwhelming thanks to the dominant part played by Dublin in Irish tourism affairs. That situation reflects the important role played by any capital city as a magnet for foreign visitors, but exaggerated in Ireland's case by the high proportion of ethnic tourism.

However, as Ogilvie noted as long ago as 1933, the origins of tourism revenues, whether foreign or domestic, are of no consequence to the beneficiaries of those receipts. Once other tourists are included, the regional distribution pattern is significantly improved as home tourists dominantly comprise Dublin residents seeking change elsewhere in the country so that the overall tourist expenditure quotient falls to its lowest value of 0.48. Cork and Kerry is clearly the pre-eminent region among home visitors and the west (viz. Mayo and Galway), too, stands out, although the southeast, containing the Wicklow Mountains and a number of seaside resorts such as Brittas Bay, also receives considerable patronage.

While domestic tourism figures are less reliable than those for overseas visitors, those for Northern Ireland are even less so, but it is worth noting that in their case Donegal, Leitrim and Sligo receive disproportionate support as one might expect given the resources and accessibility particularly of Donegal. All in all the initial revenue flows in favour of western Ireland are identifiable, with only the north-western counties suffering from their peripheral position with respect to everywhere except Northern Ireland. Even so they do record the second highest Tourist Expenditure Quotient (2.14). Thus, when compared with the distribution of personal incomes within the country, tourism revenues do seem to indicate some movement in favour of the west of Ireland, but it is also clear that not all western counties receive equal treatment, to say nothing of impoverished parts of the Midlands which obtain little benefit either.

Of course, revenue is not the same as income, and there is no guarantee that money spent in the more deprived parts remains there to boost incomes of local inhabitants on any scale. Controversy has always surrounded the development of the tourism industry particularly in the Third World and in lesser developed parts of a single economically more advanced country in that it can represent economic dominance of the periphery by the core in an almost pure form (Pearce, 1981). This can occur when the clientele originates either in metropolitan countries or in central conurbations with ownership of the tourism industry's assets and expenditure patterns ensuring an outflow of monies to those same centres. In Ireland's case the possibility exists for income flows both from west to east as well as from Ireland as a whole to centres of capital and administration in North America, Britain, and Continental Europe.

Of the various items of tourism expenditure, accommodation accounts for almost one quarter. It is both the largest item and possesses considerable potential for income outflow depending on hotel and other accommodation ownership and policy with respect to treatment of profits. However, the presence of a high proportion of locally based family hotels, guest houses, and farmhouse type accommodation - common facets of the Irish tourism industry - will also serve to retain the maximum amount of income in the areas of expenditure. Moreover, accommodation providers, whatever their origin, will incur substantial wage and salary payments given the labour intensity of this operation, thus ensuring a further flow of essentially local income.

Similar arguments may be raised with respect to the other items of tourist expenditure although, in general, they may be regarded as being less beneficial to the peripheral areas than payments for accommodation. While some fresh foods will be of local origin in these agricultural areas, most food and drink will be manufactured and thus purchases thereof, apart from wholesale and retail margins earned locally, ultimately benefit more industrialised parts of eastern Ireland and, indeed, abroad. The weak manufacturing base of the west will similarly limit the income flows from expenditures on clothing and gifts, although demand for some items of local origin, notably knitwear and tweeds, will be stimulated whereas, in an all Ireland context, home-produced cut glassware is highly popular among visitors.

In the absence of detailed studies, such comments must remain speculative. Knowledge of the distribution of economic gains from Ireland's tourism industry is very much a grey area requiring basic research into the geographical distribution of incomes deriving from it. As part of that research it is also necessary to examine inter-industry linkages and so proceed to calculate the multiplier effects of tourists' expenditure at both the regional and national levels. Without such calculations there can be no understanding of the full value of tourism to Ireland's economy or the regional economy of the west. In 1969 the national multiplier effect of tourism was estimated to be between 1.6 and 1.8 (Keane, 1971), but the derivation of those figures is obscure, and may not in any case apply in the mid-1980s given the changes that have occurred in both Irish tourism and the Irish economy.

Table 9:   **Tourism and the Republic of Ireland's balance of payments, 1970-84**

		(IR £ M)		
	1970	1975	1980	1984
All Credits	694	1,985	5,808	11,435
All Debits	771	2,017	6,846	12,272
Tourism and travel[a]	74	118	282	442
Passenger fare/				
Receipts[b]	22	34	87	149
Tourism and Travel/				
as % Credits	10.7	5.9	4.9	3.9
Pass. fare receipts/				
as % Credits	3.2	1.7	1.5	1.3

Source : Central Statistical Office. Irish Statistical Bulletin (Dublin).
a. Including passenger fare receipts.
b. From persons resident outside Ireland.

Annual estimates of tourism's contribution to the national economy are also provided by Bord Fáilte, which in 1985 considered that 6.2% of Gross National Product derived from foreign and domestic tourism combined, with a further 2.5% being generated through the multiplier mechanism. However, there does seem to be some exaggeration built into the former figure in that 6.2% appears to be the tourist revenue figure expressed as a proportion of GNP without any allowance for leakages due to foreign sources of goods and services consumed by tourists to Ireland. The initial revenue figures are, however, relevant when considering Balance of Payments effects, and in that respect tourism earnings have for many years been an important factor in keeping a negative current account balance within manageable proportions (Table 9). Both 'Tourism and Travel' and 'Passenger Fare Receipts' (i.e. the earnings of Irish shipping companies and airlines from non-Irish travellers), show significant contributions to the credit items of the Balance of Payments although there has been a steady proportional decline over recent years resulting from the relatively faster growth of overseas sales from an expanding Irish economy.

Finally, some mention should be made of the economic contribution of tourism to Northern Ireland, although the position of the region as part of the United Kingdom dismisses the possibility of calculating equivalent Balance of Payments effects. All that can be said in that regard is that Northern Ireland's overseas tourists contributed £16M to the United Kingdom's goods and service exports in 1984 (NITB, 1985). In terms of regional expenditure, visitors are combined with home holidaymakers, but the indications are of total dominance of the more affluent east of the province, Belfast alone receiving almost one quarter of all expenditure, followed by Coleraine and Down with 12% and 10% respectively. The counties of the west with their lower incomes and higher unemployment levels benefit to a far smaller degree, so that tourism does nothing towards ironing out income inequalities within the province in a manner akin to the process acting in the Republic. However, this is not to decry tourism for, without it, even fewer revenue-earning opportunities would exist for the less-favoured areas of Northern Ireland and, at least on the north and south-east coasts, it does help to provide some counterbalance to the otherwise overwhelming dominance of Belfast. The precondition for any more even spread of the benefits overall must inevitably be the return of peace to the province, for the geographical distribution of rural violence must act to the detriment of the western counties and thwart the realisation of the undoubted potential of their moorlands, mountains, river and lake systems.

CONCLUSION

No matter what the future holds in respect of demand for any products of manufacturing industry, one can be certain that the market for tourism will continue to expand. That alone is no guarantee that Ireland can take advantage of the general trend towards higher incomes and longer leisure hours, for international venues competing for that trade will grow apace, while tourism also remains highly sensitive to alternative destination prices, and even more so to political and security conditions prevailing. Ireland has suffered more than most from post-war terrorist activity as well as the trend in favour of warm-water, guaranteed sunshine resorts, without having the compensating lure of winter sports facilities of many other European nations. These events have conspired to relegate Ireland towards the sideline in the battle to catch the would-be foreign visitor's eye. Fortunately, however, tourism comes in many guises, allowing Ireland to exercise her own individuality and appeal, while the more discerning and knowledgeable visitor is either perfectly capable of seeing the highly localised and intermittent violence of parts of the North as of no real threat, or dismisses it in the same way as the visitor to Israel or Cyprus as an event of singularly unfortunate timing should he be involved. So despite generally adverse circumstances, Ireland has maintained its presence among European tourist markets and, furthermore, has retained its unique natural and cultural qualities, which are indeed the principal foundations of its tourist appeal. In this Ireland has had to accept the common tourism trade-off: for, although there may be no Irish equivalent of the Costa de Birmingham, Dusseldorf or Rotterdam, there is equally no comparable economic gain. Ireland's character and selective market is maintained by its own detractions, namely its peripherality, poverty of access and cost. The Irish tourist industry consequently cannot compete in the high density tourism of the mass resorts of Southern Europe. Rather, it needs to make the most of a relatively small number of tourists, capitalising on high income, high spending visitors. At present, with the dominant part played by ethnic visitors and self-catering holidaymakers, such tourists are relatively few in number, but the scope for expansion is there, building on Ireland's prime initial advantage of North American and Australasian nostalgic connections, while simultaneously promoting activity holiday opportunities for the European market. Many parts of the historically deprived and depressed west remain relatively untouched, and tourism clearly has its part to play along with the Industrial Development Agency's policies for redressing the income imbalances and population decline. In those respects there has indeed been some success particularly from Galway southwards to Cork. Continuing that success and extending it into northwestern Ireland and Northern Ireland itself which undoubtedly has natural resources for tourism development quite comparable with anything elsewhere in north-western Europe, is a responsibility less for Bord Fáilte and the NITB than the people of Ireland in terms of the image of the island they project beyond its shores.

REFERENCES

Bord Fáilte (annually), *Trends in Irish Tourism*, Bord Fáilte, Dublin.

Bord Fáilte (annually), *Tourism Facts*, Bord Fáilte, Dublin.

Central Statistical Office (annually), *Irish Statistical Bulletin*.

Clarke, W. and O'Cinnheide, B. (1981) *Tourism in the Republic of Ireland and Northern Ireland*, Co-operation North, Paper V, Belfast and Dublin.

Keane, E. F. (1972) *Irish Tourism: Industry in Strategic Change*, Bord Failte, Dublin.

Mowat, P. (1984) *The Administrative Factor in the Development of Tourist Resources and Markets in North-West Ireland*, D.Phil. thesis, New University of Ulster, Coleraine.

Northern Ireland Tourist Board (annually), *Annual Report*, NITB, Belfast.

Northern Ireland Tourist Board (annually), *Tourism in Northern Ireland*, NITB, Belfast.

Northern Ireland Tourist Board (annually), *Tourism Facts*, NITB, Belfast.

Ogilvie, F.W. (1933) *The Tourist Movement*, King, London.

Patmore, J.A. (1983) *Recreation and Resources*, Blackwell, Oxford.

Pearce, D. (1981) *Tourist Development*, Topics in Applied Geography Series, Longman, London.

United Nations (annually), *Statistical Yearbook*, UN, New York.

Whittow, J.B. (1975), *Geology and Scenery in Ireland*, Penguin, Harmondsworth.

13 IRISH ENERGY : PROBLEMS AND PROSPECTS

Palmer J. Newbould

INTRODUCTION

In a perfect world, there would be a world energy policy and
each nation state would have its own energy policy conform-
ing with the world energy policy. In the real world, it is
doubtful if any country has a well-developed and coherent
energy policy. There are many good reasons for this. Consensus about the relative weighting of political, social, economic
and environmental objectives is continually changing. Events
outside the country over which it can have no control also
change. Few people in the 1960s predicted the steep rise in
oil prices in 1973 and few in the later 1970s predicted that oil
prices would ever fall as low again as they did in 1986. It is
difficult to adjust supply to demand in a big energy sector.
Thus if oil production is running at a certain level and there is
a drop in demand, no-one wishes to cut their production level,
storage is very limited and there can be a sharp and unreal fall
in price.

Politicians and administrators have a vested interest in
predicting economic growth, often unrealistically; this leads
to over-projection of energy demand and distorts sensible and
frugal energy planning. The system, like a supertanker at full
steam ahead, is slow to react to change. An example of over-
projection, even on a fairly short time scale is shown in Table
1. One moral might be that if you have the luxury of four scenarios, the lowest, if not negative, should at least be level

demand.

Energy planning usually centres on electricity. This is partly because electricity supply systems tend to be monopolistic, centralized and pretty much under direct government control. Developed countries become heavily dependent upon electricity to the point where they require a high degree of security of supply. Power cuts cause chaos and suffering. This leads to costly over-investment in capacity, as insurance against power cuts. Even so it is no insurance against industrial action (stoppages) by electricity workers.

Table 1: Projection of energy demand for 1985

	Million tonnes oil equivalent
Energy Ireland, Dept of Industry, Commerce and Energy, July 1978	13.2
NBST Energy Forecasts for Ireland, Feb. 1980 Lowest of 4 projections	9.10 (Highest 11.05)
NBST revised forecasts Dec. 1981	8.97 (Highest 10.40)
Actual consumption 1985 (Energy in Ireland, Dept. of Energy, Dublin).	8.3

Once an electricity utility has over-invested in capacity it has a vested interest in selling more units of electricity, thereby reducing the average unit cost to the consumer and justifying its investment. Eventually this affects peak demand, thereby creating a need for more capacity, i.e. there is positive feedback. This runs counter to sensible energy planning. There is fair consensus that energy conservation is a good thing. Yet separate utilities, such as electricity, gas, oil and coal spend money trying to persuade their customers to consume more of their particular fuel. The message is not put in such blatant terms, each utility may stress economical use, but it is left to government to promote energy conservation.

To build a large power station or to develop a new energy resource takes at least 10 years during which time many external pressures change. It also represents a large investment of money. There is often political reluctance to enter into this and even more reluctance to change plans or withdraw from a development once embarked upon. Once built the investment requires a high rate of return which means high utilization and the hard sell.

The consequences of not having an energy policy are well exemplified by Ireland. Northern Ireland is geographically part of Ireland and politically part of the United Kingdom. In energy terms it is isolated; attempts to maintain an electricity inter connector between Northern Ireland and the Republic of Ireland were frustrated by terrorism. Electricity prices in Northern Ireland are subsidised by the UK Government. The Republic of Ireland, distinguished by having the highest electricity prices in the EC, is equally isolated. Indeed Ireland and Greece are the only two EC countries not linked to the European grid. Both parts of Ireland are relatively poorly endowed with indigenous energy resources so both are considerable net importers of energy.

While energy use *per capita* is not greatly different, the mix of energy sources, pricing structure, environmental impact etc. are very different between Northern Ireland and the Republic. Thus in terms of fossil fuel resources the Republic has a substantial natural gas field, and has been exploiting peat as a major energy resource since 1946. Northern Ireland seems set to make considerable use of lignite. Significant hydroelectric power and pumped storage schemes have been developed in the Republic but not in the North, where plans for a pumped storage scheme foundered on land acquisition problems. The North has looked seriously at wave power and at tidal power, whereas the Republic has put more emphasis on wind power. None of these so-called alternative sources has been developed commercially to any extent in either country.

In both countries there has been research on energy production from biomass using short-rotation deciduous forestry (willow coppice), although work on this promising but minor energy source now seems to be in abeyance.

The nuclear power option was carefully examined in the Republic in the 1970s and then abandoned, not necessarily for the right reasons. It was never seriously considered in the North.

Northern Ireland built a large oil-fired generating station at Kilroot and then decided to convert it to dual coal/oil firing. The Republic built a large coal-fired station at Moneypoint near Limerick. In neither country were the environmental effects of these coal-fired stations taken seriously at a time when the EC was strenuously seeking reductions in SO_2 emissions. The coal fires of Dublin make it one of the most air-polluted cities in Europe (see Chapter 16).

There has already been mention of the failure to maintain interconnection of the electricity grids across the border despite the clear economic advantages to both countries of doing so (New Ireland Forum, 1984). Plans to extend the Kinsale gas pipe line from Dublin to Belfast also foundered, mainly on economic grounds. These problems and events will be reviewed in greater detail below.

THE PRESENT ENERGY SITUATION

The energy situation in 1984, the latest year for which full figures are available, is set out in Tables 2 and 3. Table 2 contrasts the diverse sources of energy in the Republic with Northern Ireland where oil represents more than three-quarters of the energy supply. Also the Republic of Ireland is shown as 38% self-sufficient for energy. Interestingly energy use per head is the same in both parts of the island. This figure, 2.3 TOE/caput/year can be compared with the EC average of 3.5; other figures are Greece 1.5, Italy 2.4, UK 3.7 and USA 8.3.

Table 3 compares the two electrical utilities for the same year, that is the Northern Ireland Electricity Service (NIES, more recently simply NIE) and the Electricity Supply Board (ESB) in the Republic. It appears that those living in Northern

Ireland take more of their 2.3 TOE in the form of electricity than those living in the Republic. Again the diverse sources of electricity in the Republic contrast with the oil-dependence of Northern Ireland. Both have ample spare capacity. In Northern Ireland the gas turbines are very expensive to run, and are kept for topping up at times of peak load. Conversely in 1984, coal was markedly cheaper than oil, so that despite the age of Belfast West power station it contributed more than its share to the electricity supply. In the Republic maximum use is made of the quota of Kinsale gas allocated to the ESB. Availability of peat and hydro limit the amount of electricity that can be generated from them, with a wet summer favouring hydro and a dry summer favouring peat.

Table 2:   Primary   energy   consumption 1984

	Thousand tonnes oil equivalent			
	Republic of Ireland		Northern Ireland	
		%		
Peat	1,421	17.5	<20	0.5
Coal	991	12.2	830	22.5
Oil	3,891	47.8	2,807	75.9
Hydro	173	2.1	-	-
Gas	1,660	20.4	39	1.1
Total	8,136		3,696	
Population	3,550,000		1,580,000	
TOE/caput	2.3		2.3	

Sources:Energy in Ireland 1984. Dept. of Energy, Dublin. Northern Ireland Annual Abstract of Statistics No. 4, 1985, Policy Planning and Research Unit, Dept. of Finance and Personnel, Stormont, Belfast.

Table 3: **Electricity supply, 1984**

	ESB	NIE
Units sold, 10^6	8,940	4,663
Units/Caput	2,500	3,000
Capacity, MW	3,260	2,200
Peak load, MW	1,994	1,100
Peak Load % capacity	61	50

	ESB		NIE	
	% units generated	% capacity	% units generated	% capacity
Oil	20	27	78	78
Gas	54	41	0.2	11
Peat	18	15	-	-
Hydro	7	16	-	-
Coal	1	1	22	11

Sources: NIE Annual Report and Accounts 1984,
year to 31 March.
ESB 57th Annual Report for year ended 31
March 1984.

No single year is typical. In 1986 the Moneypoint coal-fired station came on-stream adding even more diversity to the ESB. The year before was more oil-dependent than usual because of the miners' strike in Britain. Cheaper oil in 1986 allowed a reduction in the real cost of electricity in Northern Ireland. The next two years (1988-1989) will see the completion of Moneypoint and the conversion of Kilroot to dual oil/coal firing, while the 1990s may see a lignite-fired power station (perhaps privately owned) in Northern Ireland, the restoration of the cross border link and the establishment of an interconnector between Ireland and Britain.

Electricity prices are shown in Table 4. The prices to the consumer in Northern Ireland are lower than those in the Republic, but this is mainly due to a special UK government subsidy designed to keep Northern Ireland prices in line with

those in England and Wales. Making allowance for this subsidy, and for the gap between the Irish pound and sterling, the
difference is narrow. The comparatively high cost of electricity in Ireland may be due in part to the smallness and isolation
of the systems, dependence on imported fuel, low population
density and extensive rural electrification. O'Boyle (1987)
indicates that one reason for the high cost of electricity in the
Republic of Ireland relates to Government levies on the ESB,
both via rates and fuel tax.

Table 4: **Electricity prices**

	1984	1985	1986
NIE unit price, pence (p)	5.207	5.281	5.413
+ government subsidy	1.284	2.211	1.313
Real price	6.491	7.492	6.726
ESB unit price, Irish p.	7.205	7.510	7.736
Sterling/Irish £ conversion*	79.81	82.94	83.36
Sterling equivalent	5.750	6.228	6.449

*Information kindly provided by Central Bank of Ireland
Sources: Annual reports of NIES and ESB

Fossil Fuel

The traditional view was that Ireland was more or less bereft
of fossil fuel resources, despite an abundance of Carboniferous strata throughout the island. There were formerly two
small coalfields in Northern Ireland at Ballycastle and Coalisland. Both closed in the 1960s, with the Ballycastle coalfield
having been worked for over 200 years mainly in association
with a glassworks. For a brief period in the early nineteenth

century the Ballycastle adit and drift mines supplied coal to
Dublin, but the supply was erratic and the trade soon ceased.
Also in Northern Ireland, thin seams of 'coal' (in reality lignite hardened by contact with lava) was worked locally at
various times. The small coalfield in the Republic at Arigna is
still active and there are known proven resources of coal in
Irish territorial waters off the east central coast (see Chapter
15).

Peat

Peat has been a traditional hand-won fuel, but the Turf Development Board (TDB, later Bord na Mona) was set up in 1934,
and developed large-scale peat-winning machinery. The first
peat-fuelled power station, Portarlington, was commissioned
in 1950. The arguments that raged between the TDB and the
ESB in that period are well reviewed by Manning and McDowell (1984, pp.100-122) and need not be repeated here. Manning and McDowell make the point that 'it (the controversy)
did at least resolve the question of the ESB's status *vis-à-vis*
government by showing very clearly that in the final analysis
the ESB was not a free agent and that its ability to formulate
its own policies independent of government supervision was
limited.' The controversy may be relevant to the proposed lignite development in Northern Ireland (see below).

The peat-fuelled power stations came to make a substantial contribution to electricity supply. Over much of their life,
the social benefits, especially the employment provided by
Bord na Mona, probably outweighed the economic disbene-
fits. With the 1973/4 oil price rise, peat-powered electricity
became economic. Environmentally, and with hindsight, it is
tragic that a small area of midland raised bog, perhaps 5% of
the total area, was not set aside for conservation. If this had
been done early in the life of the TDB, prime conservation
sites could have been selected on ecological criteria and
maintained in prime condition at very little expense. As it is
the few small sites which various bodies (An Taisce, Irish
Peatland Conservation Council, Forest and Wildlife Service)
with the cooperation of Bord na Mona are now trying to conserve have all been damaged in some way and are not necessarily the best examples of this internationally unique

ecosystem.

The remaining questions are how long will the peat last, and what will happen to the land thereafter. One option canvassed for the land is the production of energy from biomass crops, notably short-rotation deciduous coppice. This is discussed below. Other options involve mainly farming or forestry and lie outside the scope of this chapter. The answer to how long the peat will last is 20 years, but this has also been the answer for the past 25 years. The Bord na Mona 40th Annual Report for 1985/6 says 'A production plateau is projected up to the year 2000 after which slow decline will begin.' Flood (1983) estimates that fuel production by Bord na Mona will continue up to the year 2030. It seems unlikely that further peat-fuelled power stations will be built, so that as the older ones begin to need major overhaul, they will probably be decommissioned. Of the 487 MW of peat-fuelled plant, 317 MW is over 20 years old. Bord na Mona, under the Turf Development Act 1981, are obliged to grant aid individuals or groups setting up private bog development schemes. Coupled with the availability of comparatively small turf-winning machines, this has generated considerable peat production in the private sector. Currently private bog development schemes are contributing around 700,000 tonnes per year, just under 10% of Bord na Mona production. The annual subsidy is running at about IR£1.25 million. These schemes are more appropriate to smaller bogs, and may also operate in blanket bog areas unsuitable for Bord na Mona. Most of the peat produced will be used as domestic fuel, although the ESB have three small (5 MW each) power stations which are fuelled by hand-won peat, some of which is indeed now machine-won.

Gas

The next major fossil fuel to come on stream was gas from the Kinsale gas field, operated by Marathon. The amount of gas was estimated in 1974 as 25 MTOE, but in 1981 it was re-estimated as 35 MTOE. The potential flow rate was about 1.1 MTOE/year giving, with the earlier estimate a life of about 20 years. The later estimate would allow an extension of duration, or of annual flow or some combination of the two. The gas consumption in Ireland in 1974 was about 0.2 MTOE/year. So the problem was to assess potential markets and to decide what uses were most appropriate for this find. The three major options were a piped gas supply, generation of electricity or use as a feedstock for the chemical industry and in particular for the manufacture of nitrogenous fertilisers. The gas is a high quality, relatively pure hydrocarbon. There is general agreement that using it to generate electricity in a 'peak-sharing' mode, i.e. to top up the system at times of peak demand, is a good thing. A gas turbine can easily be switched on and off. But using it to provide base-load electricity can be regarded as an extravagance, not necessarily economically since the price of gas is artificially fixed, but in the sense of squandering a valuable resource. Convery *et al.*, (1983) suggest that more useful heat (by a factor of two) could be obtained from the gas by piping it directly to consumers than by converting it into electricity prior to distribution.

If a relatively cheap piped gas supply substitutes directly (i.e. in the home) or indirectly (in the power station) for oil or coal there would be a reduction in air pollution, and this is especially crucial to Dublin.

In practice 190 million therms/year were allocated to the NET (Nitrigin Eireann Teoranta) plant at Marino Point, a minimum of 260 million therms/year to the ESB and gas supply, especially to Cork and Dublin, was thought likely to rise to 100 million therms/year. This total consumption of 550 million therms/year would amount to 1.26 MTOE/year and implies an increase in the annual flow in preference to extending the duration.

Table 5: The market for natural gas

	1985	1985 (projected)
ESB	65%	35%
NET	19%	25%
OTHERS	16%	40%

Source: O'Boyle, 1987

ESB usage of natural gas has fallen from a peak generation by gas of 53% of electricity units in 1984 to a level of 47% of units in 1986. This may be reduced further as the coal-fired Moneypoint station takes up more of the load. O'Boyle (1987) suggests a likely change in the market for natural gas as indicated in Table 5. Even using 35% of the gas to generate electricity may represent a serious misuse of a precious resource which just happens to be underpriced (see Convery *et al.*, 1983).

Despite the availability of this underpriced feedstock, NET continue to make substantial annual losses, even to the point of near-bankruptcy. Current overproduction of agricultural produce in Europe, and talk of the possible imposition of nitrogen quotas suggest that the allocation of Kinsale gas to NET could be greatly reduced. While the use of gas as a domestic fuel is generally a good thing, it requires careful installation and good maintenance. Events like the two explosions, in Raglan House and Dolphin House, both in Dublin, early in 1987, are clearly a setback to the case for gas.

There were negotiations between the Irish and UK governments to extend the gas pipeline from Dublin to Belfast. Belfast town gas supply, manufactured from naphtha, had become grossly uneconomic and was due to be closed down. It appeared at one stage as if the UK government had undertaken to buy a quota of Kinsale gas. However, negotiations broke down and the pipeline was not extended. It is difficult to discover to what extent the problem was economic or political. Gas consumption in Belfast was very low, and to justify the extension of the pipeline, gas sales would have had to increase greatly. If this increase had been at the expense of oil, well and good, but if it had been at the expense of electricity, already heavily subsidised by the UK government, there was a problem. Any decline in electricity consumption would lead to an increase in unit costs and a need for further subsidy.

Lignite

The next fossil fuel to come to prominence was lignite, this time in Northern Ireland. At the time of writing there is

341

plenty of speculation and a minimum of hard fact about lignite. It is certain that there are three substantial deposits of lignite (Griffiths *et al.*, 1987). In the Crumlin area there are two major lignite seams, the lower from 17-24 m thick, the upper from a centimetre or so to 25 m. The estimate is that about 120 million tonnes of lignite are recoverable from the onshore area and a further 220 million tonnes lie under the present bed of Lough Neagh. Burnett and Hallamshire plc planned to work this deposit but subsequently sold their interest to BP Coal. On the west shore of Lough Neagh, in the area of Coagh, there is a considerable extent of lignite. BP Coal hold a licence for further exploration in this area. It seems unlikely that this deposit extends under Lough Neagh. The third deposit, possibly including the thickest seams of all, is in the vicinity of Ballymoney, and is being explored under licence by an Australian Company, Meekatharra Minerals. The Northern Ireland Economic Council (NIEC, 1987) suggest that this deposit may provide the best option for development but their economic analysis pays little attention to environmental factors.

Lignite is mined and used in some other countries of the European Community (see Table 6). It is mainly used for power generation, and is regarded as competitive with other fuels. Given the absence of other fossil fuel resources in Northern Ireland, the presumption would be in favour of mining the lignite with the aim, in particular, of diversifying the energy resource and reducing dependence on imports. Raw lignite is a wet and bulky fuel and in general is not exported or traded between countries because of transport costs, though lignite briquettes are tradeable. The proposal to mine lignite in Northern Ireland probably requires a lignite power station to justify it, i.e to use enough of the lignite to be worth while. Once the mine is operational then some sales of lignite fuel, as briquettes or crushed, can be expected. The NIEC (1987) indicate that the economics of briquette production depend upon scale with a 200,000 tonnes/year plant looking more favourable than one half that size. Since about 2.25 tonnes of raw lignite are required to produce 1 tonne of briquettes, this implies a consumption of 450,000 tonnes/year for such a briquette plant.

Table 6: **Lignite production in the European Community, 1984**

	million tonnes
Federal Republic of Germany	126.7
Greece	31.5
Spain	24.3
France	2.4
Italy	1.8
Total	186.7

Source: Energy in Europe, 4, 1986, 31-3.

BP Coal have outline planning permission to develop the Crumlin deposit. The suggestion, supported by Northern Ireland Electricity, is that there should be a mine-mouth power station. Two consortia, the American Bechtel Corporation and the British Costain-Wheeler group have applied to build and operate this station, and have set up companies in Northern Ireland to further their cases. NIE have also applied to build and operate the power station themselves.

It appears as if the present government may extend their privatisation philosophy to the building and operation of this lignite power station. This would represent a radical departure from present practice, and is seen by some as a trial run for the subsequent privatisation of the Central Electricity Generating Board in Britain. It has introduced a major political dimension into what would otherwise be an economic and environmental issue. Understandably it has aroused Trades Union opposition. It would be unfortunate if the lignite development was blocked for political reasons.

The mine itself raises environmental issues and will probably be the subject of a public enquiry. The power station is likely to require a full Environmental Impact Assessment under European Community Directive No. 337 of 1985, though the UK response to this directive is not yet clear. The Crumlin mine will displace relatively few landowners and has not aroused much local opposition. However, the Coagh deposit, although less well known or publicized is thought likely to displace many more people and to disrupt tight-knit communities, Lough Neagh fishermen and farmers among them. This has aroused articulate and well-organised opposition, and currently BP Coal seem reluctant to exercise their option to explore further. At Ballymoney indications are that in addition to 250 m tonnes directly under the town and not available, there is another 275 m tonnes on the immediate south and southwest outskirts of the town (NIEC, 1987).

If, as seems likely, Crumlin proceeds first, what is known of its effects? The power station is usually projected as 400-500 MWe. Assuming 450 MW, 33% conversion efficiency, 50% load factor and 15 GJ/tonne/year as the energy value of lignite, annual consumption might amount to 2.8 million tonnes of wet lignite or 1.5 million tonnes of lignite at 13% water content. Therefore the 120 million tonnes available on land should ensure a 30-year life for the power station and allow some marketing of lignite fuel on the domestic or commercial market.

Without precise information about the mining plan or the power station, it is difficult to predict likely environmental impact (see Newbould, 1987a). The mine itself will have substantial impact on the local environment but could ultimately be restored to form a country park or a nature reserve, with some open water, islands and fringing vegetation. The visual impact of the Crumlin mine should be small since the site is inconspicuous. Dust and noise will be local pollutants. The lignite is fairly soft and no blasting will be required. Given a good settlement pond, it should be possible to pump the site water into Lough Neagh. The major impacts of the power station will be visual, noise, cooling water, gaseous effluent, disposal of ash and transmission lines. Once-through cooling using Lough Neagh water might appear attractive but would probably have a significant and deleterious effect on Lough Neagh which is shallow and eutrophic. Thus cooling towers will probably be needed.

The lignite is said to contain 0.2% sulphur and 10% ash, both low values compared with bituminous coal, but it will take twice as much lignite as coal to produce a unit of electricity. While a change from a coal or oil-fired power station to a lignite one would result in a reduction in SO_2 emissions, it will probably be unacceptable in the 1990s to build a new fossil fuel power station without SO_2 scrubbers in the chimney.

An alternative to a mine-mouth power station would be to transport the lignite, after some processing and preferably by rail to Kilroot Power Station. A rail line passes within 1 km of the Crumlin site while another line actually crosses the Ballymoney deposit, so movement of lignite within the Province would be relatively easy. This would solve the cooling water problem, and would mean that no extra transmission lines would be needed. At present prices there would be no economic advantages in burning lignite rather than coal at Kilroot; it is difficult to quantify the advantages of diversifying fuel sources, and of using indigenous rather than imported fuel.

I believe that none of the technical problems of mitigating adverse environmental effects are insoluble. They do present a challenge to the designers and operators of the mine and the power station and will also need careful supervision by the Department of the Environment and other government agencies involved.

Oil

Finally there is the ever-present prospect of oil, and prospecting continues both on land and around the coasts (see Chapter 15). From time to time there are rumours of an economic find, but so far there are no definite plans to bring oil ashore from any of the more promising offshore oilfields around Ireland. There may be a better find, but more likely, perhaps around the turn of the century, a rise in world oil prices consequent upon oil becoming increasingly scarce will make some of these present finds more economic to work. However, this is really gazing into the crystal-ball.

Nuclear Power

In 1971 when Dáil Éireann approved the Nuclear Energy Act the prospects for a nuclear power station in Ireland seemed good. Nuclear power was perceived to have numerous advantages, and also to be a trendy new technology. It diversified fuel supply, substituting uranium for rapidly depleting fossil fuel resources. Its proponents claimed economic benefits. The easy ability to store three years fuel supply on a power station site gave security of supply. The Nuclear Energy Board (NEB), which would advise on the licensing of a nuclear power station, was set up, under the Act, in 1973. It should be stressed that the NEB had no role in formulation of policy, but merely in examining the safety and environmental aspects of proposals involving nuclear power. Contrary to the belief of its many critics, the NEB had no pro- or anti-nuclear stance. It was, however, required in the Act to provide information about nuclear power. The ESB set up a nuclear team and proceeded to select the most favourable site, which turned out to be Carnsore Point in Wexford, selected from a short list of five sites, in turn whittled down from an initial list of 30.

The OPEC oil price rise of 1973-4 reinforced the nuclear argument. It was predicted that Ireland would be 83% oil dependent by 1980/1. But there were also doubts. The technology was alien to Ireland, there was little expertise in the country. The design and much of the engineering and construction would have to be bought 'off the shelf'. The minimum viable size for a nuclear power station, for which estimates varied between 600 and 1000 MWe, was large in the context of the ESB system; nuclear power was acknowledged to be primarily base load supply. In this respect the Turlough Hill pumped storage scheme was an excellent complement to a nuclear station. However, an unforeseen breakdown could have had bad effects on the system. Opposition to nuclear power, both informed and otherwise, burgeoned in a country which is basically anti-authoritarian.

The case for a nuclear power station was based on wildly optimistic projections of demand; when there was a sudden collapse in the growth of demand in 1974-5, the nuclear power station was necessarily deferred. Also in 1974 the Government allocated the ESB 60% of the expected Kinsale Gas flow. Towards the late seventies, when demand was again rising the ESB decided on a joint nuclear/coal strategy, Moneypoint to be followed by Carnsore. Desmond O'Malley, then Minister for Energy in early 1978 described nuclear power as 'the only source of energy that could make a significant contribution to the country's electricity needs for some time to come'. A Government discussion paper, *Energy Ireland,* published in July 1978 stated 'Coal-fired and nuclear-powered generating plant appear as the only realistic alternatives (to oil) now available towards meeting our likely future demands for electricity. To opt for one to the exclusion of the other would be merely exchanging our existing dependence on imported oil for an equally heavy dependence on another source of energy.' The Institution of Engineers in Ireland felt nuclear power was essential and the Confederation of Irish Industry supported it.

However, once again the demand for electricity did not grow as fast as anticipated. The benefit of nuclear power was being questioned in many countries, and the world nuclear power programme was not developing as fast as expected. European Community energy policy, however, remained firmly pro-nuclear and considerable loan finance would probably have been available from that source. Desmond O'Malley was replaced by George Colley as Minister for Energy, and this change signalled a change in policy and a great weakening in the desire to build a nuclear power station.

Meanwhile fossil fuel prices had eased in real terms, natural gas had become a reality and the ESB mix of oil, peat, hydro, pumped storage and gas seemed reasonable. Given Government hesitation about nuclear power, and foreseeing a need for additional generating capacity by about 1985/6 the ESB went ahead with plans for the large coal-fired power station at Moneypoint on the Shannon estuary. In their annual report for 1985/6 they say 'It is the Board's view now that a nuclear power station should not be considered an option in its next long-range plan which will cover a period beyond the year 2000.' It seems to me far more likely that by interconnection with Britain and Europe, Ireland will enjoy the benefits, if any, of nuclear power vicariously.

The ESB still own land at Carnsore Point and may use it for other purposes. Unless there really is an economic bonanza in the near future, Ireland has time to wait and watch the success or failure of nuclear power in other countries.

RENEWABLE ENERGY SOURCES AND ENERGY CONSERVATION

Official pronouncements usually acknowledge the value of so-called alternative energy sources but go on to say that their potential contribution is small to negligible in the overall energy scenario, that they are expensive and that the technology is unproven. Proponents of alternative energy, environmentalists, conservationists, often with particular hobby-horses, take a contrary view. They point to successful wind generators, solar panels, to all the old watermills around the country; they scale up small projects without always appreciating the economic or engineering problems involved.

The Domestic Scale

How can we reconcile these two extreme viewpoints. It is partly a matter of scale. It seems to me that insulation, wood-burning stoves and boilers, biomass plantations, solar panels, small wind generators and heat pumps could in theory go a fair way to meet the space and water heating requirements of houses and farms, especially in rural areas.

Wind generators are noisy, dangerous and unsightly in an urban environment. Used for space and water heating for rural houses and farms they have two functional problems, energy storage and back-up during calm periods. Storage can be by batteries, hydrogen, hot water or hot stones. Batteries, the traditional energy store, are expensive, but essential if it is necessary to store electricity rather than heat. Hydrogen is probably unsuitable on the domestic scale. Bogle *et al.* (1979) examined the energy potential of the wind at Aldergrove, based on hourly weather data for wind speed and air temperature collected over the period 1949-75. They constructed a theoretical simulation model consisting of a wind generator, a house and a heat store, and by varying the size and performance of these three components, they examined the potential of wind power for domestic heating. The better the insulation and the larger the heat store, the less supplementary electric heating is required. The economics depend a good deal on the cost of this back-up. Systems involving substantial heat stores are best installed in new houses, and if this type of solution is to make a significant impact on energy demand, public sector housing must show the way. However one of the problems (or advantages, according to your viewpoint) of alternative sources of domestic energy is that they tend to need love and care from the householder. Electricity, and to a lesser extent oil-fired central heating diminish the responsibility and role of the householder. We have undergone a change of life style. Hypothermia, for example, among the aged has become a social problem because it is now perceived as the role of the welfare state to keep us warm in the winter, a role it is ill-equipped to discharge.

Solar panels, though environmentally more benign than wind generators, share some of the problems of energy storage and back-up. In Ireland they are more marginal than in the Mediterranean. However Lawlor (1975) has shown that there is potential here for space and water heating by solar panels, The practical problem is that once again to make an impact public sector housing would need to adopt the idea. The Solar Energy Society of Ireland (1978) have developed a plan which involves the fitting of solar water heating into 60% of new houses and solar space heating into 5% of new houses (O'Rourke *et al.*, 1976). Both systems would require supplementary electrical heating in the winter, but around the year would supply 25-50% of the heat energy required.

Heat pumps similarly have looked a good investment for the past 10 years or more, but are still very seldom installed even in new housing schemes.

In all these cases substantial demand would reap benefits in allowing mass production and lower unit costs. One is forced to conclude that environmental energy sources will not play a significant role until governments give them a push. This could be done now, perhaps by taxing electricity heavily, and fossil fuels slightly less heavily, and diverting this tax revenue to developing and installing the environmental energy sources mentioned. Such directed policies are regarded as inappropriate in a Western democracy, although there is an element of hypocrisy since government measures already affect energy policy considerably.

A more likely outcome is that little will happen until oil and gas really begin to run out, and coal and nuclear power are both found to be environmentally unacceptable, when there could be a drastic switch to environmental energy.

Another energy source which may become more important at the domestic level is biomass. Wood, after all, was the original fuel and is still the main fuel for the vast majority of the world's population. Its popularity as a fuel was a major reason for the decimation of native woodland in Ireland so that by 1905 only 1.4% of the country was still under forest (O'Carroll, 1984). Ireland shares with Iceland the distinction of being the least-wooded country in Europe.

Biomass as an energy source involves growing, harvesting, conversion and combustion. There has been much research, north and south, on growing willow coppice (see McElroy, 1986; Neenan and Lyons, 1977) harvested on a 3-5 year rotation. The problems of peat soils are not yet all solved, but on wet mineral soils research plots may yield 15-20 tonnes/ha/year dry matter, and a general level of 10 tonnes/ha/year should be attainable in practice. It is desirable to diversify from the present over-dependence on one willow clone to include other willows, alders, poplars and other hardwood species. The 10 tonnes of dry matter has roughly half the calorific value of coal and is therefore equivalent to 5 tonnes of coal. So one hectare should be sufficient for water and space heating for a house. If, for example, a quarter were harvested each year this could be done by hand.

However, there is a problem of conversion. The thicker (>3cm) stems can be cut into minilogs but this leaves a lot of thin branches which need to be chipped and perhaps made into briquettes. Chips can be burnt in a commercial boiler, as has been done at Loughgall for heating a greenhouse, yet for domestic use a briquette perhaps similar to peat or lignite briquettes (both now available) would be more suitable. The willow rods, as harvested, are bulky, so what is really needed is a mobile chipper-briquetter going round the farms in the winter. The benefit of this is that it could also handle hedge cuttings and any lop and top resulting from woodland management (Newbould, 1987b).

Combustion, whether of chips or briquettes, on a commercial scale is no problem, but on a domestic scale, in a wood-burning stove for example, there are problems of smoke and tarry deposits in the chimney. It is difficult to keep the combustion temperature high enough to avoid these. The smoke pollution would tend to make this fuel unsuitable for urban use.

One way to avoid this would be to convert the biomass into charcoal, which is a much cleaner and more transportable fuel, tending to burn at a higher temperature. This needs further exploration, as does conversion into ethanol or methanol.

The final domestic-scale energy ploy is energy conservation, especially by improved insulation and draught proofing. The techniques are well known and highly cost-effective.

However, everything has snags. Draught proofing of houses in high radon areas may allow build-up of significant radon levels in these houses. (Radon is a gaseous decay isotope of uranium, which tends to accumulate in buildings and may be inhaled causing irradiation, especially of lung tissue.) This affects only a small proportion of the housing stock but is potentially serious where it occurs (McLaughlin, 1987; National Radiological Protection Board, 1987). There can also be a increase in condensation. The preferred solution is a full air-conditioning scheme with a heat exchanger extracting heat from the exhaust vent and using it to warm the air intake. Another problem is that some insulating materials give off toxic fumes in the event of a house fire.

The National Scale

Moving up from the domestic scale, both wind and biomass can be scaled up to contribute, for example, to electricity generation. Solar farms, that is large arrays of solar panels, are probably not a good idea in Ireland. Energy conservation takes the form of increased energy efficiency in industry, and the development of combined heat and power schemes and district heating (Byrne, 1987). Other environmental energy sources become feasible on a larger scale including wave, tide, hydro-, geothermal and biogas.

Countries like the Netherlands and Denmark are planning greatly increased dependence on wind power. The economics of wind power depend not only on the availability of wind but also on the cost of providing back-up capacity and units during calm periods. Individual wind generators tend to vary between about 0.1 and 1 MW, and are often grouped in wind farms which tend to be unsightly, noisy and dangerous.

Once built the fuel is essentially free. Tinney (1981) estimates that by the year 2000 there might be as much as 1,000 GWh of wind-generated electricity in the Republic, this is about one-tenth of present electricity demand. However, the trial programme is already slipping behind schedule. Of the four wind generators purchased by the ESB, only two are still operable. The most successful is a vertical axis 60 kW unit of the Darrieus design. A new programme, using more modern design, may be launched soon.

Biomass for electricity production seems to me less viable, because of the large area of plantation required. A typical milled peat power station has an electrical capacity of 40 MW. At an annual production of 90 tonnes/ha/year of milled peat (a good year) this would require about 3,200 ha of productive bog. To fuel the same power station on wood chips, assuming a production of 10 tonnes/ha/year (readily attainable on mineral soil but not necessarily on peat) and assuming the same conversion efficiency, would require an area of 14,200 ha. Neenan and Lyons (1977) make more optimistic assumptions and find an area of 11,700 ha is needed to fuel a 50 MW power station. Assuming the most favourable configuration, a circle with the power station at the centre, the maximum haul would be 6-7 km for these two scenarios, and the realistic maximum haul is likely to be double that. Transport costs for this bulky fuel start to become significant. The intermediate scenario of large-scale biomass plantations producing chips or briquettes for heating in commerce or industry or district heating schemes seems more feasible.

Wave power has always looked promising because the energy available is so great. Thus Mollison (1982) finds that the overall resource around the west and south coasts of Ireland is estimated at 25 GW mean of which over 10 GW might be landed by cost-effective devices. He suggests that 10 km of Salter ducks, the most productive of the devices currently under development in the UK, could provide nearly one-quarter of the Irish electricity demand. Wave energy suffers from the same problem as wind energy, back-up capacity is needed for calm periods.

It is not clear that the engineering problems associated with extracting wave energy from rough seas, converting it into electricity and bringing it ashore have yet been solved, let alone solved at a reasonable price. Research and development in this area is expensive and all Ireland can really do is wait and watch other countries' successes and failures (see also Chapter 15).

Tidal power requires a special coastal configuration, usually an estuary or sea lough. The most thorough study made in Ireland was for Strangford Lough. Wilson (1965) made a feasibility study of Strangford and Carlingford Loughs and updated this with reference only to Strangford Lough in 1980-1981. The economic viability of the scheme was examined by the Northern Ireland Economic Council in 1981. It was estimated that a tidal barrage could contribute about 10% of present (1980) electricity production in Northern Ireland and could be just about economically viable on the basis of fuel saving, since once again electricity generation would be discontinuous and back-up generating capacity would be needed. NIES was unenthusiastic and a preliminary examination of the environmental impact of the scheme suggested that this would be large and mainly unfavourable (Carter and Newbould, 1984). The characteristics of massive tidal circulation in a relatively sheltered environment that make Strangford Lough suitable for tidal power also make it ecologically unique. It also has massive overwintering populations of waders and wildfowl. At present the project has lapsed, but it may reappear in the future.

There has been some prospecting for geothermal energy in Ireland without conspicuous success. There are two major hydroelectric schemes on the Shannon and the Erne. There is scope for smaller hydroelectric schemes but in general the configuration of the country is not favourable. Thus Lough Neagh is only 12.5 m above mean sea level and also too far from the sea. There are possibilities for smaller schemes but they would mainly be environmentally disruptive and are unlikely to be introduced.

This leaves only one renewable source of energy for discussion and that is biogas (methane) produced in the main from animal slurry, silage liquor, sewage and other forms of waste. A very successful anaerobic biogas digester is currently operating at Bethlehem Abbey at Portglenone, producing ample gas to heat the Abbey and its farm buildings, and also producing an odourless organic manure which is marketed under the name Dungstead (Gornall , 1987). One interesting finding is that the addition of silage liquor gives a rapid boost to gas production. There is considerable scope for reclaiming energy from solid and liquid wastes which helps to reduce pollution at the same time.

Alternative energy sources and energy conservation could make a major contribution towards domestic energy requirements and a minor contribution to national energy requirements (see also a review by Robinson, 1983).

CONCLUSIONS

Energy planning is extremely difficult. Energy policy is largely determined by politicians who have not yet come to terms with the post-industrial society. Unexpected events, outside national control, can change the whole picture. The developed countries are hooked on the high-energy lifestyle. Major energy sources have long lead times. There are questions about the future of nuclear power. Fossil fuel resources are finite; oil and gas are the fuels of the twentieth century. Fossil fuels tend to have massive environmental impacts, e.g. acid rain and increasing CO_2 levels. The so-called environmental sources of energy are diffuse, small and insufficiently developed. No country, with the possible exception of Norway, has invested the proceeds of the fossil fuel bonanza in creating infrastructure for a future sustainable society. Certainly the British oil and gas revenues and the Irish gas revenues appear to have conferred no lasting benefit on their respective economies.

What suggestions can be made about future energy planning? In such an uncertain world *diversity* seems crucial. Energy remains a precious commodity so that *conservation* is desirable, both in terms of making optimal use of the energy that is converted and in extending the useful life of fossil fuel resources. The environmental energy sources in some cases have the benefit of being decentralised and encouraging local self sufficiency.

In Ireland there is great benefit to be gained by basing energy planning on the geographical rather than the political entity. This does not require political unity, simply coordinated planning by good neighbours. Physical links in both electricity and gas systems are crucial. While there may also be benefit in links with the European gas and electricity grids, the first step should be to establish (for gas) or re-establish (for electricity) the cross border links.

Finally short-term economic objectives are usually held paramount in energy planning. Long-term economic objectives are not workable because of uncertainties in the projections on which they are based, e.g. a small change in discount rate can turn many scenarios on their head. In future energy planning much more weight must be attached to social and environmental objectives even where they cannot be translated into economic terms. The course of action preferred on economic grounds is not necessarily (some would say seldom) the best.

ACKNOWLEDGEMENTS

I am most grateful to Mr W.Foley of Bord na Mona, Mr E.F.O'Mahony of ESB and Mr M.Fagan of the Central Bank of Ireland for supplying information requested. They are in no way responsible for my use of it.

REFERENCES

Anon (1986) The development of lignite and peat in the European Community *Energy in Europe*, 4, 31-3.

Bogle, A.W., McMullan, J.T., Morgan, R. and Murray, R.B.(1979) Modelling of a domestic wind power system including storage. *International Journal of Energy Research*, 3, 113-27.

Bord na Mona (various) *Annual Reports*.

Byrne, P. (1987) Combined Heat and Power, a premium use for natural gas. *Technology Ireland*, 19(5), 25-9.

Carter, R.W.G. and Newbould, P.J. (1984) Environmental

impact assessment of the Strangford Lough tidal power barrage scheme in Northern Ireland, *Water Science Technology*, 14, 455-62.

Convery, F.J., Scott, S. and McCarthy, C. (1983) Irish Energy Policy, *National Economic and Social Council Report*,No. 74, Dublin

Department of Energy (1985) *Energy in Ireland*, Dublin.

Department of Industry, Commerce and Energy (1978) *Energy Ireland*, Discussion document on some current energy problems and options, Stationery Office, Dublin.

Electricity Supply Board (various) *Annual Reports*.

Flood, H.C. (1983) Current and future development, *The Engineers Journal*, 36, II and III, v-x.

Gornall, L. (1987) Slurry processing for fibre and fuel, in *Energy in Northern Ireland*, Northern Ireland Conservation Society, Belfast, 41-6.

Griffiths, A.E. (1987) The geological occurrences of lignite in Northern Ireland, in Nevin G. (ed.) *Proceedings Northern Ireland Conference on Environmental Impact Assessment*, University of Ulster, 128-31.

Griffiths, A.E., Legg, I.C. and Mitchell, W.I. (1987) Mineral Resources, in Buchanan, R.H. and Walker, B.M. (eds) *Province, City and People: Belfast and its Region* Greystone Books, British Association, 43-58.

Lawlor, E. (1975) *Solar Energy for Ireland*, National Science Council, Dublin.

McElroy, G.H. (1986) Biomass from short-rotation coppice willow on marginal land, in *Forestry and the Community*. A symposium of the Institute of Biology, Northern Ireland Branch.

McLaughlin, J.P. (1987) An assessment of indoor radon exposure in Ireland. *Paper to Second International Specialty Conference on Indoor Radon*, New Jersey.

Manning, M. and McDowell, M. (1984) *Electricity Supply in Ireland: the History of the ESB*, Gill and Macmillan, Dublin.

Mollison, D. (1982) *Ireland's Wave Power Resource*, National Board for Science and Technology, Dublin.

NBST (1980) *Energy Forecasts for Ireland.* NBST, Dublin.

NBST (1981) *Energy Forecasts for Ireland*, NBST, Dublin.

National Radiological Protection Board (1987) *Exposure to Radon Daughters in Dwellings*, NRPB-GS6, NRPB, HMSO.

Neenan, M. (1984) Short rotation forestry as a source of energy and chemical feedstock, in *Energy from Biomass using Short Rotation Coppice*, Institute of Biology Symposium, New University of Ulster, Coleraine.

Neenan, M. and Lyons, G. (1977) *Energy from Biomass*, An Foras Taluntais, Dublin.

Newbould, P.J. (1982) Energy Resources, in Cruickshank, J.G. and Wilcock, D. (eds) *Northern Ireland: Environment and Natural Resources.* Queen's University of Belfast and The New University of Ulster, 241-64.

Newbould, P.J. (1987a) Environmental aspects of lignite, in *Energy in Northern Ireland*, Northern Ireland Conservation Society, Belfast. 16-21.

Newbould, P.J. (1987b) Energy from biomass, *Energy in Northern Ireland*, Northern Ireland Conservation Society, Belfast. 38-40.

New Ireland Forum (1984). *Sectoral Studies. Opportunities for North/South Cooperation and Integration in Energy*, Stationery Office, Dublin.

Northern Ireland Economic Council (1981) *Strangford Lough Tidal Energy Report 24*, Belfast.

Northern Ireland Economic Council (1987) *Economic strategy: impact of lignite*, Report 65, Belfast.

Northern Ireland Electricity Service (various) *Annual Reports*.

O'Boyle, A. (1987) Energy - A key to industrial success, *Technology Ireland*, 19(5) 13-7.

O'Carroll, Northern (1984) *The Forests of Ireland*, Turoe Press, Dublin.

O'Rourke, K., Lewis, J.O. and O'Connell, D. (1976) Solar energy for domestic space heating, *Technology Ireland*, 8(6), 23-5.

Robinson, K. (1983) Renewable energy in Ireland: how much and of what kind?, in Blackwell, J. and Convery, F. (eds) *Promise and Performance: Irish Environmental Policies Analysed*, Resource and Environmental Policy Centre, University College Dublin, 249-59.

Solar Energy Society of Ireland (1978) *Towards Energy Independence*, SESI, Dublin.

Tinney, S. (1981) Power Generation, paper presented at a seminar on *Technological Support for Industrial Development*, sponsored by An Foras Forbartha, Dublin.

Wilson, E.M. (1965) Feasibility study of tidal power from Loughs Strangford and Carlingford with pumped storage at Rostrevor, *Proceedings of Institute of Civil Engineers*, 32, 1-29.

14 WATER RESOURCE MANAGEMENT

David N. Wilcock

INTRODUCTION

Water is a plentiful resource in Ireland and management problems result not from any inadequacy of the overall supply, but from the variety of conflicting demands on that supply and the technical problems of satisfying these demands simultaneously (Figure 1). Agriculture's need for fertilizers and good drainage, for example, may conflict with the use of rivers for fisheries. Industrial and urban effluent, if improperly treated and discharged into rivers and lakes, may cause water pollution. An upland river valley may be seen by water authorities as an ideal site for a water supply reservoir but by local residents as a valuable amenity to be preserved untouched. Although such conflicts have always been with us they have become more intense in the last two decades with rising populations, more intensive agriculture, the increase of leisure time and the consequent use of water for recreation. Added to these recent pressures have come new ones of a political character. Both the Republic of Ireland and Northern Ireland are in the European Community (EC) and subject to its environmental policy which seeks to introduce consistent management standards throughout the Community.

359

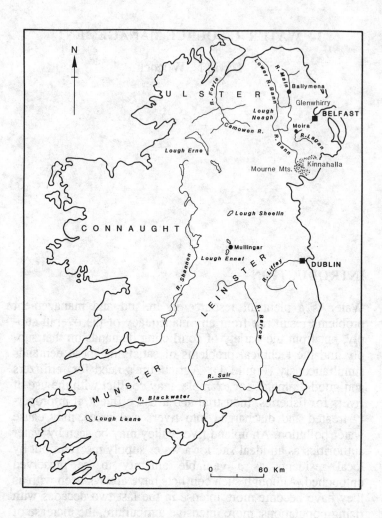

Figure 1 Place names quoted in the text

SUPPLY AND DEMAND

The total water resource of a country may be assessed as the difference between precipitation and evapotranspiration. In Ireland, annual precipitation is about 1,130mm and evapotranspiration 430mm. The remaining 700mm infiltrates into the soil and groundwater, eventually entering the rivers and discharging into the sea. As groundwater or surface runoff, water is available for direct abstraction or storage.

A figure of 700mm represents a colossal surplus of water in relation to demand. In the Republic of Ireland, for example, total surface water abstractions for public bodies and industry amount to only 6.97mm per year and groundwater abstractions to only 1.75mm per year. Agricultural stock requirements and domestic use in rural areas account for another 2.51mm (Cabot, 1985), bringing the total to 11.23mm per year. Precisely equivalent figures are not available for Northern Ireland but a total annual use of 16mm has recently been estimated (Wilcock,1982). These figures imply that available resources are some fifty times in excess of demand, although a more realistic estimate made by An Foras Forbartha for the Republic of Ireland is that total exploitable resources are some twenty times greater than present demands. Yet, whichever estimate is used, there is clearly no overall resource availability problem. This is not to say that local water supply problems do not exist. Both Dublin and Belfast will require additional water supplies before the year 2000 and about half of the rural communities in the Republic do not yet have piped supplies. Rural supply is better in Northern Ireland where less than 4% of all households are without piped water.

Interesting contrasts in water use between the Republic of Ireland and Northern Ireland emerge from a more detailed examination of water use statistics. For example, industry accounts for 31% of water abstractions in the Republic but only about 19% in Northern Ireland. This latter figure may be an overestimate since it relates to water supplied by meter, not all of which is for industry (Water Service (NI), 1986). Another difference between the two countries is in the amount of water supplied from groundwater. In the Republic, groundwater supplies some 22% of all water abstractions while in Northern Ireland it supplies only about 7%, mainly in the remoter parts of the west and north-east. Elsewhere in Northern Ireland, rivers and impounding reservoirs or lakes provide the bulk of water consumed. The largest single source is Lough Neagh which supplies 28% of all water consumed in the Province (Figure 2).

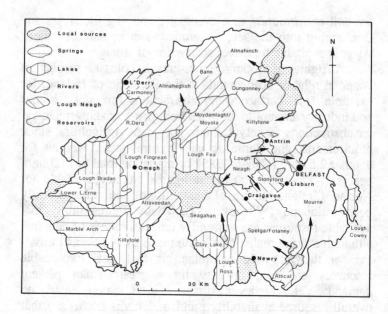

Figure 2 Northern Ireland Water Supply Zones (Source: Northern Ireland Water Statistics, 1985)

One noteworthy difference between the two countries is the apparently larger consumption of 650 litres per person per day in the Republic of Ireland as opposed to 442 litres per person per day in Northern Ireland (Cabot, 1985; Water Service (NI), 1986). This difference is probably accounted for by inclusion in the Republic's figures of estimates for consumption by agricultural stock and rural households. However, excluding these two categories, consumption per head per day falls to 504 litres, still 13% greater than in Northern Ireland. Perhaps the difference is best explained by the fact that the Republic of Ireland has many smaller water supply systems and a longer total length of pipeline. Given both the expense of maintaining supplies plus the relatively great age of installations throughout many parts of the country, it might be expected that losses through leakage in the Republic's system will be greater than in Northern Ireland's. Thus differences in consumption might be more apparent than real.

DRAINAGE

Agriculture plays an important role in the Irish economy accounting for 16% and 5.6% of the Gross Domestic Product (GDP) in the Republic of Ireland and Northern Ireland respectively. In the Republic agriculture employs 23% of the workforce while in Northern Ireland the equivalent figure is 9.7%. Compared with Britain and most countries in the developed world these figures are high; for example, in the United Kingdom as a whole agriculture accounts for only 2.5% of the GDP and 2.7% of the workforce.

In reaching its present level of efficiency, agriculture in Ireland has had to overcome a wide range of environmental difficulties. Because of a high ratio between rainfall and evapotranspiration, for example, each unit area of land has to discharge 45% more surface water than an equivalent area in Britain. Herein lies the basic agricultural water management problem that has confronted successive governments in Ireland for centuries; how to dispose of this surplus surface water without repeatedly flooding lowland areas downstream.

This problem is made worse by the mainly frontal nature of Irish rainfall, resulting in parts of County Donegal and other western counties experiencing rainfall, on average, two days out of every three. Hours of sunshine are correspondingly short and solar energy received at ground level is less than in many countries at similar latitudes. This significantly limits evapotranspiration, a crucial process for drying the soil.

Climatic problems are compounded by topography and soils. Relief is highest around the coast and consequently the largest rivers have long stretches in their middle and upper reaches with very gentle gradients. Water moves only slowly in such channels and overbank flooding is common. Moreover soils are impermeable over large areas and absorb and release water only slowly (Wilcock and Essery, 1984). The worst soils in this respect are probably the boulder clays. Peats and peaty soils which cover 16.2% of Ireland (Hammond,1981) also pose drainage problems. It is popularly believed that peats act as a 'sponge', absorbing water at times of heavy rainfall and releasing it steadily throughout the summer. Yet there is no evidence that peats in Ireland act in this way as such behaviour requires high storage capacity and high permeability. While peat has high storage capacity its permeability is usually very low except in the surface sphagnum. The analogy of a peat bog to a sponge is inadequate because sponges have to be squeezed to release their stored water and there is no natural squeezing mechanism for releasing water from peat.

The strategy adopted by governments in both the Republic of Ireland and in Northern Ireland to solve the drainage problem has had profound effects on natural river systems. This strategy is basically simple. A phase of arterial drainage is carried out on the lower reaches of a river. This involves channel widening, deepening and straightening to accommodate larger discharges, to restrict overbank flooding and to provide lower outlets for field drains. Arterial drainage (or channelisation) proceeds upstream (Figure 3) until all tributaries have been drained. Once the river channels have been enlarged, drainage ditches and sheughs can be dug to feed runoff directly into the rivers (Figure 4). This is the second phase of the drainage strategy. Field drainage is the third and final phase of a coordinated drainage plan and involves the laying of tile and/or mole drains about one metre below the soil surface to lead water into the sheughs and thence into the rivers (Figure 5).

Such a sequence of drainage events is rarely followed slavishly in any one catchment, although much of Ireland now has such integrated drainage systems, perhaps installed in piecemeal fashion over several decades. In the Republic of Ireland, the policy of the Office of Public Works has been to implement arterial drainage schemes on a catchment basis, completing work in one basin before starting in another. In Northern Ireland this has never been the practice, mainly for political reasons. Because funding for such work has always come via central Government and been undertaken by the Department of Agriculture, it has always been thought politically circumspect to invest approximately equal amounts in each of the six counties at any one time. Thus several arterial drainage programmes have usually been operated concurrently and individual catchments have experienced several phases of arterial drainage, sometimes spread over several decades.

Figure 3. The River Main, north of Ballymena, Co. Antrim, looking downstream. Arterial widening and realignment can be seen in the background. Natural meanders, to be removed as part of the river's arterial drainage scheme, can be seen in the foreground. The process of drainage proceeds upstream. Cutover raised bogs, part of Northern Ireland's diminishing areas of wetland, can be seen adjacent to the river in the foreground.

Figure 4. Drainage ditches (sheughs) being cleaned out and enlarged in the River Main basin, Co. Antrim

Figure 5. Field drainage being installed. This type of drainage is usually undertaken by farmers themselves, grant-aided by central government. In contrast, arterial drainage costs are all borne by central government.

There is little doubt that these drainage policies have improved the physical basis of agriculture. Growing seasons are longer in that soils now shed excess moisture sooner in Spring, thereby quickening the rise of soil temperatures, and become wetter later in Autumn. Liming and the use of fertilizers has become effective in raising the base status of soils. Fertilizers are no longer leached out of the soil as quickly as formerly and oxygen can enter the soil more easily to facilitate processes of mineralization. In the light of these improvements agricultural output has risen markedly over the past three or four decades, and quite spectacularly so in the period since 1960 (e.g. Wilcock, 1979).

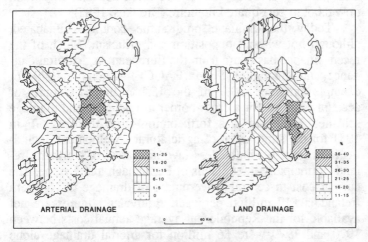

Figure 6. Agricultural land benefitting from arterial and land drainage in the Republic of Ireland (Source: Bruton and Convery, 1982)

The scale of channelisation work throughout Ireland has been impressive. In the last forty years, for example, some 4,800km of main and tributary rivers in Northern Ireland have been channelised and field drains installed on half the agricultural land area. For the Republic of Ireland figures are not available for the length of waterways affected, but rather for the area of farm land affected. According to Bruton and Convery (1982), 5.2% of farm land has benefitted from arterial drainage and 24.3% from field drainage (Figure 6). Midland and western counties have benefitted most, followed by counties in the south-west. Offaly with 23.4% of farm land under arterial drainage and Waterford (0.3%) represent the extremes of drainage activity, although there are three counties - Carlow, Kilkenny and Wicklow - in which no arterial works have been undertaken.

Arterial drainage has thus included a large proportion of the rivers of Ireland. In Northern Ireland the density of arterial drainage is 0.09km per square kilometre (if only main rivers are included) and 0.34km per square kilometre if tributary rivers are included as well. By way of contrast, the density in USA is only 0.003km per square kilometre and in Britain only 0.06km per square kilometre (Brookes, 1985).

Today the drainage programme continues unabated, although not without opposition and criticism. Much of the recent finance has come from the European Agricultural Guidance and Guarantee Fund (EAGGF) which in 1979 for example provided 50% of the 15.1 million EC units of account required for a cross-border arterial and land drainage scheme. More aid was forthcoming from the EAGGF in 1981 for the stimulation of agricultural development, including land drainage, in the 'less-favoured' areas of Northern Ireland - principally the uplands of Fermanagh and Tyrone. The current cost to central Government of drainage in Northern Ireland is £17 million per year. Equivalent figures are not available for the Republic but average annual costs between 1970 and 1979 were £6 million for arterial drainage alone (Bruton and Convery, 1982; Cabot, 1985). Since 1945, £245 million has been invested in arterial drainage by the Office of Public Works in Dublin, while over the same period in Northern Ireland accumulated actual costs have totalled about £50 million (Wilcock, 1979). The principal beneficiaries of this investment are the dairy farmers, but at a time of 'milk lakes', 'butter mountains' and cutbacks in EC quotas for dairy products generally, other users of the countryside are emerging to question the scale of this investment in drainage and to state their own priorities for the future management of Ireland's rivers.

Prominent among these other users of rivers are game fish anglers. Since the mid 1960s, salmon stocks on the Foyle system, once one of Europe's finest salmon rivers, have declined dramatically and the river now has only 25% of the spawning and only 10% of the angling catch it enjoyed twenty years ago (Wilcock, 1982). A similar story is true for rivers in the Republic (Mawhinney, 1979). Overfishing at sea and disease have both been invoked to explain these declines, but arterial drainage is believed by official angling bodies to be an important factor. For example, arterial drainage increases flood peak volumes by 60% and shortens the duration of individual flood spates (Bailey and Bree, 1981). No published figures on increased suspended sediment loads are available but ongoing work by the author on the River Main in County Antrim shows that suspended sediment loads can be increased tenfold by arterial drainage. The shorter duration of flood spates, i.e. the 'flashier' nature of the river following arterial drainage, reduces the opportunity for fish to move upstream into the spawning grounds. The effect of increased suspended sediment loads is often to convert cobble-bedded streams, essential for oxygenating salmon ova, into silty or sand-bedded streams. Natural pools and riffles on the coarse-bedded rivers are eliminated by dredging and bed morphorlgy becomes uniform, destroying the diversity of food supplies necessary for healthy fish populations. Removal of bankside vegetation removes shade and water temperatures may rise. It is argued that such effects are especially damaging to game fish and derive from arterial drainage (Figure 7).

How long the effects of drainage works persist has been a point of conjecture among fish ecologists. One study on the Camowen (Kennedy et al., 1983), identifies several interesting processes controlling the recovery of fish populations after drainage. Recovery appears to take place in a downstream direction and is related to the time taken for bedload in the channel to become consolidated following bed excavation. The presence of an upstream section of river, unaffected by arterial drainage works, appears important as a reservoir from which recruitment and re-establishment can take place in the drained section following cessation of engineering work. Over a seven kilometre stretch of upland river on the Camowen, recovery took six years and the authors point out that longer stretches of modified river may recover more slowly. Furthermore, channel relocation on the Camowen was minor compared to other schemes and this would favour relatively rapid recovery. Much appears, therefore, to depend on local circumstance, although ecological damage is real enough and may last for a decade or more on some of the larger and more radical channel relocation schemes. In this connection, the Northern Ireland practice of phasing arterial drainage on individual rivers for political reasons is bound to have serious long-term environmental implications in that recovery is continuously interrupted.

Within the last ten years, arterial drainage has also been criticised on purely economic grounds (e.g. Kelly, 1980; Bruton and Convery, 1982; Armstrong, 1982). Some economists, for example, argue that the catchments deriving the greatest economic benefit from investment in drainage were drained long ago, and that public money would now be better spent on farm amalgamation and on improved management and marketing, which would enhance agricultural output per unit of investment (see Armstrong, 1982).

An interesting historical dimension to the role of drainage in Irish society is that when the very first, modest, arterial drainage schemes were introduced in the nineteenth century, rural population densities were often 300 per square kilometre, landholdings were small and thus many individuals shared the benefits of government investment. In addition, the drainage work employed large numbers of unskilled, local people. Today, rural population densities are much lower, often under 75 per square kilometre, and fewer people benefit directly. As a means of redistributing income, investment in drainage no longer has the social relevance it had in the last century, or even 40 years ago. Local employment from drainage is now minimal and a total professional workforce of about 1,500 can service all the machinery required to implement drainage works throughout Ireland.

For all these reasons - environmental, economic and political - it seems likely that government in Ireland will soon have to reassess its attitude to arterial drainage. This may have started in the Republic of Ireland; the 1982-1987 Fine Gael government announced important institutional changes regarding arterial drainage, including a plan to levy modest charges against beneficiaries. If implemented, such charges would no doubt reduce, perhaps quite radically, much of the clamour for arterial drainage in the Republic of Ireland. Notwithstanding, the impact of arterial drainage on the hydrology, geomorphology and ecology of Irish rivers will remain with us for a very long time to come.

Figure 7 The River Main at Dunminning, Co. Antrim; (a) before and (b) after arterial drainage. Note the straighter, deeper channel, the loss of bankside vegetation and the reduction of morphological diversity in the channel

SURFACE WATER QUALITY IN IRELAND

Surface waters supply 62% of water consumed in the Republic and 92% of that consumed in Northern Ireland. They also sustain valuable game and coarse fisheries and are a principal means by which sewage and effluent is biodegraded (broken down into harmless material). Such breakdown requires high dissolved oxygen content which is therefore one of the commoner measures of water quality. Rivers are also a valuable visual amenity as well as a recreational resource generating a considerable amount of revenue both locally and nationally. For all these reasons it is important to maintain high water quality throughout Ireland's rivers and lakes.

Other pressures to encourage good water quality are fuelled by the recent spate of EC legislation on the subject. Much of this legislation attempts to prevent individual member states from unfairly attracting foreign industrial investment. An individual state could easily do this, for example, by not insisting on the same high environmental standards as in other member states. Pollution control costs industry money and raises industrial prices, especially if industry has to bear some of these costs. Acceptance of low environmental standards by a member state might reduce production costs and act as an unfair inducement for industrial investment. To ensure that this is not the case, the EC attempts to establish uniform standards throughout member states, often through the issuing of Directives. No less than nine of these since 1975 relate to water quality in inland waters and the number is set to double before 1990.

The various Directives usually identify two types of standard; 'I' values are the less strict of the two but are mandatory. 'G' values are higher standards than 'I' values and although member states are always encouraged to reach these values with regard to all water quality characteristics, they are not mandatory. The exact manner of 'I' standard implementation and enforcement, and to what parts of the national resource a particular Directive relates (lakes, rivers, groundwater, bathing beaches etc.), must be communicated to Brussels, often within three years of a Directive being issued. European Community Directives pressurise member governments to monitor water quality systematically, and monitoring programmes have therefore increased dramatically in Ireland over the last fifteen years.

Largely as a result of water quality monitoring (e.g. An Foras Forbartha, 1983), we now have a good idea of the state of surface water and groundwater resources in Ireland and more than half of the Republic's 12,000km of river have now been classified by biological survey. According to An Foras Forbartha, 84.2% of rivers in the Republic are unpolluted, 13.8% are slightly to moderately polluted and 2% are seriously polluted (Cabot, 1985). In the Republic, the national classification of river quality is based on biological field surveys of macro-invertebrates. The sampling is undertaken on riffles in river beds, because riffles are well-oxygenated and tend to show pollution levels more clearly than pools (Clabby *et al.*, 1984, p.164). The other means of classifying water quality is by means of its physicochemical properties (dissolved oxygen (DO), biochemical oxygen demand (BOD), pH, ammoniacal nitrogen, suspended sediment, and nitrite and nitrate levels). However, the collection of such data is expensive and different physico-chemical properties have different degrees of significance for different water uses. Sampling, moreover, needs to be repeated at very short time intervals because physico-chemical properties can fluctuate very rapidly as a result of changes in river flow. A large flood, for example, lasting only a few hours, can dilute the effects of a pollution incident leaving little trace in the water chemistry, despite damage or even obliteration of animals and plants. Therefore, biological indicators show effects and extent of pollution better than physicochemical properties which are perhaps best employed in measuring causes of pollution rather than its effects.

In the Republic of Ireland, physicochemical surveys are routinely taken by An Foras Forbatha only on rivers likely to have been polluted. In Northern Ireland, however, as throughout the UK, physico-chemical characteristics form the basis of the official river classification into five categories: Class 1A (unpolluted), Class 1B (high quality), Class 2 (doubtful quality), Class 3 (poor quality) and Class 4 (grossly polluted). Class 1 rivers can support game fish and have dissolved oxygen levels equal to or exceeding 60% of saturation levels for 95% of the time. They also have low BOD values. Class 2 rivers can only support coarse fish and have DO levels exceeding 40% of saturation for 95% of the time. Rivers in Classes 3 and 4 cannot support any form of fish life and have very low DO levels (less than 10% of saturation for 95 % of the time). Rivers in Class 3 have low amenity value, while those in Class 4 are officially classified as a nuisance from the amenity point of view.

In Northern Ireland, periodic water-quality surveys are undertaken on 1,291km of river officially designated for fish life under the 1978 EC Directive (78/659/EC). Of these 1,291km so designated, 1,062km are salmonid (game fish) rivers and only 29km are cyprinid (coarse fish) rivers. Overall, 38.2% of Northern Ireland rivers suitable for fish life are offically designated as unpolluted (Class 1A), 48% are of high quality (Class 1B), 13.1% of doubtful quality (Class 2) and 2.7% of poor quality (Class 3) (NI Water Service, 1986). None of the rivers designated for fish are experiencing gross pollution (Class 4).

Despite differences in classification procedures, the state of surface waters appears similar in both countries. Overall water quality is satisfactory, and much better than in Britain. Nonetheless, there are small, serious pockets of pollution and water quality in many areas is deteriorating (Figure 8). In the Republic of Ireland, for example, the condition of three lakes, Lough Ennal in Westmeath, Lough Leane in Kerry and Lough Sheelin in Cavan has caused concern for a decade or more. In Northern Ireland most attention has focused on Lough Neagh, mainly because it forms the largest water supply, and the costs of treating poor quality water are considerable. However, in Lough Erne oxygen levels are declining and algal growths excessive (Downey and Ni Uid, 1977; Wood, 1982).

In the case of all these loughs the main cause of poor water quality is nutrient enrichment. This enrichment can emanate from domestic sewage and/or intensive agriculture. Much domestic sewage is untreated or receives only primary level treatment, involving removal of coarse solids by screening and of fine solids by sedimentation. Removal of nutrients is one of the most sophisticated and expensive forms of waste water treatment, removing nitrogen and phosphorus compounds, the main causes of eutrophication and oxygen depletion in Ireland's rivers and lakes. Many Irish lakes have become progressively enriched over the last century (e.g. Wood and Gibson, 1973; Battarbee, 1986), due to increasing use of agricultural fertilisers and domestic detergents (Wilcock, 1982; Patrick, 1987).

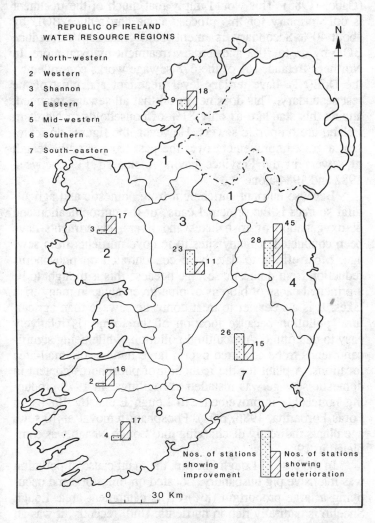

REPUBLIC OF IRELAND
WATER RESOURCE REGIONS

1 North-western
2 Western
3 Shannon
4 Eastern
5 Mid-western
6 Southern
7 South-eastern

Nos. of stations showing improvement

Nos. of stations showing deterioration

0   30 Km

Figure 8 Results of the 1982/83 Water Quality Survey (Source: An Foras Forbartha, 1983)

375

The national picture relating to sewage treatment provision is difficult to quantify in the Republic because so many local authorities are responsible for this service. One recent estimate puts the amount of post-1975 BOD reduction at about 40% for industrial waste and 19% for domestic waste (Cabot, 1985). This would imply that much of the treatment is only primary for this process alone can reduce BOD by about 40%. Secondary treatment (biodegradation) can reduce BOD by up to 70% and tertiary treatment by even more. In Northern Ireland, 161 of the 678 sewage works controlled by the DoE(NI), have tertiary level treatment and most of the rest secondary. This does not mean that all sewage treatment attains this standard as only 81% of households in Northern Ireland are on public sewers. However, the figures do represent a great improvement over the past ten years, the length of sewers in the Province having increased 13% between 1982 and 1984 alone.

Because nutrient enrichment from domestic and agricultural sources is the principal cause of the eutrophication and de-oxygenation of Irish lakes and rivers, experiments have been conducted at many sites to remove nutrients from sewage. Most efforts to date have concentrated on phosphorus reduction from domestic sewage because this is thought to be a principal cause of blooms of blue-green algae in many Irish lakes. It is also easier to treat domestic sewage than agricultural pollution because location of the former is relatively easy to determine. Agricultural pollution, while being significant, tends to be scattered over a large number of small-area locations. A plant for the reduction of phosphorus content in domestic sewage was installed at Mullingar in 1979 producing noticeable improvements in Lough Ennal by 1984 (An Foras Forbartha, 1986, p.51). Phosphate removal at 10 sewage plants indirectly discharging into Lough Neagh was completed in 1982.

In the case of Lough Sheelin, the main cause of pollution was intensive pig husbandry. Located on limestone and occupying a large proportion (8%) of its catchment area, Lough Sheelin is naturally rich in nutrients. Until recently, it was an excellent trout fishery generating considerable tourist income. In the 1970s, however, the number of pigs reared in intensive units throughout the catchment increased substantially and production of pig manure quadrupled. Slurry spreading is inadvisable in winter because water tables are high and surface runoff dominates, so that slurry is transferred quickly into water courses, raising nutrient levels dramatically. Increased nutrient loadings in Lough Sheelin during the 1970s were largely a result of poor management of pig slurry, and led to extensive algal crops. These crops reduced penetration of sunlight into the water column, in turn curtailing production of submerged plants. As such plants provide the favoured habitat for the invertebrates on which trout feed, their progressive elimination produced an associated decline in trout numbers from 1970 onward. A strategy to transport pig slurry out of the catchment has recently been implemented and this has produced a gradual improvement in the water quality of the lough. The growth of submerged plants resumed in 1982-83 though no restoration of fish stocks had taken place by that date (Dodd and Champ, 1983).

Problems akin to slurry disposal can affect rivers as well as lakes and the short-lived impact of such incidents may well go undetected by physicochemical monitoring at the usual two week intervals. Some pollution incidents in rural Ireland are disastrous. Silage effluent of the type described above for Lough Sheelin may have a BOD value of 60,000 mg /litre and is often dumped directly into rivers. Given that the EC standard for the BOD value of cyprinid rivers is 6mg/litre, the colossal impact of such discharges on fish life is only too apparent. In 1983 alone silage discharges killed one million fish in rivers of the Fisheries Conservancy Board in Northern Ireland and one such discharge eliminated a two-year cohort.

The current deterioration in Ireland's surface waters often derives from geographical factors; the growth in size of the population living in towns (Bannon, 1979; Bannon, 1983) and the consequent increase in pressure on rivers to dispose of urban effluent; the intensification of agriculture and its wastes, many with potentially serious impacts on river water chemistry; the expansion of mining, quarrying and industrial activities since the mid-1960s creating high suspended sediment concentrations and toxic materials; the drift into the countryside of isolated owner-occupiers, often with their own water supply and septic tanks for effluent disposal. One common feature of such developments is the highly localized nature of their environmental impact. This is especially true of water pollution in Ireland making control difficult for any central authority.

Because of the isolated point source nature of waste material, it is difficult to quantify how much is contributed by different types of activity. In the Republic of Ireland, industry produces 49% of all wastes, domestic sources 39.5% and agriculture 11% (Toner and O'Sullivan, 1977). Most waste is disposed of into the sea (81% in the Republic) (see Chapter 15), but of the total load contributed to inland waters, industry accounts for 30% more than domestic and agricultural sources put together. The rivers Suir and Barrow in southeast Ireland have perhaps the highest industrial loads of all rivers in the Republic and, with the Munster Blackwater and Shannon, are among the four rivers in the Republic with the highest waste loads (Toner and O'Sullivan, 1977). Most industrial loadings come from older, established industries related to the processing of agricultural products - milk, meat, and sugar beet.

In Northern Ireland a report by the Department of the Environment in 1978 shows that effluent loadings on the Lagan can sometimes almost equal low flows and that this river can suffer prolonged periods of low water quality in summer (Wilcock, 1982). Parts of the Lagan are officially classified as of poor quality (Class 3) and the whole of the river downstream of Moira has recently been officially classified as of doubtful quality (Class 2).

One pollution problem that has developed over the past twenty years concerns the increasing concentrations of nitrates in surface and groundwaters. The use of nitrogen as a fertilizer in grass production has increased dramatically over this period. In the Lough Neagh basin nitrogen usage increased from 42kg/ha/year in 1967 to 106kg/ha/year in 1982 (Smith *et al.* 1982). Excess nitrogen is leached out of the soil and accumulates in runoff into rivers and lakes. The significance of nitrogen in surface waters relates to its role in algal growth. Increased use of nitrogen has increased the nitrogen to phosphorus ratio (the N/P ratio) in rivers and lakes, an increase which is accentuated by simultaneous efforts to reduce phosphorus loadings from domestic sewage. On Lough Neagh, where this problem has been studied for twenty years, the increased N/P ratio has been associated, over the same period, with a change in the species composition of algae growing in the Lough. Since 1969 there have been no extensive green surface blooms other than on one occasion in 1975. Instead algae now tend to be suspended throughout the total water column. Surface blooms are nitrogen-fixing and related to abundance of phosphorus. Non-surface blooms are not nitrogen-fixing. It would appear that these latter blooms do not need to fix nitrogen because of its greater abundance from agricultural runoff. Despite the change in species, however, the total crop of algae has steadily increased throughout the 1970s. The non-appearance of surface blooms tends to divert public awareness away from the algal pollution problem. This on-going, but more subtle form of pollution, is a matter of great concern, since it continues the process of de-oxygenation in the Lough.

Another concern surrounding the increased use of nitrogen in Ireland's rivers results from the association between nitrate levels in drinking water and the infant disability methaemoglobinaemia. World Health Organisation (WHO) standards permit 11.3 mg N/litre in infants' drinking water and whereas mean levels of nitrate in Irish rivers appear to be much lower than this, maximum levels may get unacceptably close to WHO limits. Studies on the River Main in County Antrim have demonstrated a rise in the mean level of N between 1969 and 1979 of only 1.0 mg N/litre, from 1.41 mg N/litre to 2.41 mg N/litre. Maximum concentrations, however, rose from 2.9 mg N/litre to 7.5 mg N/litre over the same period (Smith *et al.*, 1982).

The rise of nitrate levels in groundwater has not been systematically monitored, although the growing significance of groundwater sources for public consumption both in the Republic of Ireland and in Northern Ireland, might suggest that this be done. In the Republic, groundwater aquifers account for 20% of total public water requirements (Irish National Committee for the International Hydrological Programme, 1982), although only 6% in Northern Ireland. Groundwater is generally more diffcult to pollute than surface waters because of the natural filtration processes that take place as water percolates down to the water table. However, once polluted, it may prove difficult to purify groundwater because of the slower rates at which fluids move through the ground. In view of the growing importance of groundwater sources, especially in remoter areas, plus the widespread potential for pollution by nitrates, it is imperative that this problem be monitored carefully. It needs to be stressed that present indications are that groundwater pollution in Ireland is only a local problem, with the principal sources being farm effluents and leachates from landfills. The problems of acid rain in Ireland are dealt with in Chapter 16.

RECREATION

Quite apart from the very important amenity provided to local residents by clean, healthy rivers, the recreational use of water in Ireland can often generate significant local income, especially in rural areas where the only other source of revenue is farming. The principal recreational use of rivers is for fishing and two types of angling are recognised - game fishing and coarse fishing. Recreational boating is important on some lakes and rivers, especially the Erne system, Lough Conn and the River Shannon.

The sensible management of Ireland's inland fisheries alongside agriculture represents a most important and delicate water management problem. Yet it is a problem that requires resolution, for at stake is a potentially large economic return. In Scotland, for example, the salmon industry's total earnings exceed £120 million per year. Although anglers take only 15% of this catch, rod-caught salmon account for about 95% of the industry's total earnings (Conference communication, Ulster Angling Federation, 1984). Equivalent figures for Ireland are difficult to come by, but Mawhinney (1979) shows that salmon anglers in the Republic spent about £1 million in 1970, equal to 42% of the salmon industry's gross output. However, as in Scotland, the rod-caught catch, at 9%, was only a small percentage of the total catch. Total income in 1970 from trout and coarse fisheries was £1.13 million and £1.79 million respectively, giving a total income from inland fisheries of nearly £4 million.

In Northern Ireland the commercial catch is currently about 160,000 kg, worth about £750,000 (Department of

Agriculture (NI), 1985). If this represents about 58% of the total income from salmon fisheries, as in the Republic, total income from salmon rod fishing is about £523,000. If income from commercial salmon fisheries were, or could be made to be, only 5% of the total income from salmon fisheries, as in Scotland, revenue from salmon anglers would be £10 million per year in Northern Ireland alone! The present ratios of about 1.4:1 between commercial and tourist income from salmon fisheries in Ireland will only be improved if salmon and trout angling is protected more aggressively.

Coarse fishing may not have the same allure or economic potential as game fish, but it is nonetheless extremely important, perhaps more so than game fish at the present time. The eel fishery of Lough Neagh, for example, is believed to be the largest in western Europe and accounts for about 85% of Northern Ireland's total catch, currently valued at about £2.25 million per year (Wood, 1982). Perch was another source of local income on Lough Neagh until about 1980 when it had a market value of £1 million per year. Unfortunately, perch have now disappeared, possibly through the inadvertent introduction and rapid growth of roach, an aggressive competing species without any commercial value.

Large bodies of standing water, natural or artificial in origin, are not always regarded as fine local amenities. In 1978, the DoE (NI) put forward a scheme to build a reservoir at Kinnahalla on the Upper Bann in the Mourne Mountains. In support of the scheme, the DoE (NI) outlined how Belfast and its surrounding districts needed to find an additional 110 million litres per day water supply before the year 2000. The proposed scheme was said to provide 88 million litres per day and was to be sited at Kinnahalla in the upper Bann valley. The new dam and its surrounding area would have been landscaped to merge in with the surrounding countryside.

Opposition to the scheme came from a group calling itself KOMRADE (Kingdom of Mourne Revolt against the Destruction of Environment). The scheme was criticized on various grounds. It would damage upper Bann fisheries, destroy access and amenity to a popular part of the Mournes, diminish river flows, and cause a consequent deterioration of water quality. Alternative supplies from Lough Neagh were cheaper or only marginally more expensive, even on the

Department's own figures. Most of these claims were contested by the DoE's (NI) own experts, but in an expensive and sometimes acrimonious Public Inquiry the Department was accused of witholding relevant information from its critics and of suggesting that the new storage provided by the scheme was greater than it actually was. The scheme did not go ahead and the debate continues today. Are Belfast's water needs to the end of the century and beyond to be provided from a gravity flow reservoir in the Mournes or the Antrim Plateau, or should Lough Neagh provide it? Upland reservoirs provide clean water, while flow through the delivery systems is gravity-controlled and relatively cheap. On the other hand, few catchments are large enough in the upland areas to provide sufficient inflow, and flow from neighbouring streams may have to be diverted thus perhaps damaging the local enivronment. The Lough Neagh water supply is enormous relative to demand but it is polluted and involves pumped storage to transfer it from the low-lying Lough to Belfast. The issue is unresolved, and at the time of writing proposals for a new reservoir at Glenwhirry in County Antrim are being aired.

LEGISLATIVE FRAMEWORK

Large amounts of widely available water, the small and isolated nature of demand outside of a few large urban centres and the jealously guarded rights of local authorities have all conditioned legislation on water supplies in Ireland. This is particularly true of the Republic of Ireland, although less so of Northern Ireland since 1973 when the existing local authority structure of counties, county boroughs, and urban and rural districts was abolished after the introduction of Direct Rule from Westminster. Local authority control for water supplies, sewerage, and drainage in the Province was abolished by the Water and Sewerage Services Order of 1973 and by the Drainage Order of the same year. Under these Orders in Council, the Departments of Agriculture and the Environment assumed responsibility for all those water functions which had previously been the responsibility of local authorities.

This legislation reflected UK government thinking on water resource management at the time and in many ways these two Orders represent the equivalent, for Northern Ireland, of the UK Water Act of the same year which created large Regional Water Authorites (RWA's), managed on a catchment basis, and subsuming within them (there are only ten such RWA's for the whole of England and Wales) the entire local authority system. This Act is widely regarded as one of the most effective pieces of resource management legislation in Britain, since it provides for the integrated management of all water resources within large catchments by one single authority. Individual RWA's are responsible for such diverse functions as water supply, sewerage, licensing, drainage, recreation, flooding, coast protection and the preparation of management programmes. Northen Ireland does not have its own RWA but the Department of the Environment in effect fulfils many of the same functions, although the Department of Agriculture retains responsibility for drainage.

Northern Ireland also has its own Water Act (1972) which created the Northern Ireland Water Council to advise government on water policy. This anticipated the creation in 1973 of the National Water Council for England and Wales, although this was abolished in 1982 as part of a UK government purge on quangos! The Northern Ireland Water Act made the conservation of water resources a major responsibility of both the Departments of Agriculture and the Environment and outlined the nature and content of water management programmes. The need to protect groundwater and surface waters and the means whereby this should be done were specified in the Act, which also encouraged the setting up of what is now the Water Data Unit in the Department of Environment. This Unit produces annual statistics relating to all aspects of water use in Northern Ireland and liases closely with Civil Engineering Branch in the Department of Agriculture which is responsible for supervising the streamflow monitoring network set-up in the wake of the recommendations of the 1972 Water Act (Common, 1982; Wilcock, 1977).

In the Republic of Ireland, responsibility for water resources still resides in the hands of local authorities, although the Department of Environment has theoretical overall control (Drew, 1979). The Offices of the Regional

Fisheries Board are the main enforcers of pollution legislation. In many countries there is a conflict between the rights of local and central authority regarding the management of water resources, and even where the greater authority resides with central government, as in Britain, local authority representation on RWA's has to be delicately provided for. However, most experts now agree that the impacts of modern society on water resources are so diverse and potentially damaging and that the transfer of those impacts out of their source region so easy and rapid, that management should be based on areas larger than those covered by most local authorities.

This principle has not yet been implemented in the Republic although the Local Government (Water Pollution) Act of 1977 does provide for the setting up of water quality management plans on a catchment basis. Agencies may also be set up to take over responsibility for water quality control from local authorities, although the latter comprise the regional units within the Act. Local Authority units, however are small and their boundaries seldom coincide with drainage basin watersheds. Many water resource scientists view this administrative framework as a disadvantage (Toner and O'Sullivan, 1977, p.33).

Under the 1977 legislation, An Foras Forbartha (1986) has prepared several water quality management plans for local authorities or Regional Development Organizations, including ones for the Shannon and Liffey. These plans usually contain details of all abstractions from, and discharges to, a particular river together with details of water quality along its length.

The effect of local authority control in the Republic of Ireland is to diversify the number of authorities with responsibility for water supply, sewerage and water pollution. Whereas in Northern Ireland there is only one such authority, in the Republic there are 87 (Drew, 1979). Nor is there a centralized inland water survey in the Republic. Instead, streamflow measurement and data collection is undertaken by the Office of Public Works, which has 240 streamflow gauges on those rivers for which it has arterial drainage responsibility, and the Electricity Supply Board which has 45 gauges, mainly on streams in the west with a realized or potential capacity for the development of hydroelectricity. The balance of about 700 stations are maintained by local authorities (An Foras Forbartha, 1986).

An anomalous feature of legislation in both countries relates to the licensing of effluent discharges. Under the 1977 Local Government (Water Pollution) Act of 1977, local authorities in the Republic of Ireland were made responsible for licensing effluents, although they themselves are major dischargers of sewage effluent and exempt from licences (Toner and O'Sullivan, 1977). A similar situation exists in Northern Ireland. The DoE (NI) has to give its consent to all effluent discharges, yet is itself a principal discharger of sewage effluent. Figures relating to the scale of these two respective activities are difficult to establish for the Republic of Ireland but in Northern Ireland the DoE(NI) issued 3,016 effluent discharge consents in 1983 alone. At the same time it investigated 1,515 pollution incidents, 323 (21.3%) of them resulting from sewage! Altogether, since 1972, when the Department was given responsibility for sewage treatment under the NI Water Act, the DoE (NI) has issued 29,516 consents for effluent discharge. Throughout this period sewage has been the second greatest source of pollution after agriculture, with industry in third place (Northern Ireland Water Statistics, 1986). In a technical sense the principal agency responsible for monitoring pollution is thus itself the second largest polluter!

CONCLUSION

Conflicts between alternative users of countryside, such as that illustrated by the Kinnahalla Public Inquiry, are more likely to increase than to decrease in the next few decades as pressure on water resources increases. An Foras Forbartha has said that present demands for water in the Republic could double by the year 2010 and that the most probable source for supplies will be direct abstraction from rivers. The predictions relating to demand may seem excessive given the already high *per capita* level of consumption in the Republic (650 litres per person per day), but this figure is some 30% higher than was predicted in 1973 for the year 2000. In an effort to reduce *per capita* consumption in Northern Ireland, the DoE (NI) operates an intensive waste water detection programme, which appears to be economically viable in terms of savings in capital investments (Holland, private communication).

Use of rivers and lakes as potable supplies requires that surface waters are kept clean and monitored as to their physicochemical properties. Despite its emphasis to date on biological classifications of pollution, An Foras Forbartha recognises the need for more physicochemical measurements in the Republic, to establish base-line data for future management. In Britain, water quality monitoring costs £20 million per year. Notwithstanding such substantial costs An Foras Forbartha would have liked to move to measurements of physicochemical properties at time intervals of 25 days, but as we have seen such a time interval is too coarse to pick up agricultural pollution which is the principal source of pollution in the country.

One increasingly common method of assessing water quality in the Republic of Ireland is by remote sensing, and tests are progressing to establish if trophic levels in isolated lakes can be quantified from satellite data. Preliminary results are encouraging and light reflectance from lake surfaces has been correlated with chlorophyll levels and amounts of algae. Yet costs are high, and alternative surveys from light aircraft may provide a cheaper means of collecting data of the same quality. The irregular return periods of light aircraft surveys is one disadvantage if trophic level changes rapidly, but as a means of establishing base-line data for the future the method has much to commend it.

Further improvements in cross-border cooperation on water management are needed. Much of the existing data, for example, are in different forms on both sides of the border. While the type and range of data collected may reflect governments' attitude to public spending on environmental and natural resource analysis, there seems little to justify the same types of data being stored, analysed and presented in different ways. In this day of data storage facilities, it should be easy and beneficial to agree on a uniform mode of data handling throughout the British Isles.

In the final analysis, the future management of water resources in Ireland depends on the priorities attached to it by the Governments. In this respect there are some recent signs that Northern Ireland is moving ahead of the Republic. Centralisation of water resource management within DoE (NI) has led to noticeable improvements since 1973, although much remains to be done. Recent legislation in Northern Ireland (The Nature Conservation and Lands Order, 1985; The Access to the Countryside Order, 1983; and The Wildlife Order, 1983) has been coupled with internal re-organization of the DoE (NI), and conservation interests appear to have been awarded a higher priority within government policy than they have ever enjoyed before (Wilcock and Guyer, 1986). Much depends on finance, the only accurate indicator of where government priorities really lie. Twenty-five years ago Sir Robert Matthew, author of the Belfast Regional Survey and Plan, recognised that Northern Ireland (and he would no doubt have agreed Ireland as a whole) has 'a type of environment of great potential value in the world of tomorrow, a type of environment becoming increasingly rare, and therefore more valuable, in Europe.' Much of the Irish environment is aquatic in character. Lakes, for example, cover proportionately three times as much land area in Ireland as in Britain. Government policies in the future throughout Ireland will need to acknowledge this fact and to accord a higher priority to the proper management of water resources for economic as well as environmental reasons. These two objectives are linked inextricably.

ACKNOWLEDGEMENTS

I would like to thank Professor Frank Convery, Department of Environmental Studies, University College, Dublin, and Mr P.G. Holland of the Water Data Unit, DoE (NI), for their helpful comments on the first draft of this manuscript.

REFERENCES

An Foras Forbartha (1983) *A Review of Water Pollution in Ireland*, Compiled on behalf of the Water Pollution Advisory Council.

An Foras Forbartha (1986) *An Foras Forbartha Review*, Dublin.

Armstrong, W.J. (1982) An introduction: the economic perspective, in Cruickshank, J.G. and Wilcock, D.N. (eds) *Northern Ireland: Environment and Natural Resources* The Queen's University of Belfast and The New University of Ulster, 1-8.

Bailey, A.D. and Bree, T. (1981) The effect of improved land drainage on river flood flows, *Flood Studies Report - five years on*, The Institution of Civil Engineers, London, 131-42.

Bannon, M.J. (1979) Urban land, in Gillmor, D.A. (ed), *Irish Resources and Land Use*, Institute of Public Administration, Dublin, 250-269

Bannon, M.J. (1983) Urbanization in Ireland: growth and regulation, in Blackwell, J. and Convery, F.J. (eds) *Promise and Performance: Irish Environmental Policies Analysed*, University College, Dublin, 261-286.

Battarbee, R.W. (1986) The eutrophication of Lough Erne inferred from changes in the diatom assemblages of 210-Pb- and 137-Cs-dated sediment cores. *Proceedings of the Royal Irish Academy*, 86B, 141-68.

Brookes, A. (1985) River channelization: traditional engineering methods, physical consequences, and alternative practices. *Progress in Physical Geography*, 9, 44-73.

Bruton, R. and Convery, F.J. (1982) *Land Drainage Policy in Ireland*. Economic and Social Research Institute, Dublin.

Cabot, D. (1985) *The State of the Environment*. An Foras

Forbartha, Dublin.

Clabby, K.J., Lucey, J. and McGarrigle, M.L. (1984) *The National Survey of Irish Rivers 1982-83*, Water Resources Division, An Foras Forbartha, Dublin.

Common, R. (1982) Water supply and demand, in Cruickshank, J.G. and Wilcock, D.N. (eds) *Northern Ireland: Environment and Natural Resources*, The Queen's University of Belfast and The New University of Ulster, 73-86.

Cunnane, C. (1982) *Hydrology in Ireland*, Irish National Committee for the International Hydrolological Programme, Dublin.

Dodd, V.A. and Champ, W.S.T. (1983) Environmental problems associated with intensive animal production units, with reference to the catchment area of Lough Sheelin, in Blackwell, J. and Convery, F.J. (eds) *Promise and Performance: Irish Environmental Policies Analysed*, University College, Dublin, 111-30.

Downey, W.K. and Ni Uid, G. (1977) Lake pollution: eutrophication control, *National Science Council*.

Drew, D.P. (1979) Water, in Gillmor, D.A. (ed.) *Irish Resources and Land Use*, Institute of Public Administration, Dublin, 39-63.

Hammond, R.F. (1981) *The Peatlands of Ireland*, An Foras Taluntais, Soil Survey Bulletin No.35, Dublin.

Kelly, P. (1980) *Time to Reconsider Arterial Drainage*, Institute of Fisheries Management, Dublin.

Kennedy, G.J.A., Cragg-Hine, C. and Strange, C.D. (1983) The effects of a land drainage scheme on the Salmonid populations of the River Camowen, County Tyrone, *Fisheries Management*, 14, 1-16.

Mawhinney, K.A. (1979) Recreation, in Gillmor, D.A. (ed.)

Irish Resources and Land Use. Institute of Public Administration, Dublin, 196-225.

Northern Ireland Department of Agriculture (1985) *Northern Ireland Agriculture, Forty-third Annual Report*, HMSO, Belfast.

Patrick, S.J. (1987) The *per capita* output of phosphorus from domestic detergents in Ireland, 1950-1982, *Irish Geography*, 20, 89-94.

Smith, R.V., Stevens, R.J., Foy, R.H. and Gibson, C.E. (1982) Upward trend in nitrate concentrations in rivers discharging into Lough Neagh for the period 1969-79, *Water Research*, 16, 183-8.

Toner, P.F. and O'Sullivan, A.J. (1977) *Water Pollution in Ireland*, National Science Council, Dublin.

Ulster Angling Federation (1984) The Angler's view, paper presented to a conference on Lough Neagh and its rivers at the Freshwater Biological Investigation Unit, Department of Agriculture, Northern Ireland.

Water Service Northern Ireland (1986) *Northern Ireland Water Statistics*, Department of the Environment for Northern Ireland.

Wilcock, D.N. (1977) Water resource management in Northern Ireland, *Irish Geography*, 10, 1-13.

Wilcock, D.N. (1979) Post-war land drainage, fertilizer use and environmental impact in Northern Ireland, *Journal of Environmental Management*, 8, 137-49.

Wilcock, D.N. (1982) Rivers, in Cruickshank J.G. and Wilcock, D.N. (eds) *Northern Ireland: Environment and Natural Resources*, The Queen's University of Belfast and The New University of Ulster, 43-72.

Wilcock, D.N. and Essery, C.I. (1984) Infiltration

measurements in a small lowland catchment, *Journal of Hydrology*, 74, 191-204.

Wilcock, D.N. and Guyer, C.F. (1986) Conservation gains momentum in Northern Ireland, *Area*, 18, 123-9.

Wood, R.B. (1982) Lakes, in Cruickshank J.G. and Wilcock, D.N. (eds) *Northern Ireland: Environment and Natural Resources*, The Queen's University of Belfast and The New University of Ulster, 87-99.

Wood, R.B. and Gibson, C.E. (1973) Eutrophication and Lough Neagh, *Water Research*, 7, 173-87.

15 RESOURCES AND MANAGEMENT OF IRISH COASTAL WATERS AND ADJACENT COASTS

Bill Carter

'Thank God we're surrounded by water..'

(Dominic Behan)

INTRODUCTION

Ireland is an island located on the northwest European continental shelf, in the eastern Atlantic Ocean. As with all islands, Ireland is vulnerable to the vagaries of geopolitics, the dangers of economic exploitation and the constraints of an insular culture and society. This essay examines the wealth and health of Irish coastal waters, considering both the natural resource base and our attempts to locate, allocate and exploit these resources. In some cases the surrounding seas must be viewed as a buffer, in others as a barrier or frontier or even as a link.

PHYSICAL AND BIOLOGICAL FRAMEWORK

In the tectonic sense, Ireland comprises a platform of largely Palaeozoic rocks situated on the rifting eastern Atlantic margin of the European plate. Extensional tectonic spreading has created the Irish Sea and Celtic Sea basins, which now separate Ireland and Great Britain (Figure 1). This epi-continental

sea is relatively shallow, little more than 50m deep. To the
north, west and south of Ireland the continental shelf extends
about 350km offshore.

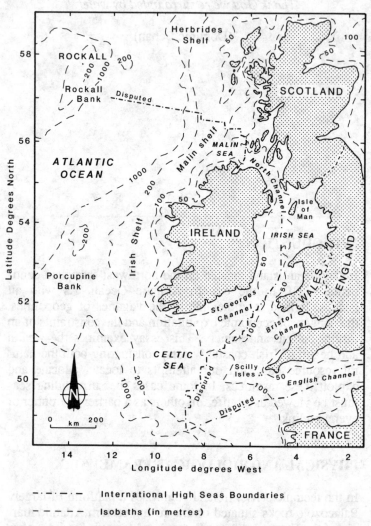

**Figure 1 Bathymetric and territorial map of Irish waters,
indicating the Continental Shelf and the Atlantic Banks.**

The coastline of Ireland is approximately 6,300km in length. There are major environmental contrasts between the east and west coasts, largely reflecting the adjacent wave climates. The eastern Irish seaboard, bordering the Irish Sea, is dominated by short sea waves (rarely more than six metres high), generated by regular west to east moving cyclonic depressions, although residual Atlantic swell enters the Irish Sea through both the North Channel and St Georges Channel (Figure 1) and is important in determining beach shape (Carter, 1983). Much of the east coast shoreline consists of eroding cliffs, cut into glacial material. There is extensive longshore drifting of sediment. On the west coast the Atlantic wave climate is both more energetic and dominated by far-travelled swell waves. Here the coastal sediments have been driven into rock-bounded embayments, allowing little or no alongshore sediment transport. In many places sand has been blown ashore to form high and wide dune fields.

Sea-level changes over the last 10,000 years have reinforced many of the physical contrasts around the Irish coast. In the south, sea-level has risen monotonously, drowning river valleys and creating wide estuaries. In the north, sea-level rose to a peak between 4000 and 6000 years ago, falling thereafter, and leaving extensive beach ridge plains and infilled estuaries. Rising sea-level was also responsible for destroying the last land bridge between Britain and Ireland, probably about 10,500 years ago (Devoy, 1985).

The predominance of strong westerly winds, relatively warm sea temperatures (winter - 7 to 9 degrees centigrade; summer 13 to 15 degrees) and generally low relief, lead to marked maritime climatic amelioration across much of Ireland (Rohan, 1975).

The biological productivity of Irish coastal waters is reasonably high, supporting commercial pelagic, demersal and shell fisheries. Of particular importance are the Irish estuaries, as these provide food, shelter and breeding grounds for many coastal water species. Nutrients, usually derived from the land, serve to enrich estuaries and coastal waters, boosting productivity. Such sources are particularly important in the shallow Irish Sea. The extensive wildfowl, wader and seabird colonies around the Irish coast form part of a significant nutrient cycle, catching fish and return faecal material to the sea.

Further offshore, productivity rises in the vicinity of oceanic fronts - comparatively turbulent zones where water masses mix. Important frontal zones occur in the Irish Sea, the Malin Sea and the Celtic Sea, where colder Atlantic waters meet warmer inshore waters (Simpson and James, 1986).

The natural coastal ecology of Ireland has been largely destroyed by Man, through reclamation, drainage, grazing and recreation. Virtually no native salt marsh remains, and all bar one (Lady's Island Lake) of the once-extensive coastal lagoons have been drained. The vegetation of the sand dunes has been altered by farming, and more recently, recreation. Only the inaccessible sea-cliff habitats remain unscathed.

It is against this dynamic backdrop that the Irish coast must be studied.

IRISH COASTAL WATERS

The Political and Strategic Value of Irish Waters

The geopolitical significance of Irish waters is perhaps less today than it was thirty years ago. Both pre-partition Ireland in the First World War, and Northern Ireland in the Second, were strategically important in the protection of Atlantic convoys and in the surveillance of enemy shipping and aircraft. Today, Ireland's allegiances are split; Northern Ireland, as part of the United Kingdom, is within NATO, while the Republic of Ireland remains neutral, and is not. By-and-large this disparity is inconsequential, as Ireland as a whole would shelter an umbrella of western security. However, Irish coastal waters, particularly in the Irish Sea and St Georges Channel may be used by warships and submarines en route to or from the Atlantic, or overflown by military aircraft.

Jurisdiction and ownership of the seas around Ireland is disputed not only by the Republic of Ireland and the United Kingdom, by also by Denmark, Iceland and France. The Irish territorial sea baseline was confirmed in 1959. Outside these three mile (5.5km) territorial waters, the neighbouring countries have contrived to argue about the drawing of maritime boundaries both to the north and south of Ireland (Symmons, 1979). At the heart of these disputes are the largely unexploited fish and mineral resources of the continental shelf (see Figure 1). The outer edge of the shelf is usually defined by the 200m isobath, although distance and geological criteria are also used in some definitions. In the case of Ireland, the continental shelf is widest (200 to 350km) to the south and southwest of the island, and narrowest (150km) to the northwest.

The maritime boundary in the Irish Sea (Figure 1) is relatively simple, comprising an equidistant line between Ireland and Britain (including the Isle of Man). To the north, in the Malin Sea, the line leaves the international land frontier at the mouth of Lough Foyle and runs north and west between Donegal and Jura (Scotland). It is beyond these politically acceptable median lines that the disputes have arisen (Prescott, 1985) (see Note on p. 419.) In the Celtic Sea, the line running south and west from St Georges Channel is disputed, largely on account of the Scilly Isles, west of Cornwall. Britian claims the median line from the Scillies, while Ireland claims from Lands End on the British mainland. The disputed zone subtends an arc of five degrees, almost 50km across at the edge of the continental shelf. France also disputes jurisdiction with Britain in the same area.

A potentially more serious disagreement has arisen over Rockall, an isolated uninhabited rock some 400km northwest of Co. Donegal (Figure 2). In 1972, the UK Government passed the Rockall Act, annexing Rockall as part of Invernesshire. In 1974, the Irish Government extended its continental shelf limit to include the Rockall Trough (southeast of Rockall), and, at the same time, declared, at the Caracas LOS Session, opposition to isolated islands being used to extend maritime jurisdiction (Symmons, 1975; Brown, 1978). Britain responded by issuing an Order in Council claiming 135,000km^2 of seabed around Rockall for exploration, on the somewhat untenable basis that it comprised 'a natural prolongation of the Scottish landmass'. Based on the 1982 UN LOS Convention (to which Britain is not a signatory), Ireland, Denmark and Iceland have rejected the claim that Rockall may generate a territorial zone. The LOS Convention states (Article 121.3) 'Rocks which cannot sustain human habitation or economic life of their own shall have no exclusive economic zone or continental shelf'. The superficially idiotic stunt of the Scotsman, Tom McClean, who camped on Rockall for a month in 1985, was not without political significance in the battle for potential seabed wealth.

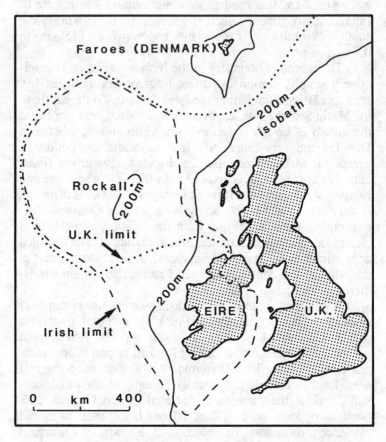

Faroes (DENMARK)

200m isobath

Rockall

200m

U.K. limit

200m

Irish limit

EIRE

U.K.

0   km   400

Figure 2 Disputed waters around Rockall, showing the most extreme national limits

Both the UK and Ireland are signatories to the 1974 Oslo and London LOS Conventions, but Ireland, alone has signed the 1982 Convention. The extensive provisions of the latter Convention cover seabed mining, marine pollution, and the designation of 200 nautical mile (nM) (about 580km) Exclusive Economic Zones (EEZs) around coastal states. EEZs allow control of fishing and mining rights, but do not restrict access. As yet, a formal EEZ has not been drawn around Ireland, but its probable limits would not differ much from the existing shelf boundary. Ireland and Britain also have the option of defining, under the EEC Common Fisheries Policy an Exclusive Fishing Zone (EFZ), within which the fish stocks would be shared on a quota basis with all EEC member states, but with the contiguous state retaining control. So far, no EFZs have been designated and accepted due to inter-Community disputes.

Resources

(i) Fisheries. Fishing has been a traditional pursuit of Irishmen for several thousand years. Irish coastal waters are relatively abundant in both pelagic (free-swimming, shoaling species) and demersal (bottom living, solitary species) fish. In addition, shellfish (lobsters, crabs, prawns, oysters and scallops) are of considerable commercial importance (Went, 1979). Recently, large-scale dredging for mussels has become important in Co. Donegal, as well as fish farming (mainly for oysters, scallops, sea trout and salmon) in sheltered sea loughs. The value of Irish fish catches is about IR£20 million a year.

The main fishing grounds are in the Irish Sea, with a lesser concentration to the north and west, particularly in spring and summer. Heavy Government investment in the fishing fleets (often with EC assistance) has enabled a recent revival in the industry, although market forces frequently depress activity.

Boaden (1982) argues that we have both neglected and mismanaged our fish resources. The crux of his argument is that exploitation has proceeded without a scientific management strategy, leading to wild fluctuations in natural fish stocks. In some cases, the fish populations have not recovered and a valuable resource has been lost. Pressure arises in two ways, first from over-fishing, and second from modification and destruction of coastal habitats that afford shelter and food, particularly for spawning. The importance of Irish waters in this context cannot be overestimated; cod, whiting, sprat and herring all spawn around the Irish coast, while other species, for example the salmon and eels, pass through the coastal zone on their way to breeding areas.

(ii) Minerals. The offshore geology of Ireland consists of a series of tectonic basins filled with Mesozoic and Tertiary sediments (Sevastopulo, 1982). These basins include salt and coal deposits, and also act as traps for oil and gas. The economic potential of the salt and coal deposits, is, at this time, relatively limited, although various proposals have been put forward for undersea coal mining off Co. Dublin (see Chapter 13). Offshore oil and gas exploration began, in earnest, in 1970, and one year later, the Kinsale Head Gas Field, off Co. Cork was discovered. This gas field is now in production, supplying natural gas for consumers as far north as Dublin. Further oil and gas finds, of commercial potential, have been located south of Waterford, and evidence of hydrocarbons has been detected in the Porcupine Trough, 150km west of Co. Kerry and in the Rathlin trough, off Co. Antrim. However, neither of these later discoveries have been declared commercially viable.

The surficial sediments of the Irish Sea are mainly sands, gravels and clays of glacial origin. In places, these materials have become sorted by tides and waves, into extensive banks of clean aggregate, suitable for commercial extraction. The most likely areas of exploitation would be the Kish Bank and Arklow Bank, south and east of Dublin, where a dredge could operate in shallow, sheltered water. Various schemes have been put forward to mine these offshore deposits, but economic, environmental and political forces have intervened to halt proposals.

(iii) Energy. Although Irish energy resources are examined in Chapter 13, it is useful to comment here upon the energy potential of the Irish coastal waters. The Irish and Malin Shelves are among the most energetic offshore wave environments in the World, with waves reaching heights of 20 to 30m at times. Mollison (1982) estimates wave power at 70,000kW/km off Belmullet, Co. Mayo, which if it could be captured (over a distance of 20 to 30km), would easily exceed Irish power demands into the forseeable future. Significant wave power generation would require a massive capital outlay, at least two to three times that required to build the equivalent conventional thermal power capacity. However, wave generator running costs would be lower and unaffected by coal or oil prices, although design life (30 to 40 years (?)) might not be significantly different. There is unlikely to be early development of Irish offshore wave power resources, although the recent (July 1987) decision of the British Government to build a prototype wave power generator (a coastal oscillating column type) on the Scottish island of Islay (55km north of Northern Ireland), might bring about a re-evaluation.

Onshore (at the coast) wave power generation potential has not been explored in Ireland, although there are a number of suitable sites along the crenellate western seaboard. Two options are feasible, one, pneumatic pumps fixed along steep, cliffed shorelines, and two, run-up collectors inside narrowing channels (see Carter, 1988).

An alternative to wave power is offered by tides, although suitable sites for the realisation of this energy source are limited to a few estuaries. Serious proposals have been put forward for Strangford Lough in Co. Down (NIEC, 1981), while other possibilities include Carlingford Lough on the border between Northern Ireland and the Republic and Lough Swilly in Co. Donegal.

Waste Disposal in Irish Waters

In Ireland, as in many countries, the coastal waters are viewed as a convenient dumping ground for the nation's waste. While Irish coastal waters are not generally polluted, the water and sediment quality of many bays and estuaries is distinctly tainted, and will require attention before too long (Cabot, 1985). Among the worst polluted bays are the Avoca, inner Dublin Bay and the Liffey estuary, Belfast Lough and outer Larne Lough (Wilson, 1980; Wilson and Jeffrey, 1986; Parker, 1982).

Dublin Bay (Figure 3) is among the most environmentally stressed of Irish waters, as it receives high industrial, domestic and agricultural pollution loads, from both the adjacent urban area (population 900,000) and the catchment of the River Liffey. Water and sediment quality, particularly in the inner Bay, are poor with high Biochemical Oxygen

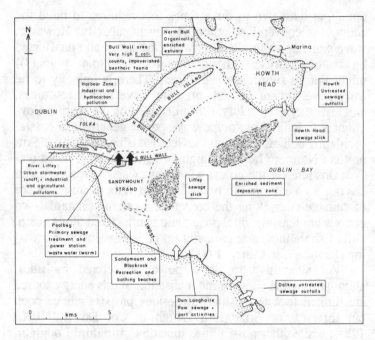

Figure 3 Labels (clockwise):
- North Bull Organically enriched estuary
- Marina
- Bull Wall area: Very high E coli counts, impoverished benthaic fauna
- HOWTH HEAD
- Howth Untreated sewage outfalls
- Harbour Zone Industrial and hydrocarbon pollution
- Howth Head sewage slick
- DUBLIN
- TOLKA
- N BULL WALL
- NORTH BULL ISLAND
- LWOST
- LIFFEY
- DUBLIN BAY
- River Liffey: Urban stormwater runoff, + industrial and agricultural pollutants
- S BULL WALL
- SANDYMOUNT STRAND
- Liffey sewage slick
- Enriched sediment deposition zone
- Poolbeg: Primary sewage treatment and power station waste water (warm)
- LWOST
- Sandymount and Blackrock Recreation and bathing beaches
- Dalkey untreated sewage outfalls
- Dun Laoghaire Raw sewage + port activities

0 — kms — 5

Figure 3 Environmental problems of Dublin Bay

Demands (BODs) and concentrations of nutrients, metals and hydrocarbons. As Keenan (1984) demonstrates, it is not easy to pinpoint pollution sources around the Bay, as at least 50 industries discharge waste, and there are eight major sewage outfalls, plus numerous storm drains. The biological status of the Liffey and the inner Bay is poor (Jeffrey *et al.*, 1978; Jones and Jordan, 1979; Magennis and Wilson, 1984; Bruton *et al.*, 1987), but elsewhere productivity is often boosted following enrichment by organic wastes. The diffusion of sewage, often as slicks, spreads effluent throughout the Bay, creating pockets of especially high biological productivity in more sheltered 'sinks', like the tidal flat behind North Bull Island. Bathing water quality, relative to EC standards, is poor, with faecal coliform counts commonly exceeding the 2,000 per 100 ml standard (Fegan, 1983). The range of problems evident in Dublin Bay are matched in other Irish estuaries, most noticeably Belfast Lough, another polluted urban embayment. Here there is a noticeable source and tidal regime related zonation of sediment and benthic fauna (Parker, 1982; Carter, 1988).

The 1974 LOS Oslo Convention lists the types of material that may be dumped into the coastal waters of the northeast Atlantic. Some substances - the 'black list' - may not be dumped, while others - the 'grey list' - may be dumped only after a permit is obtained from the adjacent country, and all other coastal states are informed of the type and quantity of waste involved, as well as the timing, method and location of the disposal.

There are many categories of waste, and various ways of discarding them. The chosen method will depend on the volume, source, toxicity and assimilation capacity of the surrounding waters. Often sub-optimal methods of disposal are employed, simply because of precedence and the prohibitive cost of upgrading existing systems. Perhaps the most emotive issues relate to domestic sewage, oil spills and radioactive waste. Each will be examined in turn.

(i) Domestic sewage. For over a century most Irish coastal towns have dumped untreated sewage into the sea. Often discharge has been via gravity outfalls located between the tide marks, so that sewage is liable to become concentrated and stranded along the shoreline. While the precise epidemiological consequencies of this process are debatable, most scientists would concede that the practice constitutes a health risk. Dispersal of organic-rich waste into water increases the BOD, which may lead to oxygen depletion (anoxia) and ultimately 'fish kills'. The degree of oxygen depletion depends on the natural aeration of the waters, so that most risk is vested in small, sheltered embayments, where wave and tide action is limited. Most satisfactory is the dumping of sewage at sea, especially if disposal is into relatively turbulent waters. Macerated sewage sludge is dumped in deep water (c. 20 metres) in the Irish Sea off Belfast Lough and Dublin Bay, and again in the Mersey Estuary. This material is dispersed by tidal currents, and biodegraded within a few days. The environmental effects of this dumping are not thought to be serious.

According to International Council for the Exploration of the Seas (ICES) data, just under 500 M m^3 of sewage is discharged into Irish waters each year. Just under half this volume is completely untreated, the remainder is either macerated (chopped), screened or chlorinated prior to discharge. Only a very small proportion (<3%) is dumped offshore, mainly from Belfast. Despite recent improvements in sewage treatment in many areas, including both Dublin and Cork, 80% of all Irish sewage still reaches the sea with little or no treatment. However, many sewage systems are being integrated to provide better treatment and longer outfalls; Cabot (1985) notes a decline of 20% in sewage outfall licences in the Republic of Ireland between 1979 and 1983.

Domestic sewage, plus agricultural waste, form a potent cocktail, capable of inititating and sustaining high levels of biological productivity. In confined spaces this may lead to enrichment (eutrophication), but more often will simply add to general ecosystem productivity. High chlorophyll *a* levels (indicating high photosynthetic activity) and depleted dissolved oxygen demand, often leading to enhanced high biomass production, have been observed in the Shannon, Suir and Blackwater (Co. Meath) estuaries (Cabot, 1985).

(ii) Oil spills. Most spills of oil occur during transhipment operations, as a result of tank cleaning, or from engine discharges. Most oil spills in Ireland are local affairs; O'Sullivan (1979) records about 20 incidents a year, of which only one or two are of major significance.

In recent years, two newsworthy oil spills have occurred, the *Betelgeuse* in 1979 and the *Kowloon Bridge* in 1986. The former exploded during offloading at Whiddy Island in Bantry Bay (Tribunal of Inquiry, 1980), while the latter was wrecked on rocks near Cape Clear in Co. Cork, following a bizarre farrago, in which the structurally and mechanically unsound ship sailed from Bantry Bay into a gale and was subsequently abandoned. In the *Betelgeuse* disaster 35,000 tonnes of light crude oil were spilt, necessitating a IR£600,000 clean-up operation. However, ecosystem damage was relatively slight, except along a one kilometre stretch of shoreline near the wreck (Cross *et al.*, 1979). The *Kowloon Bridge* incident, while involving only a few hundred tonnes of heavy fuel (bunker) oil has resulted in widespread, but largely sub-lethal, pollution along the south shore of Co. Cork. Such small disparate spills are perhaps harder to clean-up than large, confined spills, and the *Kowloon Bridge* oil has caused both logistical and financial headaches to the authorities.

(iii) Radioactive waste. The discharge of small quantities of low-level radioactive waste into the Irish Sea has been one of the most contentious environmental issues of the last decade. Many people suspect that, even small amounts of radio-activity constitute a serious health hazard.

The largest volume of radionuclides in the Irish Sea originate from the British Nuclear Fuels (BFN) re-processing plant on the Cumbrian coast at Sellafield, opposite Co. Down. Releases of radio-active caesium (^{137}Cs, ^{134}Cs), strontium (^{90}Sr) and plutonium ($^{238-240}Pu$) have all given cause for concern. Discharges of ^{137}Cs, the most common material, peaked in the late 1970s, but have since declined tenfold due to the imposition of stricter discharge standards by BFN. Except for locations within a few kilometres of the discharge zone, concentrations of these elements are relatively low, usually only one or two becquerels (Bq) per litre of seawater. (A becquerel is equal to one atomic disintegration per second; in time, these disintegrations lead to the decay of the radioactive material into inert forms of the element. The average 'background' radiation in British Isles ranges from about 25 to 70 Bq.) Animal and plant tissue samples from the east coast of Ireland contain between 50 and 100 Bq/kg (Cunningham and O'Grady, 1986; Mitchell *et al.*, 1986).

However, direct measurements of radionuclide concentrations are misleading on two counts; first, ionising effects are not proportional to simple concentrations, and second, the susceptibility of both irradiated organisms and organs is statistically variable. (Nuclear energy doses are measured in Grays (Gy), while the impact or dose equivalent is measured in Sieverts (Sv) ($1Sv = 1 Gy * K$ (a proportionality factor)).) Radiation may been inhaled, absorbed or ingested. The International Commission on Radiological Protection (ICRP) recommends a maximum annual dose of 1mSv (=0.001 Sv) for adults, so that even the most voracious consumers of Irish

405

Sea fish are unlikely to receive 10% of this dose (McAulay and Doyle, 1985). Yet it is difficult to set a limiting intake of ionising radiation, as not only does susceptibility vary from person to person, but also side-effects may take years (or generations) to become evident. The most disquieting aspects are the possible increases in carcinomas or genetic birth defects among Irish coastal communities. The evidence is both hard to unravel, as often radiation may only accelerate cancers, rather than initiate them, and in many cases, the level of incidence is very low, making it hard to test statistical significance. In an interim report to the Northern Ireland Department of Health and Social Services, Lowry (1986) finds no statistical justification for assuming ionising radiation from Sellafield has led to increased incidence of either leukaemia or Down's syndrome among Northern Ireland coastal population cohorts, although the short time scale of the study and crude aggregation of spatial data does not inspire confidence in the conclusions. Similar conclusions were reached by Black (1985), working on the Cumbrian data, although here also the statistical data and tests have been called into question and are being re-analysed.

Beyond these official reports one is left with sinister, yet unexplained, coincidences and contagions of disease, for example, the remarkably high incidence of Down's syndrome children born to former pupils of a grammar school in Dundalk, perhaps related to airborne contamination following a fire at Sellafield in 1958 (Sheehan and Hillary, 1983).

The evidence is confusing, the risks often unexplained. The arguments over radioactivity in the Irish Sea will continue for years.

COASTAL RESOURCES AND MANAGEMENT

The coastal resources of Ireland include beaches and dunes for recreation, shore protection and minerals, estuaries for fisheries, wildfowl, port and harbour development, land reclamation and waste disposal, as well as many other varied uses. Over the past two centuries, man has made various efforts to exploit these resources, with varying degrees of success. Piecemeal attempts to improve the coastlands have led to unforeseen consequences, often far removed both in time and space. For example, the origin of much of the pernicious shoreline erosion in Ireland can be traced to land reclamation or shoreline 'protection' undertaken in the nineteenth century. Man's relationship with the Irish coast has been an uneasy one; over the last 150 years the response to change has shifted from one of bewilderment and abandonment to one of partial understanding and adjustment (Carter, 1988), although whether this conceptual advance has led to more prudent allocation and use of coastal resources is debatable.

Resource Management of Beaches and Dunes

While there is a long tradition of using Irish beaches and dunes for minor agricultural purposes, their main use over the last hundred years has been for recreation. The attractions of the Irish coast for the leisure industry are manyfold, including sailing, surfing, angling and bathing, as well as more passive pursuits. However, exploitation of this leisure resource has been sadly under-managed, so that today the coast is suffering from a deterioration in aesthetic value, low water quality, shoreline vegetation damage and erosion and inadequate access control. Perhaps the most seriously affected areas have been the sand dunes, where unfettered access and trespass have led to widespread vegetation damage followed, in many instances, by soil and sediment erosion. The late-twentieth century mobilisation of the dunescapes parallels many earlier phases (Carter, 1985), but it is unmatched in its severity. Many isolated Irish dune systems are showing signs of rapid deterioration, which, will in time lead to both the elimination of wildlife habitats and the loss of amenity value.

The deterioration of the beach environment is perhaps less noticeable, but nonetheless presents serious problems. The threat comes from several directions. First, the unchecked removal of beach sands and gravels for agricultural and construction purposes, is causing accelerated shoreline erosion in many places. Although unauthorised extraction is banned under the 1940 Minerals Act in the Republic of Ireland, few attempts have been made at enforcement. Recent plans to mine, on a commercial basis, beach sand placers in Co. Wicklow, must be viewed with concern from an environmental standpoint. Second, sporadic attempts to protect Irish beaches through the construction of shoreline defences - seawalls, groynes and latterly beach nourishment - have been based on inadequate scientific knowledge and funding, leading not only to wasteful use of resources, but also, in some cases, accelerating the original problems and initiating new ones.

Many Irish beaches are 'protected' by seawalls, although these structures have, in themselves, often sowed the seeds of instability and change. At Portrush in Co. Antrim, the beach level has dropped over 1.5 metres and the beach width shrunk by 30 metres, since the completion of the seawall in the 1963. Similar problems have been reported at Bray, Co. Dublin, Youghal, Co. Cork, Lahinch, Co. Clare and elsewhere.

The effects of constructing jetties and seawalls have been less predictable. In many examples the ensuing coastal changes have been limited to minor adjustments before the establishment of a new equilibrium. At the other extreme, Portballintrae in Co. Antrim (Figure 4) has completely lost its beach in the last 80 years, due to the disruptive influence of a small jetty. Almost 100,000 m^3 of sand from this bay have been scoured into deepwater (Carter *et al.*, 1983).

The undubitable value of the Irish beaches and dunes to the economy (see Chapter 12 on Tourism) makes it paradoxical that very little is done to manage the coast. In some places, management plans have been implemented, for example by the National Trust at Murlough, Co. Down, but in far too many places the beach environment has simply been allowed to deteriorate. *Blitzkreig* tactics, as for example in the 1960s restoration of Brittas Bay, Co. Wicklow, are only successful if day-to-day management is provided to maintain the restored environment. The skills needed to manage such 'honeypot' sites (those attracting large numbers of people for only a few times a year) are very different from those associated with conventional countryside or coastal locations.

Figure 4 In 1935, Port Ballintrae, Co. Antrim, still had a wide sandy beach (a). Unfortunately the influence of a pier (visible bottom right of bay), rebuilt in 1895, led to erosion so that by 1987 (b) the sand had been completely removed.

Coastal Land Use

The coastal lands of Ireland have, and are, used for many diverse purposes. In Ireland, traditional uses for ports and harbours have expanded, so that the shoreline has become a mosaic of industrial, commercial, residential, agricultural and recreational land (and water) uses. In many places, the coast is under considerable 'pressure' form one, two or even several of these uses, and increasingly the remaining coastland is being designated for preservation or conservation. In addition to conflicts between competing activities, there are others between activities and the natural processes. In some cases these cause and effect links are clear, in others they are obscure and cause widespread consternation to local communities, leading to the incorrect application of remedial measures.

Carter (1987) has summarised the main conflicts and compliments of coastal land use in Ireland. Often these have changed through time; in the nineteenth century many of the problems related to agricultural expansion into marginal coastlands, and attempts at 'reclaiming' (more truthfully 'claiming') estuary shores. At this period almost all the coastal lagoons in Ireland (there may have been 50 to 60) were drained and many of the estuaries, including the Shannon, the Slaney (Wexford Harbour), the Foyle and the Swilly, were partly enclosed and dewatered. These activities extracted their toll. Much of the fresh and salt water coastal wetlands were lost, and tidal regimes irrevocably altered. Luckily, perhaps, several of the more grandiose schemes, either failed for financial and/or engineering reasons or never reached fruition. Early attempts at coastal problem solving were very limited, partly because of a poorly developed, poorly funded political structure, and partly because of the reduced pressure after the Great Famine in the 1840s (Carter, 1985). Notwithstanding, the Congested Districts Board (later the Irish Land Commission) sponsored a number of land restoration schemes on the coast in attempts not only to boost local economies, but also, in the case of sand dunes, to stop drifting sand blowing over nearby settlements.

Although estuary reclamation has provided some of the most fertile land in Ireland. Such reclamation has had serious side-effects, mainly through the blocking and narrowing of tidal channels, causing a reduction in the tidal volume (prism) in estuaries. In time, this reduction has led to a hydraulic closing of the estuary mouths and eventually to major coastal changes, including shoreline erosion (Carter *et al.*, 1987). One of the best Irish examples of this type of change, is in Co. Wexford where nineteenth century reclamation in Wexford Harbour has led to a baffling (to local people) sequence of spit beheading and shoreline erosion (Orford, 1988). However, there are numerous similar cases; Mulrennan (1986) has described the problems at Portmarnock, Co. Dublin, Carter and Mulrennan (1985) at Donabate, Co. Dublin and Shaw (1985) at Horn Head in Co. Donegal. An intriguing, but as yet undocumented, example comes from the southern end of Lough Swilly in County Donegal (Figure 5), where artificial closure of the tidal channel between Inch Island and the mainland in the early twentieth century has led to a long-term readjustment of the shoreline. There is something quintessentially Irish about the situation where the golf club is losing ground to the nearby yacht club, which is silting-up. Unfortunately there is very little that can be done about this and other similar problems. Turning the clock back is not a management option.

The loss of land through marine erosion is to some extent inevitable given the 'soft' nature of much of the Irish coast, formed of glacial and glacigenic sediments. Carter and Johnston (1982) estimated that 130 to 160 hectares of land (worth over IR£1,000,000) are lost every year, mostly along the eastern Irish coasts of Counties Down, Dublin, Wicklow and Wexford. This land loss has created problems of adjustment; near Killiney the Great Southern Railway line had to be re-sited inland in 1915 following cliff collapses, near Rosslare and Blackwater, both Co. Wexford, and at Kilkeel, Co. Down houses have been undermined and abandoned as the shoreline has retreated, and in numerous places agricultural land has been lost and roads disrupted. However, much of the coast erosion in the last 100 years has been caused, directly and indirectly by man. From small-scale instances, like the run-off from house roofs causing cliff gullying near the Cush Gap, Co. Wexford, to large-scale problems like the extensive degradation of dunes through pedestrian trampling, man-related coast erosion in Ireland is a significant problem. The common practice of sand 'winning' (extraction) also leads to erosion, if unchecked. These problems can be controlled, by tighter regulations, reduced access provision and, especially, by better education.

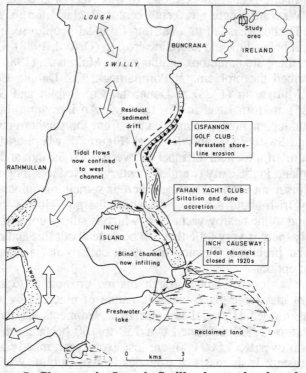

Figure 5 Changes in Lough Swilly due to land reclamation

One of the most pervasive impacts on coastal land in Ireland has been the post-war spread of caravans (Carter, 1982a, 1987). This almost unplanned and unrestricted growth has opened up many coastal sites to a new and significantly different use. Caravan sites form semi-permanent 'holiday home' settlements at often isolated locations, where the facilities and general socio-economic infrastructure is weak. Thus such settlements tend to put great pressure on small areas, perhaps without offering much in the way of benefits to the indigenous population. There are around 300 coastal caravan sites in Ireland, constituting about 35,000 'pitches'. As Carter (1982a) notes, the caravan economy in Ireland (at least in Northern Ireland) is directed more at the richer, owner-occupiers, rather than at the less well-off, hiring population or at the touring market.

AN INTEGRATED APPROACH TO COASTAL AND MARINE MANAGEMENT IN IRELAND

The complex problems of the Irish coast demand an integrated approach to decision-making and problem-solving. In the Northern Irish response to the World Conservation Strategy, Boaden (1983) put forward a strong case for a Government Department of Marine Resources, to both administer and coordinate activities in the marine environment. The World Conservation Strategy has not been accepted by the United Kingdom, but interestingly an almost parallel initiative has been realised in the Republic of Ireland as part of the incoming (February 1987) Fianna Fáil Party's policies. Fianna Fáil have set up a Department of Marine Affairs (Roinn na Mára) to embrace all existing coastal and maritime institutions and activities, including transport, natural resources, planning and recreation. At this moment (Summer 1987), this fledgling Department is just finding its wings; axiomatically its coordinating role and success will depend on the level of funding, the transfer of responsibilities from existing authorities, and the general commitment of both the designated Minister and the Government. This innovation in the Republic of Ireland parallels moves in Canada and to some extent in the USA, but is contrary to established practice in Europe (Carter, 1988). Northern Ireland lags behind. Some reorganisation of the Department of the Environment to form the Countryside and Wildlife Branch has led to a better and more sympathetic understanding of the need for coastal and marine resource studies, especially the need for site specific management plans. However, the general lack of integration between government departments has changed little in the last forty years. One significant difference between Northern Ireland and the

Republic of Ireland is the heavy involvement of non-governmental organisations in coastal matters in the former (Carter, 1982b). Both The National Trust and The Royal Society for the Protection of Birds own and manage coastal lands and waters, while, increasingly, local naturalist trusts and the Conservation Volunteers are becoming more active - and more vociferous. One particularly meritorious scheme, administered by The National Trust, is the Strangford Lough Wildlife Scheme, which tries to defuse and avoid conflicts in and around the Lough, through regular meetings of interested parties, both within and without government.

The role of the EC in Irish coastal and marine affairs, is at the moment, relatively limited. The Common Fisheries Policy has done little, apart from re-finance many fishing fleets, in evening-out the wild fluctuations in supply and demand, which bedevil the more marginal activities in this economic sphere. Special Directives, like the Bathing Water Directive of 1975, have either been largely ignored (as in Northern Ireland), or not complied with (as in the Republic of Ireland). (Recently some movement has been made on this matter, but real progress awaits large-scale EC funding for the improvement of sewage treatment.)

It is clear that the sea and the coastline are important resources for the economic, social and environmental well-being of Ireland. These resources need careful management if they are to be maintained for future generations. New objectives from both the Irish and British Governments, as well as from the EC and the international community as a whole may have an important bearing on our future use of the seas around us, but as with so many proposals, only time will tell.

REFERENCES

Black, D. (1985) *Investigation of the Possible Increased Incidence of Cancer in West Cumbria.* HMSO, London.

Boaden, P.J.S. (1982) Estuarine and inshore waters, in Cruickshank, J. and Wilcock, D.N. (eds) *Northern Ireland: Environment and Natural Resources*, The Queen's University of Belfast and The New University of Ulster, Coleraine, 101-18.

Boaden, P.J.S. (1983) *Conservation and development of the marine and coastal resources. Response in relation to Northern Ireland to United Kingdom*, Report No. 4 of the World Conservation Strategy, unpublished, Department of the Environment (NI).

Brown, E.D. (1978) Rockall and the limits of national jurisdiction, *Marine Policy*, 2, 181-211 and 275-302.

Bruton, M., Convery, F.J. and Johnson, A. (eds) (1987) *Managing Dublin Bay*, Resource and Environmental Policy Centre, University College, Dublin.

Cabot, D. (1985) *The State of the Environment.* An Foras Forbartha, Dublin.

Carter, R.W.G. (1982a) Coastal caravans in Northern Ireland 1960-1980, *Irish Geography*, 15, 107-111.

Carter, R.W.G. (1982b) The coast, in Cruickshank, J. and Wilcock, D.N. (eds.) *Northern Ireland: Environment and Natural Resources*, The Queen's University of Belfast and The New University of Ulster, Coleraine, 119-38.

Carter, R.W.G. (1983) Raised coastal landforms as products of modern process variations and their relevance to eustatic sea-level studies, *Boreas*, 12, 167-83.

Carter, R.W.G. (1985) Approaches to sand dune conservation in Ireland, in Doody, P. (ed.) *Sand Dunes and their Management.* Nature Conservancy Council, Peterborough, 29-41.

Carter, R.W.G. (1987) Man's response to change in the coastal zone of Ireland, *Resource Management and Optimisation*, 5, 127-64.

Carter, R.W.G. (1988) *Coastal Environments*, Academic Press, London.

Carter, R.W.G. and Johnston, T.W. (1982) Ireland - the shrinking island, *Technology Ireland*, 14(3), 22-8.

Carter, R.W.G. and Mulrennan, M. (1985) Coastal Environments of Co. Dublin, *Geographical Society of Ireland, Field Guide*, No.1.

Carter, R.W.G., Lowry, P. and Shaw, J. (1983) An eighty year history of erosion in a small Irish bay, *Shore and Beach*, 52 (3), 34-7.

Carter, R.W.G., Johnston, T.W., McKenna, J. and Orford, J.D. (1987) Sea-level, sediment supply and coastal changes: examples from the coast of Ireland, *Progress in Oceanography*, 18, 79-101.

Cross, T.F., Southgate, T. and Myers, A.A. (1979) The initial pollution of shores in Bantry Bay by oil from the tanker *Betelgeuse, Marine Pollution Bulletin*, 10, 104-7.

Cunningham, J.D. and O'Grady, J. (1986) *Radioactivity monitoring of the Irish marine environment during 1982-1984*, Nuclear Energy Board, Dublin.

Devoy, R.J.N. (1985) The problems of a late-Quaternary land bridge between Britain and Ireland, *Quaternary Science Reviews*, 4, 43-58.

Fegan, L. (1983) How dirty is Dublin Bay? *Technology Ireland*, 15(6), 30-3.

Jeffrey, D.W., Pitkin, P.H. and West, A.B. (1978) The intertidal environment of northern Dublin Bay, *Estuarine and Coastal Marine Science*, 7, 163-71.

Jones, G.B. and Jordan, M.B. (1979) The distribution of organic materials and trace metals in sediments from the River Liffey estuary, Dublin, *Estuarine and Coastal Marine Science*, 8, 37-44.

Keenan, T. (1984) Sources of pollution in Dublin Bay, in *Dublin Bay - An Outstanding Natural Resource at Risk?*, Dublin Bay Environment Group, Dublin, 29-46.

Lowry, S. (1986) *Investigation into Patterns of Disease with Possible Association with Radiation in Northern Ireland*, Department of Health and Social Security, Northern Ireland.

Magennis, B. and Wilson, J. (1984) Impacts of pollution on the biology of the Bay, in *Dublin Bay - An Outstanding Natural Resource at Risk?*, Dublin Bay Environment Group, Dublin, 47-64.

McAulay, I.R. and Doyle, C. (1985) Radiocaesium levels in Irish Sea fish and the resulting dose to the population of the Irish Republic. *Health Physics*, 48, 333-7.

Mitchell, P.I., Sanchez-Cabeza, J.A., Clifford, H., Vidal-Quadras, A. and Font, J.L. (1986) The distribution of radiocaesium, radioiodine and plutonium around the Irish coastline using *Fucus vesiculosus* as a bio-indicator, in Richardson, D.H.S. (ed.) *Biological Indicators of Pollution*, Royal Irish Academy, Dublin, 1-12.

Mollison, D. (1982) *Ireland's Wave Power Resource*, National Science Council, Dublin.

Mulrennan, M. (1986) *The Coastal Environment of Portmarnock, Co. Dublin*, Unpublished Report to Dublin County Council.

Northern Ireland Economic Council (NIEC) (1981) *Strangford Lough Tidal Energy*, Report Number 24, 56

O'Sullivan, A.J. (1979) Oil pollution on the coast of Ireland 1978 and 1979, *Irish Journal of Environmental Sciences*, 1, 85-7.

Orford, J.D. (1988) Alternative interpretation of man-induced shoreline changes in Rosslare Bay, Southeast Ireland, *Transactions of the Instituite of British Geographers*, New Series,

417

13, 65-78.

Parker, J.G. (1982) Structure and chemistry of sediments in Belfast Lough, a semi-enclosed marine bay, *Estuarine, Coastal and Shelf Science*, 15, 373-84.

Prescott, J.V.R. (1985) *The Maritime Political Boundaries of the World*, Methuen, London.

Rohan, P.K. (1975) *The Climate of Ireland*, Irish Meteorological Service, Dublin.

Sevastopulo, G.D. (1982) Economic geology, in Holland C.H. (ed.) *A Geology of Ireland*, Scottish Academic Press, Edinburgh, 273-301.

Shaw, J. (1985) *Holocene coastal evolution, Co. Donegal, Ireland*, Unpublished D.Phil Thesis, University of Ulster, Coleraine.

Sheehan, P.M.E. and Hillary, I.B. (1983) An unusual cluster of babies with Down's Syndrome born to former pupils of an Irish boarding school, *British Medical Journal*, 287, 1428-29.

Simpson, J.H and James, I.D. (1986) Coastal and estuarine fronts, in Mooers, C.N.K. (ed.) *Baroclinic Processes on Continental Shelves*, American Geophysical Union, Washington, DC, 63-93.

Symmons, C.R. (1975) The Rockall dispute, *Irish Geography*, 8, 122-6.

Symmons, C.R. (1979) British offshore continental shelf and fishery limit boundaries: an analysis of overlapping zones, *International and Comparitve Law Quarterly*, 28, 703-33.

Tribunal of Inquiry (1980) *Report on the Disaster at Whiddy Island Co. Cork, 8th January 1979*. Government Stationery Office, Dublin.

Went, A.E.J. (1979) Fisheries, in Gillmor, D. (ed.) *Irish*

Resources and Land Use. Institute of Public Administration, Dublin, 152-70.

Wilson, J.G. (1980) Heavy metals in the estuarine macro-fauna of the east coast of Ireland, *Journal of Life Sciences, Royal Dublin Society*, 1, 183-9.

Wilson, J.G. and Jeffrey, D.W. (1986) Europe-wide indices for monitoring estuarine quality, in Richardson, D.H.S. (ed.) *Biological Indicators of Pollution*, Royal Irish Academy, Dublin, 225-42.

Note: In November 1988 the Irish and British Governments signed a Treaty resolving many of the disputes and claims over territorial waters.

16 AIR POLLUTION PROBLEMS IN IRELAND

John Sweeney

In Ireland, environmental priorities rank lower with most people than in any other EC country (Table 1). At least in part, this reflects the fact that Ireland is inherently less susceptible to pollution-related problems than some of its European partners. A vigorous westerly wind regime facilitates the dispersal of indigenous air pollutants on most occasions, and minimises episodes of acid-rain-bearing easterlies from the UK and Europe. Having most of its large urban centres on the east coast is a further advantage. However, growing stress on the dispersal capabilities of the atmosphere is apparent and seemingly intractable air quality problems now exist in some locations.

A small, energy-deficient, newly industrialising, country such as Ireland inevitably faces a dilemma between fostering growth and imposing pollution abatement costs on its exporting industries. This tension manifests itself in the setting of standards which reflect the judgement of society as to what is, and is not, an 'acceptable' amount of environmental deterioration commensurate with particular economic benefits. Although such standards have their origins ostensibly in dosage-response relationships, they are essentially politically determined manifestations of this social consensus. This consensus changes, often radically, in response to a heightening of awareness of risk due to individual events. The accident at Chernobyl demonstrated this clearly, as did the London smogs of the 1950s. Sensitisation of the electorate determines whether or not the political will to solve environmental problems exists. It may be asserted that the political will to tackle air quality problems does not exist at present in Ireland. Indeed in a currently static economy, adequate resources will not be diverted easily into even maintaining existing levels of air quality.

Table 1: Position of the EC member states on 'trade-off' questions

	Protection of the environment or price control			Protection of the environment or economic growth		
	Envir-onment %	Prices %	Don't know %	Envir-onment %	Prices %	Don't know %
Community	60	19	21	59	27	14
Belgium	50	30	20	50	30	20
Denmark	74	9	17	75	14	11
Germany	54	12	34	64	21	15
France	63	19	18	58	30	12
Ireland	34	53	13	29	58	13
Italy	66	18	16	67	20	13
Luxembourg	69	16	15	64	26	10
Netherlands	72	13	15	56	34	10
United Kingdom	57	28	15	50	36	14
Greece	67	17	16	56	26	18

Source: Cabot (1985)

THE PROBLEM OF URBAN SMOKE AND SULPHUR DIOXIDE POLLUTION

Smoke, suspended particulates less than 15 μ m in diameter, consists mainly of unburnt carbon, grit and ash and results principally from the incomplete combustion of coal or peat. National emissions for 1985 are of the order of 117,000 tonnes (Bailey *et al.*, 1986), some 56% above the comparable figure for 1975. This is almost entirely due to a large-scale switch to coal combustion in the urban domestic sector, particularly following the increase in oil prices of the late 1970s. Domestic consumers were actively encouraged to switch to solid fuel burning appliances by installation grants at this time. Today, almost 80% of smoke emissions are attributable to domestic emissions and only small contributions to total smoke emissions originate from commercial/industrial sources and from power stations (Figure 1). Smoke is not a particularly buoyant effluent, and furthermore is emitted from sources only a few metres above ground level. This means that it is a localised problem, particularly during calm conditions, or when a low level temperature inversion exists to restrict dispersion. The latter often occurs during nocturnal cooling on calm, clear, frosty nights in winter, conditions which also tend to provoke maximum coal burning.

Smoke pollution is a minor problem in many small towns, but a major problem in Dublin city where 58% of households burn coal (Brady, 1986). Smoke emissions mirror closely housing density, being greatest in the inner city residential areas, where stationary sources are also augmented by mobile sources arising from vehicular traffic. These latter sources amount to about 15% of the total. Several suburban areas in Dublin, with relatively low housing densities nevertheless are also important contributors, with values of greater than 100 tonnes/km^2/year. Overall Dublin produces an average of 55 tonnes/km^2/year, some six times the comparable figure for London. More coal in total is burned in Dublin than in London, although the latter is six times larger, and considerable commercial interests are mobilised to downplay the city's smoke problem (Coal Information Services, 1985).

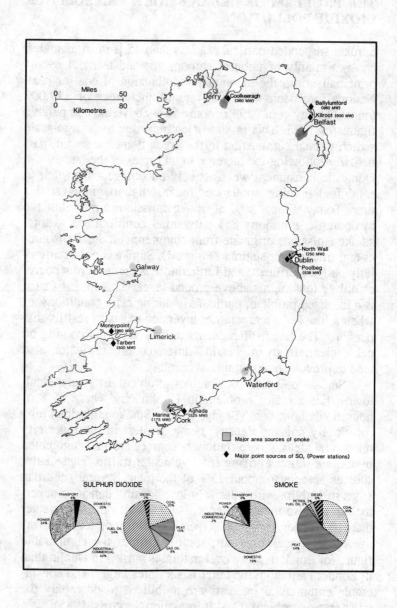

Figure 1 Major sources of industrial air pollution in Ireland

Inevitably these emission figures mean that ground level concentrations of smoke are comparable to those of British cities in the 1960s. Figure 2 shows a rather indeterminate trend for smoke concentrations until 1980/81, with city centre averages of 40-60 μ g/m³. The following winter, smoke values more than doubled to a mean of 90 μ g/m³. Oscillations around this higher level since then reflect favourable or unfavourable winters for dispersion of smoke before it reaches the ground level monitors. Winter smoke levels in Paris and London are also shown on Figure 2, and it is clear that the considerable improvements in winter smoke levels which have occurred widely in Europe in the past two decades have not occurred in Dublin. Indeed the deterioration in air quality in Dublin now places it in the invidious position of having probably the most smoke polluted atmosphere of any EC capital.

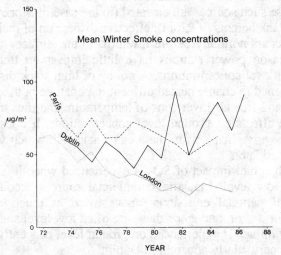

Figure 2 Winter smoke concentrations in Dublin, London and Paris

National emissions of sulphur dioxide are estimated by Bailey *et al.* (1986) at 138,000 tonnes, about one-third of which is accounted for by the power stations (Figure 1). These have, over the past few years, burned natural gas in large amounts, and this has contributed to a substantial drop in national emissions from a high of 234,000 tonnes in 1979. However, this is officially a temporary expedient, and projections of primary energy demand imply a doubling of national SO_2 emissions by the turn of the century (Kavanagh and Brady, 1980). This is an order of magnitude greater than the likely average in the rest of Europe. It is attributable partly to a non-nuclear electricity generation policy, necessitating a reliance on oil, coal, and peat for the bulk of electricity supplies, and entailing the construction of large fossil fuel burning power stations. The completion of the Republic's 900 megawatt coal-burning power station at Moneypoint on the Shannon Estuary will alone result in an emission of 67,000 tonnes of SO_2. Further fossil fuel burning power stations are likely to be necessary in the medium term to cater for a growth in electricity demand which averaged 8.5% per annum for the two decades before 1980, and 5% since. Because such sources are elevated (in the case of Moneypoint with stack heights of 220m), effective dispersion of their effluent occurs normally before it returns to the surface, and for this reason power stations have little impact on trends in ground level concentrations. A policy of high stack dispersal is designed to enable heated effluent to penetrate both surface based and free air inversions of temperature and thus minimise the effects on ground level concentrations. It does however have implications in the area of acid rain which will be examined later.

The chief impact of SO_2 is experienced when it is emitted at low level. Industrial/commercial sources account for 42% of national emissions, about twice as much as the domestic sector, and since these are often low level sources a greater impact is experienced on ground level concentrations. This is particularly apparent in Dublin.

Emissions of SO_2 in the Dublin area, 32,000 tonnes, correspond to an emission rate of 110 tonnes/km^2/year (Bailey *et al.*, 1986). This is very similar to the comparable figure for London (Schwar and Ball, 1983). As polluting industry has relocated out of the city centre, and as cleaner alternative forms of space heating become established (electricity and natural gas) so the winter average concentrations of SO_2 have declined (Figure 3). Increases were observed during the

1981/82 winter when particularly unfavourable dispersion conditions existed. Since then, SO_2 and smoke levels appear to move in unison from year to year, a hint that coal burning is the chief influence on year to year variations in urban SO_2 also. Similar scales of reduction in SO_2 have occurred in Dublin by comparison to other European cities.

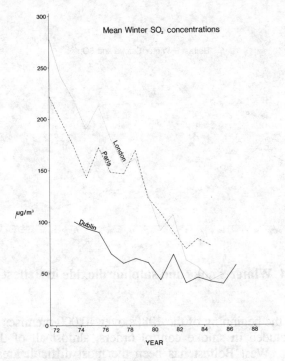

Figure 3. Winter sulphur dioxide concentrations in Dublin, London and Paris

The experience in Northern Ireland is somewhat different to that of the Republic. The provisions of the Clean Air Act (NI), introduced in 1964, have resulted in a steady drop in urban smoke levels throughout the mid 1960s and 1970s. In Belfast, winter means have fallen from 264 μ g/m^3 in 1961/62 to 38 μ g/m^3 in 1985/86 (Figure 4). Even before the implementation of the Act; however, Belfast was less smoke polluted than many of the larger British cities, despite a very high *per capita* consumption of bituminous coal. The geographical situation of the city, in a valley aligned southwest-northeast, parallel to the prevailing wind direction, undoubtedly was partly responsible for this.

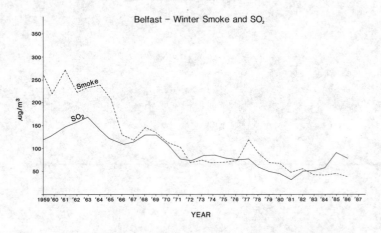

Figure 4 Winter smoke and sulphur dioxide in Belfast

By the beginning of the 1980s over 40,000 premises had been included in smoke control orders, almost all of them domestic. West Belfast has been the most difficult area to enforce and the smokeless programme has skirted clockwise around the city from the east. Conversion to smokeless systems is grant-aided by both the government and local authority. Winter smoke levels today are somewhat higher than those prevailing in other parts of the UK, and have, in recent years, stabilised at around 40 μ g/m^3 on a city average basis. The absence of a natural gas network in Northern Ireland is likely to inhibit progress towards pollution parity with other areas of the UK.

Sulphur dioxide levels in Belfast have not fallen as far or as fast as might have been hoped. The absence of natural gas has meant that conversion of open fires often resulted in the

428

installation of oil burning central heating systems, particularly in the era of cheap oil in the late 1960s and early 1970s. The 1985/86 mean winter level of 78 μ g/m^3 compares unfavourably with 55 μ g/m^3 for Dublin (1986/87), a reflection also of the greater industrial component in Belfast's emission environment.

Table 2: **EC Directive limit values for sulphur dioxide and smoke**

	Limit value sulphur dioxide 5g/m³	Assoc-iated smoke value 5g/m³	Absolute smoke value 5g/m³
Annual median	80	40	80
(Median value of daily means)	120	<40	-
Winter median (October-March)	130	>60	130
(Median value of daily means)	180	<60	-
Daily value	250*	>150	250
98 percentile of daily means	350*	<150	-

* Member states must take all appropriate steps to ensure that this value is not exceeded for more than three consecutive days.

URBAN AIR QUALITY AND THE EC DIRECTIVE

On 1 April 1983 an EC Directive specifying limit values for smoke and SO_2 became mandatory, providing a framework for air quality management throughout the European Community. Although these standards were ostensibly geared to WHO research findings on health impacts of air pollution, particularly the synergistic effects of smoke and SO_2, in reality their impetus also came from a desire to regulate competition between polluting industries among member states. Annual, winter and daily limit values are specified (Table 2).

Table 3: Dublin area breaches of EC Directive limit values

	Annual	Winter	Daily 98% of daily mean values	More than three consecutive days	Number of Daily Observations of Smoke $>2505g/m^3$
1973/74	4*	-	4*	2*	28
1974/75	2*	1*	3*	1*	6
1975/76	2*	-	1*	1*	-
1976/77	-	-	1*	2	35
1977/78	-	-	1	1	33
1978/79	-	-	-	-	4
1979/80	-	-	1	-	39
1980/81	-	-	-	-	8
1981/82	2*	-	7	7	145
1982/83	-	-	-	-	27
1983/84	-	1	5	1	101
1984/85	-	-	6	4	125
1985/86	-	1	3	1	71
1986/87	-	1	6	4	119

* Breaches due to sulphur dioxide.
The number of valid sites in the Dublin Corporation network varied between 11 and 14 during the period.

Full compliance with the provisions of the Directive is required by 1993. This is most unlikely to be achieved with respect to smoke in Dublin, where a deterioration in winter levels continues. Table 3 shows the breaches of the Directive which have occurred since 1973/74. In the early 1970s the problem was obviously SO_2, whereas since the mid 1970s it has almost always been smoke which gives rise for concern.

The number of daily observations of smoke in excess of 250 μ g/m^3 has changed substantially since 1980/81, with half the network in a typical year failing to satisfy the requirements of the Directive in this respect alone. Indeed, alarmingly high concentrations are becoming common in the western suburbs of the city, where the network is sparse. At Ballyfermot, for example, for the four months from November 1986 to February 1987, 30 days with over 250 μ g/m^3 smoke were measured, with one continuous period of 7 days above this value being registered. At the same location a value of 1,429 μ g/m^3 almost six times the daily smoke limit, was recorded on 31 January 1987. Values in excess of 1,000 μ g/m^3 were recorded at other sites in the network also during winter 1986/87, and exceedences of the daily smoke limit also occurred in the county council areas outwith Dublin city centre.

An incipient smoke problem also exists in Cork city, where the daily smoke limit is occasionally exceeded. possibility of short duration high concentrations exist here because of katabatic drainage of cold air into the Cork basin. However, the problem is as yet not serious, and the longer availability of natural gas in the city may safeguard its future air quality somewhat. Outside of these two cities there has been no record that the Directive's provisions have been exceeded.

Links between SO$_2$ and mortality and morbidity in Dublin were suggested for the 1970s by Kevany et al. (1975), Bailey et al. (1987) and Sweeney (1982). Indeed, Ireland had the highest mortality rate from respiratory disease in a survey of twenty two European countries carried out during the mid 1970s (WHO, 1974). However, directly attributing health impacts to air pollution has proven elusive. Explicit cause-and-effect relationships are difficult to show conclusively in this area, although even this may be more feasible if present trends continue.

One episode of high smoke and SO$_2$ concentrations has been accompanied by a mortality peak. Figure 5 shows the radio-sonde ascent data for Valentia (Co. Kerry) for 10 January 1982. An inversion of temperature at about 950mb fell close to the surface in the following 8 hours, accompanied by light winds and bitterly cold temperatures. The inversion effectively trapped the greatly enhanced emissions of smoke and SO$_2$ in the Dublin area, causing concentrations of smoke to reach 1,812 μ g/m^3 at the Cornmarket monitor on 13 January. One week later a peak in mortality of 120 was recorded at the largest city hospital, about twice the normal monthly level (Kelly and Clancy, 1984). Though the 56 excess deaths involved may be linked to a multiple causation hypothesis, the extremely high levels of atmospheric pollution preceding them are undoubtedly major contributory factors.

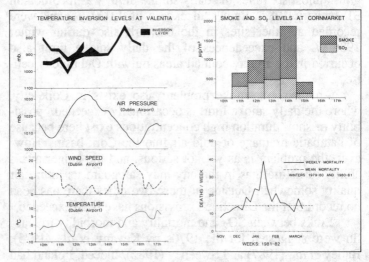

Figure 5 Meteorological information relating to a pollution incident in Dublin, January 1982

The Problem of Long Range Transport of Air Pollution

Ireland has largely escaped the consequences of the 'emit hot and high' philosophy which has rendered much of Europe susceptible to the effects of trans-boundary movements of sulphur and nitrogen oxides. Extensive forest damage, the acidification of water bodies, and the mobilisation of toxic heavy metals from soil, have become widely publicised by research findings in the past two decades, particularly in Scandinavia and central Europe. A location 'upwind' of the principal exporters of these gases, however, ensures that only for the 3.5% of the year when easterly winds blow in Ireland is importation of pollutants from these sources possible. In addition, the geological configuration of Ireland could hardly be more favourable, with over 40% of its surface underlain by Carboniferous limestone and an even greater amount veneered with limestone-derived glacial drift. Considerable buffering against the effects of acidification of surface waters may be envisaged over much of the island.

Monitoring of the pH value of Irish rainfall has been carried out since 1960 by the Irish Meteorological Service, and this indicates a fall in median pH values from 5.9 in the period 1960-64 to 5.7 in 1980-84. Values less than 5.6 are usually considered acidic. A marked geographical variation in the results is apparent and probably any trend reflects mainly local and national influences (Mathews *et al.,* 1981). As part of the European Monitoring and Evaluation Programme a site was established for daily measurements at Valentia and this has suggested median values of about 5.14, still favourably comparing with France, Germany, Denmark, and Netherlands (4.50) or Sweden (4.10) (OECD, 1985). However, more intensive studies have now been completed around the Dublin area (Bailey *et al.,* 1986) and these have shown substantial increases in acid deposition with easterly winds. At Glencree in County Wicklow, for example, 5% of daily samples showed pH values of 4.0 or less, and marked enhancement of sulphate and nitrate deposition was recorded on easterly circulations. In nearby Glendalough Lake Upper, minimum pH values were occasionally observed as low as those which have been linked to damage to lake ecosystems elsewhere in Europe (Bailey *et al.,* 1986).

Long-distance transport of pollutants from continental Europe has for long been associated with visibility impairments in Ireland (Lovelock, 1972). Aerosol sulphate is probably primarily resonsible, though substantial ozone transport may occur especially in the summer months. No regular monitoring of ozone or nitrogen oxides is presently carried out, however. Figure 6 shows median summer visibility at Dublin Airport since 1970 based on hourly observations. Wind directions from easterly quadrants are clearly associated with periods of diminished visibility and since no local sources of pollution lie in this direction it may be hypothesised that long-distance transport of air pollutants from beyond Ireland is responsible.

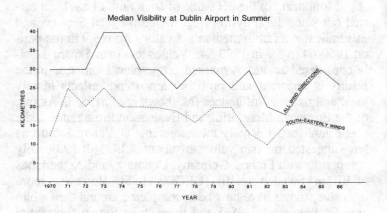

Figure 6 Summer visibility at Dublin Airport

The meteorological background to one such episode is shown in Figure 7. On this occasion elevated ozone concentrations were measured throughout England, and back trajectory analysis by Simpson and Davies (1985) indicated sources in the Ruhr and northern France. To this burden would have been added emissions from industrial parts of England prior to the air mass reaching Ireland. More than 50% of sulphate deposition within Ireland probably is as a result of long-distance transport (Highton and Chadwick, 1982).

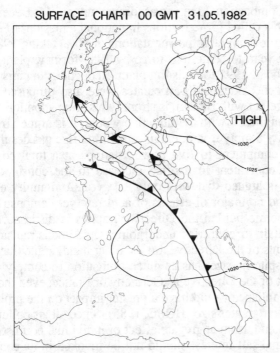

SURFACE CHART 00 GMT 31.05.1982

Figure 7 Synoptic conditions associated with European air pollution on 31 May 1982

Long Range Transport and the Draft EC Directive

The EC includes western Europe's most prolific producers of sulphur dioxide: the United Kingdom (4.2Mt), West Germany (3.5Mt), France (2.9Mt) and Italy (3.1Mt). Radical progress towards tackling the problem has thus been inexorably slow with member states divided in their attitudes towards what are inevitably extremely expensive control options. Divergent national strategies, however, imply unequal competition, and the Commission has sought to remove hidden 'subsidies' from countries who solve their local pollution problem by exporting it to their neighbour. From the early 1980s onwards West Germany became an enthusiastic proponent of emission controls, possibly as a consequence of the sudden discovery of extensive acid rain damage to its forests and lakes. Largely at its behest the Draft Directive on Limiting Emissions from

435

Large Combustion Plants was drawn up on 15 December 1983. This called for reductions in national emissions of sulphur dioxide from large power stations of 60%, taking 1980 as the base year, to come into force by 1985 for new plant, and by the end of 1995 for existing plant. Reductions in emissions of nitrogen oxides and particulates were also proposed. Ireland, together with Britain, is strongly resisting adoption.

Compliance with the directive would, it is argued by the Electricity Supply Board, entail retrofitting flue gas desulphurisation equipment to five power stations at a total cost of £400M, equivalent to a price increase to the consumer of 20%. It is argued that with only 1.1% of Community emissions, and no major ill-effects of acid rain yet demonstrated, the economic burden the Directive implies is disproportionately unfair to countries undergoing rapid industrialisation. Proponents of the Directive, on the other hand, argue that Ireland, despite its size, has as much obligation to comply with the spirit of the Directive as Germany or France, even though their compliance would have a greater impact on the problem due to their greater size (HOPE, 1985). Since all air pollution, they contend, is an aggregate effect of individual point sources, national self interest should not be invoked in its control. A slowing down in demand for electricity, however, makes it feasible that perhaps two of the larger power stations fitted with flue gas desulphurisation equipment would enable compliance. In addition, less costly estimates for installation of this equipment exists. The Central Electricity Generating Board in the UK estimate a cost of £70-80M for retrofitting a plant of similar size to Moneypoint (Dudley et al., 1985), less than half the ESB estimate. Of the EC countries, only Ireland, Britain and Greece, have not pledged a 30% reduction in sulphur emissions by 1993 and the position adopted by Ireland appears essentially to be one of sheltering behind British intransigence.

Environmental Legislation and Policy Formation

Legislative provision for environmental protection in Ireland is rooted in a plethora of statutes, some of which (such as the Alkali Act of 1906) predate independence. The overhaul of these has been proceeding, especially since EC entry, and in some respects Ireland has been in the forefront of new approaches. This is especially apparent in the requirement for Environmental Impact Statements contained in the Local Government (Planning and Development) Act of 1976, in which national provision for an EIS were incorporated in advance of most other members of the Community. The problem has always been in implementation. During the 80 years since the enactment of the Alkali Act no prosecution under it has occurred, and only one prosecution under the Control of Atmospheric Pollution Regulations (1970) has taken place (Cabot, 1985).

The most recent initiative in air pollution legislation was introduced in the Seanad on 10 February 1986 and became operative on 1 September 1987. This is a comprehensive bill which repeals all existing air pollution legislation and consolidates it under one act. A licensing system for industrial emissions is required, and provision for local authorities to create smokeless zones in their jurisdiction is provided. For the first time the problem of domestic emissions is tackled. In its slow passage through the legislature it has become apparent that the necessary resources for its successful implementation will not be made available in the medium term. Indeed, the concerns of the coal industry were a paramount consideration in one amendment moved by the Minister for the Environment. It is most unlikely that the Dublin smoke problem will be solved as a consequence of this bill.

There has never been a tradition in Ireland of explicit policy formulation with respect to environmental protection. Environmental issues have rarely been electoral issues, and the successful management of the Irish 'commons' has in the past necessitated little or no intervention. Increasingly, though, the 'hidden agenda' in environmental management is apparent to an ever-more sophisticated and articulate public. It is most unlikely that they will in the future be content with cosmetic legislative exercises designed to assuage the EC Commission rather than safeguarding environmental quality for their children.

REFERENCES

Bailey, M., Bowman, J. and O'Donnell, C. (1986) *Air Quality in Ireland : the Present Position*, An Foras Forbartha, Dublin.

Bailey, M., Kevany, J. and Walsh, J. (1987) *Air Pollution, Climate and Health in Dublin*, National Board for Science and Technology, Dublin.

Brady, J. (1986) The impact of clean air legislation on Dublin households, *Irish Geography*, 19, 41-4.

Cabot, D. (1985) *The State of the Environment*, An Foras Forbartha, Dublin.

Coal Information Services (1985) *Towards a Planned Improvement in Dublin's Air Quality*, Coal Information Services, Dublin.

Dudley, J., Barrett, J. and Baldock, J. (1985) *The Acid Rain Controversy*, Earth Resources Research, London.

Highton, N. and Chadwick, M. (1982) The effects of changing patterns of energy use on sulfur emissions and depositions in Europe, *Ambio*, 11(6), 324-30.

HOPE (1985) *Acid Rain: Ireland*, Help Organise Peaceful Energy, Bantry.

Kavanagh, R. and Brady, J. (1980) *Energy supply and demand: the next thirty years*, National Board for Science and Technology, Dublin.

Kelly, I. and Clancy, L. (1984) Mortality in a general hospital and urban air pollution, *Irish Medical Journal*, 77, 322-4.

Kevany, J., Rooney, M. and Kennedy, J. (1975) Health effects of air pollution in Dublin, *Irish Journal of Medical Science*, 140, 108-17.

Lovelock, J. (1972) Atmospheric turbidity and CCl_3F

concentrations in rural southern England and southern Ireland, *Atmospheric Environment*, 17, 151-9.

Mathews, R., McCaffrey, F. and Hart, E. (1981) Acid rain in Ireland, *Irish Journal of Environmental Science*, 1(2), 47-51.

OECD (1985) *Environmental Data Compendium 1985*, OECD, Paris.

Schwar, M. and Ball, D. (1983) *Thirty Years On - A Review of Air Pollution in London*, Greater London Council, London.

Simpson, D. and Davies, S. (1985) *Trajectory analysis of ozone episodes at a rural site in Sibton, Suffolk 1980-1983*, UK Department of Industry, London.

Sweeney, J. (1982) Air pollution and morbidity in Dublin, *Irish Geography*, 15, 1-10.

World Health Organisation (1974) *World Health Organisation EURO 4905(5)*, Copenhagen.

17 REGIONAL DEVELOPMENT STRATEGIES

James A. Walsh

INTRODUCTION

Within the context of the international space economy Ireland can be described as a problem region with significant intra-regional imbalances in the nature and extent of its economic development. The five million inhabitants of the Republic and Northern Ireland are part of a European Community which comprises twelve states and over 170 regions which have a combined population of more than 321 million people. Considerable disparities exist between the regions in relation to a wide range of demographic and economic variables. The extent of the disparities between Ireland and other parts of the European Community is accentuated by its peripherality (Keeble *et al.* 1982). The regions of the Community, excluding Greece, Spain and Portugal, have recently been ranked according to a 'synthetic index' which measures the relative intensity of regional problems on the basis of GDP measures and the rate of unemployment. Of the 131 regions examined Northern Ireland had the second lowest ranking while the Republic of Ireland was fourth lowest (Commission of the European Communities, 1984).

Within Ireland there are considerable variations in the nature and extent of economic and social development. Politically the island has been partitioned since the 1920s into the Republic (comprising 83% of the total area and just under 70% of the population in 1981) and Northern Ireland, the

441

latter being part of the United Kingdom. Hoare (1982) has shown that since 1945 Northern Ireland has constantly been the most disadvantaged economic region of the UK. The extent of its economic malaise is accounted for by reference to its peripheral location and small size *vis-à-vis* the UK mainland, its high level of specialisation in the past on industries which have been in decline since the 1960s, its more recent reliance on branch plants of transnational corporations among which there have been many closures, and finally the poor public image which it has acquired after almost two decades of civil unrest.

Partition has also had a considerable impact on the economy of the Republic of Ireland, particularly in border areas which have been transformed into inland peripheries and detached from their natural trading hinterlands. It has recently been estimated that as a result of the violence since 1969 the total direct and indirect costs to the economies of the UK and the Republic were in the region of IR£9,500 million and IR£2,300 million, respectively, in 1982 prices (New Ireland Forum Report, 1984). Since partition the level of economic performance in Northern Ireland is largely influenced by the level of economic activity on the UK mainland. However, within Northern Ireland there is some scope for policy determination particularly in relation to industrial development assistance.

Within both the Republic and Northern Ireland there are considerable regional disparities (NESC, 1981; Horner *et al* 1987). In Northern Ireland the Belfast sub-region (consisting of places within approximately 24 km of Belfast city centre) in 1981 contained approximately 730,000 persons (47% of the total population), 55% of those employed in manufacturing industries, and 53% of those with professional, technical or managerial type occupations. By contrast this sub-region contained only 42% of the total number of unemployed persons, representing an unemployment rate of 14% compared with 28% for the Strabane District Council area, for example.

In the Republic of Ireland similar problems are associated with the dominance of Dublin and the East region (Walsh, 1987). In 1986 Dublin contained over one million persons, 28.9% of the total population, while the East region as a whole contained 37.7% of the total. The regions share of those employed in manufacturing industries was 38%, while its share of white-collar employment was approximately 50%. In 1980 the average household income in the East region was 19% above the average for the State, compared with the Northwest/Donegal region where it was 28% below the average. The imbalance which exists between the East region and the remainder of the State involves more than just an imbalance in numbers. It also involves an excessive level of centralisation of political power and decision making (Barrington, 1975), and an undue influence through information flows from Dublin on attitudes and opinions affecting many aspects of life in other parts of the country (Lee, 1985; O'Tuathaigh, 1986).

REGIONAL PLANNING 1950-1972

The importance of a regional dimension in economic development was first recognised in Ireland as far back as 1891 when the Congested Districts Board (CDB) was established to improve living standards in eight western counties. The work of the Board from 1891 to 1923 represented the first attempt at comprehensive regional development. Measures were introduced to encourage agrarian restructuring, the improvement of breeds of livestock and poultry, the development of the fishing industry and other small industries. While the CDB was solely concerned with the problems of peripheral rural areas there was also growing concern over the extent of social deprivation in the larger urban areas. The need to integrate town and country planning within a regional framework was first suggested in 1911 by Geddes (Bannon, 1985a).

Following the achievement of political independence the concept of a regional dimension to economic policy in the Republic was virtually abandoned until the early 1950s when the Undeveloped Areas Act (1952) was introduced. This Act was passed to foster industrial development in the poorer areas of the country which coincided largely with the areas previously covered by the CDB. A second body, An Foras Tionscal, was set up under the Act to administer grants in order to encourage industrialisation in the undeveloped areas.

443

The Industrial Development Authority (IDA) had been established in 1950, but its role at this stage was to promote and develop industry, without any specific regional dimension to its activities. Two additional agencies were established in the late 1950s : Gaeltarra Éireann in 1958 with specific responsibility for operating and promoting manufacturing industry in Gaeltacht areas, and Shannon Free Airport Development Company (SFADCO) in 1959 with the objective of creating demand for air traffic facilities and labour at Shannon.

In 1958 the government published its first White Paper on Economic Development which proposed that, in order to maximise the return from scarce resources, new industries should be located at or convenient to the larger centres of population (Walsh, 1976). This was in contrast to Newman (1958) who proposed that six or eight towns in every county should be developed for the provision of social services and industrial employment. The concentration versus dispersion debate continued into the 1960s, with the emphasis on concentration evident in a series of reports produced for the government between 1963 and 1965 (NESC, 1975). The report prepared by a Committee on Development Centres and Industrial Estates (1965) proposed that industrial growth from both indigenous and external sources should be concentrated in a small number of development centres, each of which in turn would act as a stimulus for regional growth. The report recommended that the following criteria should apply in selecting such centres: size of town, labour availability, infrastructural facilities, availability of land, communications and existing industrial base. The report was followed up by a recommendation from the National Industrial Economic Council that both regional and national development could be promoted by focusing on a total of eight centres including Dublin and Cork, and that the development centres should be locations for both manufacturing and service industries. It also recommended that, because of the lack of large centres in the poorer northwestern counties, a range of alternative policies would have to be sought for these areas. The culmination of the first stage of the debate on development centres came in 1965 with the first official statement on regional policy (NESC, 1975, pp.77-78). While accepting that such centres could be 'an effective means of promoting the further expansion of economic activity' the government nevertheless considered that, in addition, 'the dispersal of industrial activity throughout the country, where this is economically feasible, yields important social advantages' (p.78). It was envisaged that the development centres would become the commercial, financial, educational, health, social and administrative centres of each region.

The advent of economic planning in 1958 coincided with the initiation of a relatively long period of successful economic development. In this period of growth and expansion the need for a comprehensive framework of physical planning became apparent (Bannon, 1983). A Local Government (Planning and Development) Act was introduced in 1963 which obliged each of the 87 local planning authorities in the state to prepare development plans for the 173 separate districts under their jurisdiction, within three years of a specified date. This marked a new departure in the role of local planning authorities which was expected to be of considerable significance in helping to foster economic development. In 1964 An Foras Forbartha (The National Institute for Physical Planning and Construction Research) was established as a state agency. The state was divided into nine regions for planning purposes and consultants were appointed to undertake studies of the long-term development of the Limerick (Lichfield *et al.*, 1967) and Dublin regions (Wright, 1967). In the same year eight regional tourism organisations were established to develop and promote regional tourist facilities. In the following year, 1965, a government report on investment in education identified eight centres outside of Dublin as locations for new third level colleges, thus giving further support to the concept of concentrating development in selected areas.

In 1966 the government commissioned Colin Buchanan and Partners to prepare a national planning strategy based on a survey of the resources of each of the nine planning regions. This report proposed that regional planning policy should be based on a hierarchy of growth centres involving, in addition to Dublin, two national, and six regional centres, together with four local centres in areas remote from large towns in the west and northwest (Buchanan, 1968) (Figure 1). It was envisaged that Dublin would be allowed to grow according to

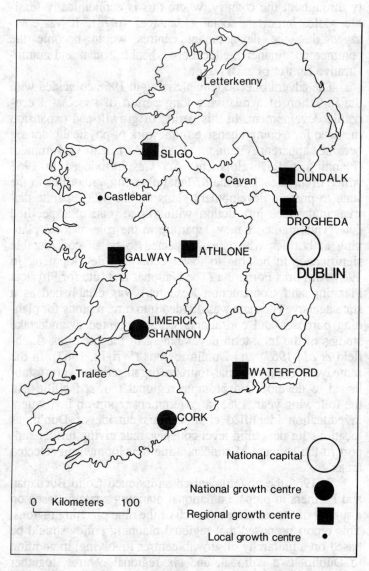

Figure 1. Growth centres in the Republic of Ireland

National capital

National growth centre

Regional growth centre

Local growth centre

0 Kilometers 100

Letterkenny

SLIGO

Cavan

DUNDALK

Castlebar

DROGHEDA

GALWAY

ATHLONE

DUBLIN

LIMERICK
SHANNON

Tralee

WATERFORD

CORK

its rate of natural increase. The principal development effort was to be concentrated on the proposed national growth centres at Cork and Limerick-Ennis-Shannon which were expected to increase their population totals to approximately 250,000 and 175,000 respectively by 1986. These centres would accommodate 70% of the population increase envisaged for the eight centres outside Dublin. In addition to the proposed hierarchy of growth centres the Report recommended that roads of motorway standard be constructed between Dublin and Northern Ireland and from Dublin towards Cork and Limerick.

The increased level of economic activity which would follow from the proposed growth centre policy would, by reducing emigration, lead to an estimated population total of approximately 3.5 million in 1986. It was envisaged that by that year the distribution of population would be: 32.2% in Dublin, 18.7% in the combined national and regional centres and 49.1% in the remainder of the state compared with 27.6%, 10.3% and 62.1% respectively in 1966. This redistribution would involve a reduction of population in half of the counties, with substantial losses of 22%, 18% and 16% forecast for Donegal, Leitrim and Kerry respectively. In order to achieve this redistribution it was proposed that the expansion of Cork, Limerick-Ennis-Shannon, Waterford, Dundalk and Drogheda would be based primarily on manufacturing employment; that of Galway on a mixture of manufacturing employment, tourism and regional services; and that of Athlone, Sligo and the four local centres on regional services. Overall it was recommended that 75% of all new industrial jobs would be located in Dublin and the eight regional growth centres. In order that each of the key centres in the overall strategy, Cork and Limerick, should not be at a financial disadvantage relative to other locations it was proposed that they should be included in the Designated Areas (formerly called the Underdeveloped Areas).

The implementation of the recommended development strategy was to be carried out under the aegis of regional planning authorities with statutory powers. Development Corporations, along the lines of the development company already established at Shannon, were proposed for each of the national growth centres to oversee their development needs.

In addition to the Buchanan report, in 1968 the government was also presented with a major report on the future of the hospital system (Fitzgerald Report, 1968). This study concluded that the best strategy for future hospital development would be to locate sixteen regional and general hospitals at twelve centres, all of which except two, coincided with those recommended by Buchanan. The Buchanan centres also corresponded closely with the locational strategy for the proposed new third level colleges (Investment in Education, 1965). Thus by the end of the 1960s there appeared to be a widely accepted (at least among government advisors) framework for developing an urban strategy, along the lines indicated in the government policy statement of 1965, which would counteract some of the strong centrifugal forces contributing to the rapid expansion of Dublin.

In May 1969 the government issued a lengthy statement on regional policy simultaneously with the publication of the Buchanan report (NESC, 1975, pp.78-83). It accepted 'in principle that growth centres can be a valuable element in a regional programme' (p.80) aimed at promoting industrial expansion and checking the tendency towards unbalanced regional development. However, it also noted that the Buchanan proposals had far-reaching implications which would require further investigation in the context of proposals for regional development generally. Regional Development Organisations were to be established in the planning regions to co-ordinate regional development programmes but they were not to be given any statuory powers, and responsibility for all planning activity was to remain in the hands of local planning authorities. This government statement also expressed the objectives of broad-based regional expansion, provision of local employment in small towns and rural areas, minimisation of population dislocation, and concern for underdeveloped and Gaeltacht areas. At the same time the IDA, which in 1969 had its role expanded to include administration of the Industrial Grants Scheme, was charged with setting up local units in each region to promote industrial development, with the exception of the Midwest region where the service was already being provided by SFADCO as agents of the IDA.

By 1972 the Regional Development Organisations had completed their regional development plans and the IDA had prepared a regional industrial strategy for the period 1973-77. In May of that year the government issued its final statement on regional policy (NESC, 1975, pp.83-86). The goals which were laid out for regional policy were consistent with the objectives contained in the 1969 statement. The policy statement proposed a regional strategy for the next twenty years which included: (1) Dublin development to accommodate the natural increase of its existing population; (2) expansion of the towns and cities identified by Buchanan as growth centres; (3) development of county or other large towns in each region, including towns in remote areas; and (4) continued development of the Gaeltacht. An important component of this strategy was the IDA's Regional Industrial Plans (IDA, 1972) in which manufacturing targets were specified for each region and for 137 towns organised into 48 town groupings. This plan envisaged that 52% of new industrial jobs would be located in the main centres identified in the Buchanan report compared with the 75% provided for in that study up to 1986. These plans, therefore, provided a strategy for regional development which was neither one of extreme concentration nor extreme dispersal but rather one specifically tailored to regional needs and potential.

The principal means of achieving this degree of spatial disaggregation of the overall national job-creation target were to be a comprehensive programme of acquiring and preparing fully serviced industrial sites throughout the country, in most cases with ready-built advance factories, a much greater degree of discretion than that applied previously in the allocation of non-repayable grants, provision of generous training grants to industrialists locating in less-developed areas, and housing for key workers (IDA, 1972). In justifying the high level of dispersion implicit in its plans the IDA argued that developments in transport and communications had conferred on industry much greater locational flexibility. Thus, in 1972 there were 271 locations with IDA grant-aided industries, of which almost half were in towns of less than 3,000 population (IDA, 1972). Of the 418 grant-aided plants which had been established in the period 1960-73 over half (237) were located in towns of less than 5,000 population, and 59% were located in the Designated Areas, compared with only 15% in county Dublin (O'Farrell, 1975). As an explanation for this phenomenon O'Farrell (1980) attached particular significance to the IDA's financial incentives, which compare most favourably with those provided by other west European countries, and to the organisation of itineraries to prospective sites for potential investors. Breathnach (1982) has argued that the high level of industrial dispersion achieved by the IDA was largely due to fundamental changes which have been taking place in the nature of manufacturing industry at the global level. In this respect he attached particular significance to: (1) developments in the realm of transport and communications technology; (2) the reduction of uncertainty concerning supplies and outlets arising from the centralisation of ownership and control of production activities; and (3) the development of infrastructural facilities in peripheral areas by national governments, thereby removing basic obstacles to the successful operation of externally orientated and controlled activities. The overall effect of these trends was to greatly weaken the basic foundation upon which a growth centre strategy might be built, namely the tendency for industry to agglomerate. The gradual abandonment of the concept of promoting growth at a small number of centres was not just confined to the industrial sphere. The 12 centre hospital strategy proposed in 1968 was initially replaced in 1974 by a proposal for an 18 centre strategy and finally in 1975 by the governments' 22 centre strategy (Horner and Taylor, 1979).

In the case of Northern Ireland both its inter- and intra-regional imbalances have been recognised for a considerable time. Attempts to overcome these problems relying mostly on industrial development legislation began in the 1930s, and were particularly successful during the late 1940s and throughout the 1960s at encouraging the establishment of new manufacturing projects, many of which were of external origin (Bull *et al.*, 1982). Attempts to reduce the extent of the imbalances between the Belfast sub-region and the less-urbanised areas in the south and west of the region were initiated in the mid 1960s. These have involved proposals aimed at altering the settlement pattern and some measures designed to encourage a redistribution of investment. Chief among the latter have been: (1) larger industrial grants for manufacturing firms setting up projects in the more disadvantaged areas, (2) provision of industrial sites and advance factories at a number of centres identified for their growth potential, and (3) the establishment of the Local Enterprise Development Unit (LEDU) in 1971 which was initially specifically empowered to aid the development of small manufacturing businesses in rural parts of the region (Bull, 1984). As a result of these measures there has been a reduction in the level of concentration of sponsored manufacturing firms in the east of Northern Ireland. Between 1945 and 1959 some 23% of sponsored firms set up within Belfast County Borough and 75% within 50 km of the city; between 1960 and 1973 when the total number of sponsored firms was double that of the previous period these proportions decreased to 11% and 65% respectively (Hoare, 1982).

The 1960s in Northern Ireland was a period of intense government concern with the locational aspects of economic development. The first report concerned with physical planning (Matthew, 1964) was confined to the Belfast sub-region, within which it proposed a 'Stop Line' around built-up Belfast to restrict its further outward growth, and a series of growth towns partly designed to accommodate the surplus that would be released from Belfast. The most important of these was to be the New Town of Craigavon, 35 km southwest of the city, which would also serve as a focus for the south and the west of the region. It was envisaged that the New Town would grow to 100,000 population by 1981. In addition six 'key' centres, where industrial development would be concentrated, were proposed for the remainder of Northern Ireland (Figure 2). Little or no attention was granted to rural areas at this seminal stage of physical planning (Caldwell and Greer, 1984). The policy of concentrating investment in particular centres was endorsed by the Wilson Report (1965) on Economic Development in Northern Ireland. While this study queried the choice of some of the 'key' centres suggested by Matthew for the southern parts of the region, it also enhanced the status of Londonderry and Coleraine to growth centres. Despite increasing concern over the lack of development opportunities in rural areas the policy of concentrating development efforts into growth centres was reiterated in the Development Programme 1970-1975 prepared by Matthew,

451

Wilson and Parkinson (1970). This report included Strabane as a 'key' centre in addition to those identified in 1964, and recommended that Londonderry and Ballymena should be centres of accelerated growth. Thus throughout the 1960s in Northern Ireland, as in the Republic, the theme of economic concentration remained dominant in physical planning considerations. In Northern Ireland the policy of concentration was not only justified on economic grounds but also for reasons relating to management of the rural environment (Caldwell and Greer, 1984).

EUROPEAN COMMUNITY REGIONAL POLICY

In 1973 Ireland and the UK along with Denmark joined the European Economic Community. This development came at a time when the EC was, for the first time, giving serious consideration to the formation of a Community regional policy. In the EC the need to work towards a reduction of regional disparities was acknowledged in the Treaty of Rome (1957) which stated that a major aim of the contracting parties is to ensure 'a harmonious development by reducing the differences existing between the various regions and the backwardness of the less favoured regions' (Sutherland, 1986, p.372). The task of promoting regional development was entrusted to the European Investment Bank which was given special responsibility for financing 'projects for developing less developed regions'. In addition to this lending agency the Community established a number of Funds to provide grants and subsidies, each of which had particular regional implications. The most significant was the European Agricultural Guarantee and Guidance Fund (EAGGF) which is concerned mainly with providing price guarantees on certain agricultural products. A second source of aid with important implications for regions with severe unemployent problems has been the European Social Fund (ESF) (Laffan, 1985).

The first enlargement of the Community in the early 1970s provided the catalyst for developing policies directly aimed at reducing differences between regions. There was also an increasing realisation within the Community that the ultimate aim of full economic and monetary union was unlikely to be achieved as long as serious regional disparities persisted. By 1973 the European Commission had put forward detailed proposals on regional policy including the setting up of a European Regional Development Fund, (ERDF), which was eventually established in 1975 (Talbot, 1977).

It was in this context that Ireland became a member of the EC. There was considerable optimism at government level that membership of the Community would contribute greatly to solving the problems of under development within the country, as reflected in its 1972 statement on regional policy (NESC, 1985, p.86). However, many problems remained in relation to the formulation of appropriate policies and strategies for comprehensive regional development in Ireland as shown by several reports prepared for the recently established National Economic and Social Council between 1974 and 1977 (e.g. NESC, 1975, 1976a, 1976b, 1977a, 1977b, 1978a, 1978b). At the same time studies were commissioned on problems relating to agricultural and rural development (NESC, 1978a, 1978b) and on relationships between urbanisation and regional development (NESC, 1979). There was a continuing concern about the problems of western areas for which a series of Development Boards were considered. However, none of these Boards were in fact established.

From the mid-1970s onwards government attention became increasingly focused on the growing level of unemployment resulting from the international recession triggered off by significant increases in the price of oil. Consequently there was a diversion of interest away from issues relating to regional development despite indications from several studies that urgent action was required if progress was to be made towards achieving the goal of an even spread of development among regions. (O'Farrell, 1978b; Ross and Walsh, 1979; NESC, 1980, 1981a). Increasingly attention was directed towards the European Community and the role which it could play in supporting the development of the economy.

In Northern Ireland the Department of the Environment in 1975 issued a discussion paper which was eventually to lead to the preparation of a Regional Planning Strategy (1977) for the period 1975-1995. Following consideration of six options the chosen strategy involved the selection of 'District Towns' which aimed at a greater dispersal of resources than hitherto by increasing the sixteen existing growth and key centres to 23 (Figure 2). Although each District Town was to be ascribed equal status, priority was to be given to the regeneration of Belfast's inner areas, to Londonderry and Craigavon as the principal complexes for industrial and population growth and to Antrim and Ballymena as major shopping centres. Extensive use has been made of development agencies to attract investment projects from mainland UK and the USA, and from indigenous sources. The funding for these agencies comes mainly from the UK Treasury and also from the European Community.

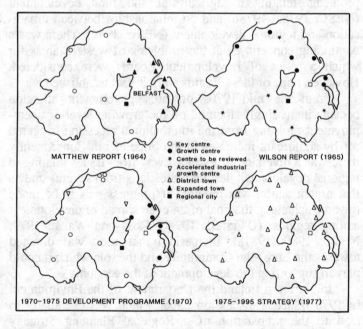

Figure 2. Northern Ireland Planning Strategies

IMPACT OF REGIONAL POLICIES

Since 1973 there have been very substantial net financial transfers from the European Community to both the Republic and Northern Ireland (Drudy, 1984; Simpson, 1984). Total receipts from the Community between 1979 and 1983 to the Republic amounted to almost 30% of the Public Capital Programme, or 3.5% of total public expenditure in 1983 (McNamara, 1984). The latter proportion for Northern Ireland was 2.1% (McGurnaghan, 1984). The principal source of aid was the Agricultural Fund, amounting to 78% of the total receipts in the Republic between 1973 and 1981 and 55.6% of the total receipts to Northern Ireland up to 1984. The bulk of these receipts, 91% in the Republic and 89% in Northern Ireland, have been in the form of price supports and other market stabilising mechanisms which have tended to discriminate in favour of the more prosperous farmers especially in the southeast and southwest (Cuddy, 1986). At the same time there has been an increase in the level of marginalisation in agriculture which is particularly serious in the west and northwest where small farms predominate (Kelleher and O'Mahony, 1984; Horner *et al.*, 1984). In Northern Ireland where similar problems have arisen the volume of agricultural output declined by an average of 2.2% per annum between 1973 and 1983 and as a result of falling incomes some downward convergence was achieved between the agricultural sectors in both parts of the country (Cuddy and Doherty, 1984)

The European Regional Development Fund is the only Community instrument set up with the specific purpose of helping to reduce regional imbalances. Despite its crucial role the amount of finance available to the Fund has only comprised between 4.8% in 1975 and 7.8% in 1984 of the Community Budget. The Fund is used to assist investments in industrial, craft and service activities and investments in infrastructure projects. In the first ten years of its operation the Republic of Ireland, which is designated as a single region for aid allocation, received IR£94 million, 6.2% of total payments from the Fund (Commission of the European Communities, 1985). In terms of ERDF assistance *per capita* the Republic of Ireland fared best among the priority regions of the Community at 207 ECUs (European Currency Units), while the *per capita* amount provided to Northern Ireland was slightly lower at 190 ECUs. In the Republic approximately 37% of the assistance was for infrastructural projects with multiregional impacts. Of the remainder approximately 45% was used to assist industrial projects. For the period 1975 and 1981 Drudy (1984) has shown that of the IR£226.8 million approved for the Republic only 17.6% was allocated to the Designated Areas. Furthermore, he noted a marked change in the distribution of assistance after 1977. The share of the total allocated towards projects in the Designated Areas declined from 41.6% for 1975-77 to 13.2% between 1978-81. However, the Designated Areas fared better in relation to projects assisted by loans from the European Investment Bank. Despite its limited size it has been estimated that between 1975 and 1984 the Fund assisted in the creation or maintenance of almost 70,000 jobs in the Republic (Commission of the European Communities, 1985) and approximately 22,000 jobs in Northern Ireland (Harris, 1984). In recent years special assistance has been made available from the Fund for projects in Border areas and for programmes of integrated development and urban renewal in Belfast.

The impact of European Community membership on regional development is, however, not just confined to financial transfers. The European Community is primarily based on a concept of a market economy in which fair, undistorted competition is supposed to ensure that available resources are allocated to the most productive economic sectors. The objective of free competition among member states, as well as free movement of capital, goods and labour has significant implications which can only be partially redressed by the instruments of regional policy which has secondary status to competition policy. It is in the area of manufacturing that the effects of competition policy are most evident. Participation in the European Community has coincided with an acceleration of a process of deindustrialisation which has been ongoing in both the Republic and Northern Ireland since the adoption of outward-looking free market policies in the early 1960s (NESC, 1981b). Indigenous industry has not responded well to the competitive pressures and potential export opportunities brought about by the free-trade environment (Blackwell and O'Malley, 1984). In the Republic there have been significant losses of employment in both indigenous and overseas controlled firms established in the 1960s which specialised in sectors such as textiles and clothing and footwear (NESC, 1983). Consequently between 1973 and 1980 there was an increase of only 2,000 in the number employed in indigenous industries compared with a net increase of 22,000 jobs in foreign-owned manufacturing industries (NESC, 1982), with most of the growth in the mechanical engineering, electric and electronic engineering sectors. By 1985 foreign-owned manufacturing firms accounted for 37.3% of all manufacturing employment compared with 25% in 1973. The structural reasons underlying these trends which have resulted in a dualistic structure in the Irish manufacturing sector have been analysed by Perrons (1986) who has attempted to relate events in Ireland to changes in the global geography of industrial production.

The combined effects of deindustrialisation, which has been especially serious in Dublin and Cork and the highly dispersed pattern of new industrial development achieved by the IDA has been a significant redistribution of industrial employment towards the less-developed regions (O'Farrell, 1984, 1986; Gillmor, 1985). Between 1971 and 1985 the share of total industrial employment in the East region declined from 48% to 37%, while the share in the combined Midlands, West and Northwest/Donegal regions increased from 12% to 19%. The extent of the decline in the East region has been such that total manufacturing employment in 1985 was less than in 1961 (Walsh, 1987). Within the region there has also been a significant decentralisation of industrial employment away from the inner city and port area of Dublin to recently established industrial estates on the outskirts of the city. This has resulted in a major increase in unemployment in parts of central Dublin, and more recently a high level of out migration from the East region.

The effects of European integration on the manufacturing sector in Northern Ireland have also been considerable. Deindustrialisation has been particularly severe in traditional sectors such as shipbuilding, engineering, textiles and clothing (Harrison, 1982; O'Dowd, 1986). Between 1973 and 1982 total employment in manufacturing industries declined by almost 60,000 compared with a decline of 15,500 between

1960 and 1973 (McGurnaghan, 1984). In contrast to the Republic the share of total employment accounted for by overseas controlled firms declined from 39% in 1974 to 37% in 1979 reflecting the increasing competition for foreign investment from other parts of the European periphery including the Republic of Ireland and especially the impact of the civil unrest which has affected the number of multinational companies coming to Northern Ireland. An additional factor in this region was the depressed state of the UK economy since 1974 which resulted in a large decline, 11,000, in the numbers employed in branch plants of mainland UK firms (Harrison, 1982) which had been established there in the 1960s as part of UK regional policy (Moore *et al*, 1978).

Within Northern Ireland there has also been some spatial redistribution of manufacturing employment. Following the introduction of physical planning measures in the mid 1960s government-assisted industrial projects were increasingly directed towards the growth centres, especially Londonderry which increased its share of assisted projects from 4% in the period 1960-64 to 16.2% over the next five years (Bull *et al*, 1982). In addition between the two periods the proportion of assisted projects locating in mainly rural areas increased from 18% to 32.4% (Bull, 1984). However, since the late 1970s the closure of many large multinational branch plants has undermined the whole basis of the regional strategy (O'Dowd, 1986). The deteriorating position in Belfast has resulted in a reversal of regional policy so that now attention is focused primarily on the inner city problem.

The industrial policies of the 1970s have had only a limited success in reducing the extent of regional imbalance. Particular concern has been expressed in recent studies over the level of dependence on foreign-controlled firms, the skill content of the employment they provide, the poor performance of indigenous firms, the relatively high cost and short duration of assisted employment, and the low level of linkages between the industrial sector and other sectors of the economy (O'Farrell and O'Loughlin 1980, 1981; NESC, 1982; NIEC, 1983, 1985, 1986a, 1986c).

The second major weakness in the approaches to regional development has been the absence of a policy in relation to the role of the services sector in which employment increased by almost 200,000 between 1961 and 1984, representing 77% of the net employment increase outside of agriculture. In the 1970s public sector services accounted for over three-fifths of the total growth in service sector employment (Humphreys, 1983). Despite many promises to decentralise sections of government departments and State agencies (NESC, 1977a) very little has been achieved by way of substantial relocation (Bannon, 1985b). The failure to implement a comprehensive regional policy in relation to the location of public and private services has allowed Dublin to improve its comparative advantage as a service location (Dineen, 1986) which, however, has also resulted in a wide range of complex problems in the Dublin area (NESC, 1981a; Horner and Parker, 1987).

A third area of concern in relation to regional planning has been the failure of centrally conceived development programmes to recognise the particular problems of the more disadvantaged areas (Cawley, 1986). This has been found to be of significance in parts of the West of Ireland, especially Gaeltacht areas (Breathnach, 1983; O'Cinneide,1987). In these areas community-based organisations have attempted to develop 'bottom-up' type strategies, with varying degrees of success (Regan and Breathnach, 1981; Breathnach, 1984).

In Northern Ireland the historical contrasts between districts in the east and west have persisted, especially in relation to the operation of the labour market. Within the east there has been a reduction of over one-quarter in the population of Belfast which has only been partially compensated for by increases in some surrounding towns. The level of industrial decline and the effects of civil disorder in Belfast have forced increasing numbers to emigrate. As a result the share of the total population residing in the Greater Belfast region declined from 52.7% to 49.5% between 1971 and 1981 (NIEC, 1986b).

FUTURE DIRECTION IN REGIONAL PLANNING

Since the early 1980s there have been three major developments in the European Community which have significance for regional development. First, the Community has been greatly enlarged by the inclusion of Greece, Spain and

Portugal. Each of these countries has a GDP *per capita* level lower than the levels in the Republic and Northern Ireland. Consequently, the extent of regional disparities in the Community has greatly increased. The number of regions eligible for 'absolute priority' assistance from the ERDF has increased from 9 to 42, and among these in 1985 41 were ranked below Northern Ireland and 37 below the Republic in GDP *per capita*. As a result the focus of the priority regions is now very much in the Mediterranean areas of the Community, and Ireland's claims for special assistance have been weakened.

The second development has been in the approach to regional policy. There have been increasing efforts to reform the Common Agricultural Policy, the objectives of which have at times been in conflict with regional policy. The reforms which are currently proposed are, however, unlikely to result in any improvement in the relative position of farmers in the poorer regions (Conway, 1986; O'Hara, 1986). Consequently there will be an increasing need to address the problems of rural areas in a wider context of regional development. The provisions of the ERDF have undergone some radical changes with special emphasis being directed towards assisting integrated development programmes rather than separate projects. While considerable progress has been made in implementing integrated programmes in the Mediterranean regions there have been no programmes of this type put forward by the government in the Republic of Ireland, despite the possibility of increased assistance from the ERDF and the European Social Fund. The Irish government remains opposed to the implementation of integrated programmes as this would give the European Commission a greater input into decisions concerning the use of ERDF assistance. The implementation of such programmes at a regional level would also require an appropriate administrative structure with statutory powers or alternatively that much greater powers be extended to local authorities: neither of these changes are likely to occur in the near future. The most recent reforms in this area (Department of the Environment, 1985) fall far short of the radical proposals which are considered necessary (Muintir na Tire, 1985).

The third development at Community level with implications for regional development is the Single European Act which commits the Community to establishing an internal market by 1992 which will imply free trade in goods, services, capital and labour, common indirect taxes and the elimination of special privilages and concessions which have a national basis. To achieve this aim programmes are being drawn up to help peripheral regions to compete, so that greater progress can be made towards achieving greater social and economic cohesion.

Within Ireland recent strategies for development do not reflect the renewed commitment to regional planning evident at Community level. While the government White Paper of 1984 on Industrial Policy did acknowledge that regional development will continue to be an important factor in industrial policies, it was envisaged that in future 'industry must be placed where it can make greatest progress' (White Paper, 1984, p.47). More recently new criteria for designation of areas for industrial growth were proposed which would ensure that industry would be encouraged to locate in the major centres of population where infrastructural development costs would be lower due to economies of scale (NESC, 1985).

Despite the absence of a national commitment to regional policy since the mid 1970s most of the nine Regional Development Organisations have either completed or are in the process of preparing long-term strategic plans. The usefulness of these studies is, however, severely limited by the absence of any overall national physical planning strategy which could provide a context for assessing the contribution of each region towards national development in the future. This limitation was particularly evident following the publication of a proposed settlement strategy for the East region up to the year 2011 (ERDO, 1985).

In the years ahead the problem of imbalance in regional development is likely to become more acute and also more difficult to ameliorate. This will be partly due to the limited quantity of resources available for assisting regional development programmes and the lack of political will to bring about a significant redistribution of resources from central to peripheral parts of the European Community. It will also be due to the increasing inability of nation-states, particularly small ones such as Ireland, to cope with the consequences of

461

processes which are facilitating an increasingly interdependent capitalist world economy (Johnston, 1986). In this context, with increased competition from the new member states of the Community, there will be a greater need for each region to develop a 'credible development strategy containing a programme of projects which have been carefully prepared, which relate to each other, which are cost effective and which maximize local involvement' (Sutherland, 1986, p.377).

Note Since the completion of this chapter there has been a change of government in the Republic of Ireland. A number of measures relevant to regional development have been adopted by the new administration, including the abolition of An Foras Forbartha, the Regional Development Organisations and the county development teams who operated in the less-developed counties. While no new regional structures have been proposed the government has committed itself to preparing integrated programmes on a pilot area basis.

REFERENCES

An Foras Forbartha (1987) *Review of Regional Studies*, An Foras Forbartha, Dublin.

Bannon, M.J. (1983) The changing context of developmental planning, *Administration*, 31, 112-46.

Bannon, M.J. (1985a) The genesis of modern Irish planning, in Bannon, M.J. (ed) *The Emergence of Irish Planning*, Turoe Press, Dublin, 189-262.

Bannon, M.J. (1985b) Service activities in National and Regional development: trends and prospects for Ireland, in Bannon, M.J. and Ward, S. (eds) *Services and the New Economy: Implications for National and Regional Development*, Regional Studies Association (Irish Branch), Dublin, 38-61.

Barrington, J.J. (1975) *From Big Government to Local Government; the Road to Decentralisation*, Institute of Public Administration, Dublin.

Blackwell, J. and O'Malley, E. (1984) The impact of EEC membership on Irish Industry, in Drudy, P.J. and McAleese, D. (eds) *Ireland and the European Community*, Irish Studies Series, Vol. 3, Cambridge, 107-44.

Boylan, T.A. and Drudy, P.J. (eds) (1987) *Regional Policy and National Development*, Regional Studies Association (Irish Branch), Dublin.

Breathnach, P. (1982) The demise of growth-centre policy: the case of the Republic of Ireland, in Hudson, R. and Lewis, J.R. (eds) *Regional Planning in Europe*, Pion, London, 35-56.

Breathnach, P. (ed) (1983) *Rural development in the west of Ireland: observations from Gaeltacht experience*, Occasional Paper 3, Geography Department, St Patrick's College, Maynooth.

Breathnach, P. (1984) Co-operation and Community: an aspect of rural development in the west of Ireland, in Jess, P.M., Greer, J., Buchanan, R. and Armstrong, J. (eds) *Planning and Development in Rural Areas*, The Queen's University of Belfast, 155-177.

Buchanan, C. and Partners (1968) *Regional Studies in Ireland*, An Foras Forbartha, Dublin.

Bull, P.J., Harrison, R.T. and Hart M. (1984) Economic planning for rural areas in Northern Ireland, in Jess, P.M., Greer, J., Buchanan, R. and Armstrong, J. (eds) *Planning and Development in Rural Areas*, The Queen's University of Belfast, 41-59.

Bull, P.J. *et al.*, (1982) Government assisted manufacturing activity in a peripheral region of the United Kingdom: Northern Ireland 1945-1979, in Collins, L. (ed.) *Industrial Decline and Regeneration*, University of Edinburgh, Dept of Geography, 39-64.

Caldwell, J. and Greer, J. (1984) Physical planning for rural areas in Northern Ireland, in Jess, P.M., Greer, J., Buchanan,

R. and Armstrong, J. (eds) *Planning and Development in Rural Areas*, Queen's University, Belfast, 63-86.

Cawley, M.E. (1986) Disadvantaged groups and areas: problems of rural service provision, in Breathnach, P. and Cawley, M.E. (eds) *Change and Development in Rural Ireland*, Geographical Society of Ireland, Special Publication No.1, Dublin, 48-59.

Commission of the European Communities (1984) *The Regions of Europe: Second Periodic Report on the Social and Economic Situation of the Regions of the Community*, Luxembourg.

Commission of the European Communities (1985) *European Regional Development Fund*, Tenth Annual Report (1984), Luxembourg.

Conway, A.G. (1986) Prospects for the CAP and its modifications, in *The Changing CAP and its Implications*, An Foras Taluntais, Dublin, 1-33.

Cuddy, M. (1986) *The Performance of Irish Agriculture in its First Decade under the CAP*, WP No 1, SSRC, University College, Galway.

Cuddy, M. and Doherty, M. (1984) *An Analysis of Agricultural Developments in the North and South of Ireland and of the Effects of Integrated Policy and Planning*, prepared for the New Ireland Forum, Stationery Office, Dublin.

Department of the Environment, Northern Ireland (1977) *Northern Ireland: Regional Physical Development Strategy 1975-1995*, HMSO, Belfast.

Department of the Environment (1985) *The Reform of Local Government*, Stationery Office, Dublin.

Dineen, D.A. (1986) *Strategy for the Development of the Services Sector in the mid-West region of Ireland*, Mid-West RDO, Limerick.

Drudy, P.J. (1984) The regional implications of EEC policies in Ireland, in Drudy, P.J. and McAleese, D. (eds) *Ireland in the European Community*, Irish Studies Series, vol. 3, Cambridge University Press, Cambridge, 191-214.

Eastern Region Development Organisation (1985) *Eastern Region Settlement Strategy - 2011*, Dublin.

Fitzgerald Report (1968) *Outline of the Future Hospital System*, Stationery Office, Dublin.

Gillmor, D.A. (1985) *Economic Activities in the Republic of Ireland: a Geographical Perspective*, Gill and Macmillan, Dublin.

Harris, R. (1984) Regional policy in Northern Ireland, in Simpson, J.V. (ed.) *European Community Policy in Northern Ireland*, Queen's University, Belfast, 21-35.

Harrison, R.T. (1982) Assisted industry, employment stability and industrial decline: some evidence from Northern Ireland, *Regional Studies*, 16, 267-85.

Hoare, A.G. (1982) Problem region and regional problem, in Boal, F.W. and Douglas, J.N.H. (eds) *Integration and Division: Geographical Perspectives on the Northern Ireland Problem*, Academic Press, London, 195-224.

Horner, A.A. (1986) Rural population change in Ireland, in Breathnach, P. and Cawley, M.E. (eds) *Change and Development in Rural Ireland*, Geographical Society of Ireland, Special Publication No.1, Dublin, 34-47.

Horner, A.A. and Parker, A.J. (1987) *Geographical Perspectives on the Dublin Region*, Geographical Society of Ireland, Special Publication No.2, Dublin.

Horner, A.A. and Taylor, A.M. (1979) Grasping the nettle - locational strategies for Irish hospitals, *Administration*, 27, 348-70.

Horner, A.A., Walsh, J.A. and Williams, J.A. (1984) *Agriculture in Ireland, a Census Atlas*, Department of Geography, University College, Dublin.

Horner, A.A., Walsh, J.A. and Harrington, V. (1987) *Population in Ireland, a Census Atlas*, Department of Geography, University College, Dublin.

Humphreys, P.C. (1983) *Public Service Employment, an examination of strategies in Ireland and other European Countries*, Institute of Public Administration, Dublin.

Industrial Development Authority (1972) *Regional Industrial Plans, 1973-1977*, Dublin

Industrial Development Authority (1979) *Regional Industrial Plans, 1978-1982*, Dublin

Investment in Education, (1965) Stationery Office, Dublin.

Johnston, R.J. (1986) The state, the region and the division of labour, in Scott, A.J. and Storper, M. (eds), *Production, Work, Territory - The Geographical Anatomy of Industrial Capitalism*, Allen and Unwin, Boston, 265-80.

Keeble, D., Owens, P. and Thompson, C. (1982) Regional accessibility and economic potential in the European Community, *Regional Studies*, 16, 419-32.

Kelleher, C. and O'Mahony, A. (1984) *Marginalisation in Irish Agriculture*, An Foras Taluntais, Dublin.

Laffan, B. (1985) *The European Social Fund and its Operation in Ireland*, Irish Council of the European Movement, Dublin.

Lee, J. (1985) Centralisation and Community in Lee, J. (ed.), *Ireland:towards a Sense of Place*, Cork University Press, Cork, 84-101.

Lichfield, N. and Associates (1967) *Report and Advisory*

Outline Plan for Limerick Region, Stationery Office, Dublin.

McGurnaghan, M. (1984) Northern Ireland and EEC membership: an economic analysis, *Journal of the Statistical and Social Inquiry Society of Ireland*, Vol. XXV, Part 1, 237-55.

McNamara, B. (1984) The impact of EEC financial assistance on Irish economic development since 1973, *Journal of the Statistical and Social Inquiry Society of Ireland*, XXV(1), 221-36.

Matthew, Sir R.H. (1964) *Belfast Regional Survey and Plan 1962: a Report*, HMSO, Belfast.

Moore, B., Rhodes, J. and Tarling, R. (1978) Industrial policy and economic development: the experience of Northern Ireland and the Republic of Ireland, *Cambridge Journal of Economics*, 2, 99-114.

Muintir na Tire, (1985) *Towards a New Democracy? Implications of Local Government Reform*, Institute of Public Administration, Dublin.

NESC (1975) *Regional Policy in Ireland: a Review*, Report No. 4, Stationery Office, Dublin.

NESC (1976a) *Rural Areas: Social Planning Problems*, Report No. 19, Stationery Office, Dublin.

NESC (1976b) *Institutional Arrangements for Regional Economic Development*, Report No. 22, Stationery Office, Dublin.

NESC (1977a) *Service-type Employment and Regional Development*, Report No. 28, Stationery Office, Dublin.

NESC (1977b) *Personal Incomes by County in 1973*, Report No. 30, Stationery Office, Dublin.

NESC (1978a) *Rural Areas: Change and Development*, Report No. 41, Stationery Office, Dublin.

NESC (1978b) *Report on Policies for Agriculture and Rural Development*, Report No. 42, Stationery Office, Dublin.

NESC (1979) *Urbanisation and Regional Development in Ireland*, Report 45, Stationery Office, Dublin.

NESC (1980) *Personal Incomes by Region in 1977*, Report No. 51, Stationery Office, Dublin.

NESC (1981a) *Urbanisation: Problems of Growth and Decay in Dublin*, Report No. 55, Stationery Office, Dublin.

NESC (1981b) *Industrial Policy and Development: a Survey of Literature from the early 1960s to the Present*, Report No. 56, Stationery Office, Dublin.

NESC (1981c) *The Socio-economic Position of Ireland within the European Economic Community*, Report No. 58, Stationery Office, Dublin.

NESC (1982) *A Review of Industrial Policy*, Report No. 64, Stationery Office, Dublin.

NESC (1983) *An Analysis of Job Losses in Irish Manufacturing Industry*, Report No. 67, Stationery Office, Dublin.

NESC (1985) *Designation of Areas for Industrial Policy, Report No. 81, Stationery Office, Dublin.*

New Ireland Forum Report, 1984, Stationery Office, Dublin.

Newman, J. (1958) The future of rural Ireland, *Studies*, 47, 388-409.

NIEC (1983) *The Duration of Industrial Development Assisted Employment*, Report No. 40, Belfast.

NIEC (1985) *The Duration of Industrial Development Maintained Employment*, Report No. 52, Belfast.

NIEC (1986a) *Economic Strategy: Industrial Development*

Linkages, Report No. 56, Belfast.

NIEC (1986b) *Demographic Trends in Northern Ireland*, Report No. 57, Belfast.

NIEC (1986c) *Economic Strategy: Industrial Development*, Report No. 60, Belfast.

O'Cinneide, M.S. (1987) The role of development agencies in peripheral areas with special reference to Udaras na Gaeltachta, *Regional Studies*, 21, 65-74.

O'Dowd, L. (1985) The crisis of regional strategy: ideology and the State in Northern Ireland, in Rees, G. *et al.* (eds) *Political Action and Social Identity: Class, Locality and Culture*, London, Macmillan, pp.143-65.

O'Dowd, L. (1986) Beyond Industrial Society, in Clancy, P., Drudy, S., Lynch, K. and O'Dowd, L. (eds) *Ireland a Sociological Profile*, Institute of Public Administration, Dublin, 198-220.

O'Farrell, P.N. (1975) *Regional Industrial Development Trends in Ireland 1960-1973*, Industrial Development Authority, Dublin.

O'Farrell, P.N. (1978) An analysis of New Industry Location: the Irish case, *Progress in Planning*, 9(3), 129-229.

O'Farrell, P.N. (1978) Regional planning policy - some major issues, *Administration*, 26(2), 147-61.

O'Farrell, P.N. (1980) Multinational enterprises and regional development: Irish evidence, *Regional Studies*, 14, 141-50.

O'Farrell, P.N. (1984) Components of manufacturing employment change in Ireland 1973-1981, *Urban Studies*, 21, 155-76.

O'Farrell, P.N. (1986) *Entrepreneurs and Industrial Change*, Irish Management Institute, Dublin.

O'Farrell, P.N. and O'Loughlin, B. (1980) *An Analysis of New Industry Linkages in Ireland*, Industrial Development Authority, Dublin.

O'Farrell, P.N. and O'Loughlin, B. (1981) The impact of new industry enterprises in Ireland: an analysis of service linkages, *Regional Studies*, 15(6), 439-456.

O'Hara, P. (1986) CAP Socio-Structural Policy - a new approach to an old problem, in *The Changing CAP and its Implications*, An Foras Taluntais, Dublin, 34-69.

O'Tuathaigh, M.A.G. (1986) The Regional Dimension, in Kennedy, K. (ed.), *Ireland in Transition*, Cork, 120-32.

Perrons, D.C. (1986) Unequal integration in global Fordism: the case of Ireland, in Scott, A. and Storper, M. (eds) *Production, Work, Territory, the Geographical Anatomy of Industrial Capitalism*, Allen and Unwin, Boston, 246-54.

Reagan, C. and Breathnach, P. (1981) *State and Community: Rural Development Strategies in the Slieve League Peninsula, Co. Donegal*, Occasional Paper 2, Geography Department, St Patrick's College, Maynooth.

Ross, M. and Walsh, B. (1979) *Regional Policy and the Full-employment Target*, Policy Series 1, ESRI, Dublin.

Simpson, J.V. (1984) *European Community policy in Northern Ireland*, The Queen's University of Belfast.

Sutherland, P. (1986) Europe and the principle of convergence, *Regional Studies*, 20, 371-77.

Talbot, R.B. (1977) The European Community's Regional Development Fund, *Progress in Planning*, 8(3), 183-281.

Walsh, F. (1976) The growth centre concept in Irish regional policy, *Maynooth Review*, 2, 22-41.

Walsh, J.A. (1986) Agricultural change and development, in

Breathnach, P. and Cawley, M. (eds) *Change and Development in Rural Ireland*, Geographical Society of Ireland, Special Publication No. 1, Dublin, 11-24.

Walsh, J.A. (1987) Dublin: region-state relations, in Horner, A.A. and Parker, A.J. (eds) *Geographical Perspectives on the Dublin Region*, Geographical Society of Ireland, Special Publication No. 2, Dublin, 82-95.

White Paper (1984) *Industrial Policy*, Stationery Office, Dublin.

Wilson, T. (1965) *Economic Development in Northern Ireland*, HMSO, Belfast.

Wright, M. (1967) *The Dublin Region: Advisory Regional Plan and Final Report*, 2 Vols. Stationery Office, Dublin.

INDEX

- Abortion 42, 43, 56, 78
- Access to the Countryside Order (NI) (1983) 387
- Accessibility 162-4, 288, 407
- Accidents 278, 279
- Achill 117
- Acid Rain 380, 421, 426, 435, 436
- Acidification 432, 433
- Act of Union 25
- Agricultural Foodstuffs 341
- Agriculture 10-11, 128, 19, 147, 161, 171-96, 203, 363
- Air Quality 19, 340, 345, 406, 421-37
- Airlines 295, 298
- Airports 276, 295-6
- Alcohol 351
- Algae 376, 379
- Alkali Act (1906) 436, 437
- Alternative Energy 348-54
- Amenity Lands Act (1965) 132
- An Bord Pleanala 263
- An Foras Forbartha (AFF) 8, 126, 135, 136, 361, 373, 384, 385, 386, 445, 462
- An Foras Tionscal 443
- An Taisce 126, 127, 131, 338
- Ancestors 309-10
- Angling 313, 369, 380-1
- Anglo-Irish Agreement 39
- Anglo-Irish Free Trade Agreement 173
- Anglo-Irish Treaty 27, 29
- Antrim 7, 105, 121, 148, 152, 210, 244, 258, 313, 320, 365, 369, 371, 382, 408, 409
- Arable 172, 180, 185, 188, 192
- Aran Islands 114, 117
- Architecture 119, 126, 315
- Areas of Outstanding Natural Beauty (AONBs) 132, 134, 156, 161

- Areas of Scientific Interest (ASI) 134, 156
- Areas of Special Control (ASC) 132, 134, 156
- Arigna 338
- Armagh, (Co.) 29, 33, 104, 122, 321
- Athlone 258, 447
- Autarky 211

- Ballycastle 337
- Ballymena 113, 244, 258, 452, 454
- Bangor 104
- Banks 243-4
- Bathing 407
- Beaches 313, 320, 407-9
- Belfast 3, 4, 7, 19, 28, 31, 33, 35, 36, 67, 67, 70, 71, 77, 89, 100, 104-5, 107, 114, 117, 122, 148, 156, 202, 210, 217, 223, 227, 240, 244, 246, 255, 256, 257, 260, 261, 262, 264, 265, 275, 278, 279, 282, 284, 291, 293, 294, 295, 314, 327, 335, 341, 361, 382, 401, 404, 427-8, 442, 450, 451, 454, 456, 458, 459
- Biochemical Oxygen Demand (BOD) 12, 373, 374, 376, 401-2, 403
- Biogas 351, 353-4
- Biomass (Energy) 195, 334, 339, 350-1, 352, 404
- Birth Rate 63, 73, 74
- Blackrock, (Co. Dublin) 245, 261
- Blanchardstown 68
- Boating 380
- Bord na Mona 338, 339
- Bord Fáilte 126, 312, 316, 321, 322, 327, 329
- Border 3, 5, 11, 28, 29-30, 115-6, 132, 152, 176, 238, 245, 262, 272, 289, 301, 302-3, 306, 333, 334, 335, 456
- Boundary Commission 29-30
- Branch Plant 216, 217, 221, 227, 442, 458
- Bray 7
- Burglary 93, 96, 97, 102, 105
- Bus Lanes 282
- Buses 160, 163, 262, 277, 279, 286-8, 297

- Caesium 405

- Cafes and Restaurants 260
- Caravan Parks 320, 412-3
- Carbon Dioxide 354
- Carlow 130, 152,368
- Cars 242-3, 276-7, 288
- Castles 320
- Catholic 26, 28, 31, 32, 37, 41, 42, 68, 78-9, 124, 214, 245, 265
- Cattle 12, 13, 176, 180, 185, 187, 188, 190, 192, 195
- Cavan 12, 130, 189, 192, 374,
- Celtic Sea 393
- Cement 292
- Central Heating 349, 429
- Cereals 184, 185
- Chemicals 224
- Civil Rights 37-8
- Civil War 23
- Clare 163, 288, 313, 319, 408
- Clean Air Act (NI) (1964) 427
- Clondalkin 7, 68, 244, 259
- Closures (Industrial) 206, 220-1
- Clothing Stores 246, 250, 252
- Coal 334, 335, 338, 400, 423, 426
- Coalisland 337
- Coastal Engineering 407-8
- Coastlines 393-414
- Cockcroft Report 133
- Coleraine 35, 260, 264, 317, 320, 327, 450
- Commission on Emigration 54, 65, 69, 75
- Common Agricultural Policy (CAP) 5, 146, 172, 174, 194, 455, 460
- Common Fisheries Policy (CFP) 5, 18, 399, 414
- Conacre 178
- Connemara 319
- Conservation 119-20, 121-22, 126-7, 134-6, 165, 195, 338, 348, 383, 387, 410
- Conservation Volunteers 414
- Consumers 238, 239-66, 336-7
- Consumers' Association of Ireland 242
- Continental Shelf 394, 397

- Contraception 42
- Control of Atmospheric Pollution Regulations (1970) 437
- Convenience Stores 251, 254, 266
- Cooperatives 148, 162, 249
- Coras Iompair Éireann (CIE) 160, 274, 287, 288, 298
- Core 208, 209, 214, 219, 221, 325
- Cork 89, 93, 97, 98, 114, 127, 130, 136, 152, 157, 210, 217, 222, 247, 256, 258, 259, 261, 262, 276, 278, 294, 295, 313, 316, 318, 319, 324, 329, 404, 408, 431, 444, 447, 457
- Countryside and Wildlife Branch (NI) 413
- County Management Act (1955) 157
- Craigavon 35, 36, 70, 244, 258, 264, 450, 454
- Crime Rate 93-9, 100, 104
- Crime 87-108, 265, 304
- Cross Border - see Border

- Dáil Éireann 27, 41
- Dublin Area Rapid Transit (DART) System 265, 282
- Death Rate 75, 431
- Democratic Unionist Party (DUP) 37, 38
- Deprivation 5, 95, 147-54, 160-2, 442
- Derry 5, 29, 32, 35, 104, 105, 114, 122, 124-5, 163, 210, 258, 260, 262, 264, 275, 276, 291, 306, 320, 450, 452, 454, 458
- Designated Areas 447, 456
- Development Cooperation 447
- Divorce 42, 43, 78
- Donegal 7, 12, 30, 136, 152, 155, 209, 275, 313, 319, 324, 363, 399, 401, 411, 443, 447, 457
- Down 68, 122, 128, 130, 148, 210, 317, 320, 321, 327, 408, 411
- Drainage 13, 16, 136, 193, 359, 363-71
- Dredging 369
- Drogheda 258, 447
- Droughts 378
- Drugs 90, 94, 95, 100
- Dublin 3, 4, 7, 19, 30, 31, 41, 44, 45, 67, 68, 70, 71, 72, 77, 89, 93, 96-9, 100, 107, 114, 118, 119, 121, 126, 127, 204, 217, 222, 223, 227, 239, 244, 245, 246, 247,

248, 249, 251, 253, 254, 255, 256, 257, 258, 259, 260,
262, 265, 266, 276, 278, 279, 282, 284, 293, 294, 295,
314, 316, 317, 318, 319, 323, 324, 341, 361, 404, 408,
411, 423, 426, 427, 430, 442, 443, 444, 445, 447, 448,
449, 457, 459
- Dublin (Co.) 97, 130, 210
- Dublin Bay 135, 401, 402, 403
- Dun Laoghaire 114, 245, 259, 276, 293
- Dundalk 258, 289, 447
- Dunes 395, 407-8, 410

- Eastern Regional Development Organisation (ERDO) 71
- Economic Multiplier 325-6
- Economy 9, 64-5, 79, 99, 171, 180, 183, 201-2, 301-2, 370, 445, 458
- Education 69, 74, 95, 146, 158
- Effluent 359, 372, 377, 380, 385, 401-6
- Egg Production 180
- Electrical Engineeering 224-5
- Electricity 332-55, 426
- Electricity Supply Board (ESB) 334, 335, 338, 339, 341, 346, 347, 352, 352, 384, 436
- Electronics 224, 457
- Emigration 60, 62, 63, 65-6, 80, 89, 90, 226, 230
- Employment 73-4, 146, 171, 177, 182, 187, 192, 201, 203, 206, 216, 220,
- Energy 331-55, 400-1
- Engineering 202, 224, 225, 457
- Enniskillen 7, 105, 121, 262
- Enterprises, (farm) 183-9
- Environmental Impact Assessment (EIA) 343, 437
- Erosion 193, 313, 407, 408, 411-2
- Estuaries 401-3, 410-1
- European Agriculture Grants and Guarantee Fund (EAGGF) 368, 452
- European Community 4-5, 6, 9-21, 44, 128, 131, 135, 148, 152, 161, 162, 165, 172, 174-6, 180, 183, 185, 194-6, 202, 227, 228-9, 240, 272, 274, 283, 289, 293, 333, 359, 368, 372, 374, 377, 399, 414, 421, 425, 429, 435, 436, 437, 441, 452-3, 452-62

- European Free Trade Association (EFTA) 228
- European Social Fund 452, 460
- Eutrophication 374-5, 376
- Exclusive Economic Zone (EEZ) 18, 398-9
- Exclusive Fishing Zone (EFZ) 399
- Export Profit Tax Relief Scheme 212
- Exports 202, 217, 224, 229 247, 457-8, 459

- Faecal coliforms 402
- Fair Employment Agency 39
- Famine 57, 110, 410
- Farm Improvement Programme 175
- Farmhouse Accommodation 325
- Farming 44, 146, 155, 171-76, 180, 187, 344, 455-60
- Farming (Part-time) 155, 171, 177, 183, 187,
- Farms 16, 176-8, 180, 182, 188, 190
- Fast Food Outlets 246, 251, 253
- Fermanagh 29, 31, 33, 122, 128, 132, 152, 320 321, 368
- Ferries, cross-channel 293-95, 310
- Fertility 72-3
- Fertilizers 182, 193, 292, 340, 359, 367, 375, 378
- Fianna Fáil 41-2, 46, 413
- Fine Gael 41-2, 46, 370
- Fish 12, 374, 377, 397,
- Fisheries 13, 18, 380-1, 399-400
- Fisheries Conservancy Board, (NI) 377
- Fishermen 344
- Flooding 363, 364, 369, 373
- Folklore 315
- Food Processing 209, 262, 378
- Foreign Industry 218-9, 221-2, 230, 457,
- Forestry 16-8, 161, 334, 350
- Free Trade 228-9, 456
- Freight Transport 274-5, 289-93, 298

- Gaelic Athletic Association (GAA) 115
- Gaeltacht 319, 444, 448, 449, 459
- Galway 93, 97, 135, 136, 155, 258, 276, 319, 324, 329, 447

- Garda Siochána 91, 92, 94, 297
- Geothermal 351
- Golf 313, 411
- Grass 16, 172
- Groceries 238, 242, 248, 249-50, 260, 261, 263-4
- Groundwater 361, 373, 379

- Health 69, 74, 75-7, 158-9, 163, 241
- Heat Pumps 348, 349
- Heritage 315
- Historic Monuments Act (1971) 121, 131
- Holidays 308
- Horse Riding 313
- Horses 189
- Horticulture 192
- Hotel and Guest Houses 303, 320, 325
- Housing 152, 155, 157, 163, 165, 349, 351
- Hydro-electric Power 333, 351, 353

- Imports 203, 262
- Industrial Development (NI) Act 212
- Industrial Development Authority (IDA) 71, 214-6, 223, 444, 448, 449-50, 457
- Industrial Development Board (IDB) 214-5
- Industrial location 214, 217-8, 226
- Industry 5-6, 20, 35, 69-70, 71, 201-31, 456-8
- Insulation 348, 351
- Investment, agriculture 370
- Investment, industrial 201, 204-6, 207-8, 211, 218-9, 221, 442, 461
- Investment, retailing 265
- Investment, transport 284-5
- Irish Association of Distributive Trades, (IADT) 263
- Irish Land Commission (formerly The Congested Districts Board) 178, 410, 443
- Irish Republican Army (IRA) 37, 41
- Irish Sea 8, 294, 314, 393-4, 395, 397, 399, 403
- Irish territorial waters 396-9
- Irradiation 351, 421

- Job losses 204-6, 209, 210, 216
- Jobs 206
- Joyriding 94

- Kerry 152, 157, 319, 374, 447
- Kilkenny 130, 368
- Killarney 7
- Kinsale Gas Field 335, 339-40, 341, 346, 400
- Kinship (family) 7, 163, 177, 190

- Lakes 313, 344, 359, 375, 377, 386, 433
- Land holding 13, 146, 161, 177, 178, 190, 370
- Land War, The 88-9
- Larne 276, 293, 321, 401
- Leitrim 12, 152, 324, 447
- Less Favoured Areas (LFA) 152, 153, 175-6
- Letterkenny 258
- Library (mobile) 162
- Life expectancy, 75-6
- Lignite 333, 336, 338, 341-5
- Limerick 12, 90, 93, 127, 258, 259, 262
- Lisburn 7
- Livestock Unit 178-80
- Local Enterprise Development Unit (LEDU) 223, 451
- Local Government (Planning and Development) Act (1963) 45, 133
- Local Government (Planning and Development) Act (1976) 437
- Local Government (Water Pollution) Act (1977) 384, 385
- Londonderry 5, 29, 32, 35, 104, 105, 114, 122, 124-5, 163, 200, 258 260, 262, 264, 275, 276, 291, 306, 320, 450, 452, 454, 458
- Longford, 130, 152
- Louth, 130

- Man-made fibres 225
- Manufacturing 202-32
- Matthew Plan 133, 451
- Mayo 157, 178, 324

- Meath 115
- Migration 43, 59, 62, 63, 64, 69, 72, 73, 154, 457
- Military 396
- Milk 11, 12, 16, 174, 185, 188, 192, 195, 290
- Minerals 292, 397, 400
- Minerals Act (1940) 407
- Monaghan 12, 189, 192
- Mortality 76-7
- Motorways 272-4
- Mourne 321
- Moyle 320
- Mullingar 320
- Multi-nationals 221, 222, 224, 227, 237, 458
- Multiple stores 238, 247-51, 253, 257, 261-2, 263
- Multiplier 326, 327

- National Economic and Social Council (NESC) 453
- National Grid 333
- National Industrial Economic Council 444
- National Monuments Act (1930) 130
- National Trust 121, 132, 134, 408, 414
- Nationalists 28, 36, 38
- Natural gas 333, 335, 339-41, 400, 425-6, 428
- Nature Conservation and Amenity Lands Order (NI) (1985) 387
- Navan 115
- New Towns 451
- Newry 262, 289, 321
- Newtownards 7
- Nitrification 379, 380
- Nitrogen 373, 374, 378, 433, 436
- Non-food outlets 253, 260
- Northern Ireland Economic Council (NIEC) 342, 353
- Northern Ireland Electricity (NIE/NIES) 334, 343, 353
- Northern Ireland General Consumer Council 242
- Northern Ireland Housing Executive (NIHE) 39, 112, 124
- Northern Ireland Labour Party (NILP) 38
- Northern Ireland Office 39
- Northern Ireland Tourist Board (NITB) 304, 312, 316, 321, 329

- Northern Ireland Water Council 383
- Nuclear power 334, 345-7, 354
- Nuclear waste 8, 401
- Nuclear Energy Act (1971) 345
- Nuclear Energy Board 346
- Nutrients 374, 376-7, 379, 402

- Offaly 114, 368
- Office of Public Works (OPW) 364, 368
- Official Unionist Party (OUP) 38
- Oil 331, 335-6, 345, 346, 400, 426, 428, 429
- Oil spills 403, 404-5
- Oxygen 372, 373, 374, 376, 379, 403, 404
- Ozone 433, 434

- Park-and-ride 282
- Partition 23-7, 178, 203, 204, 441-2
- Pasture 172, 180
- Peat 338-9, 363-4, 423, 426
- Periphery 5, 147, 148, 208, 211, 213, 289, 293, 301, 325, 442, 458
- Petrol stations 254
- pH 373, 433
- Phosphorus 374, 376, 378
- Pigs 12, 13, 180, 183-5, 187, 189, 374
- Placenames 114
- Planning 45, 72, 120-1, 122, 123, 126, 146, 147, 154-7, 208, 255, 264,
- Planning and Development Acts 156
- Planning regions 213, 282, 441-66
- Plantations 26
- Plutonium 405
- Police 89, 92, 102
- Pollution 7-8, 13, 77, 125, 192-3, 279, 334, 344, 372, 373, 374, 376, 382, 385, 401-6, 428-9, 432-7
- Population 55-82
- Portadown 264
- Portlaoise 114, 258
- Ports 276, 277, 294, 295, 322, 409, 457
- Post offices 162, 164

- Poultry 12, 180, 183-5, 187, 189
- Power station 19, 333, 334, 335, 336, 338, 339, 340, 342, 343, 344, 345, 346, 347, 352, 426
- Price support, agriculture 173, 187
- Price war, retail 262
- Progressive Democratic Party (PD) 49
- Proportional Representation (PR-STV) 38, 46-7, 50
- Prostitution 90
- Protestant 26, 33, 36-7, 78, 124, 214, 245, 265
- Public houses and bars 238, 252, 262
- Pumped storage 333, 346, 382

- Radioactive waste 403, 405-6
- Radon gas 351
- Recreation 380-5
- Regional Development Fund (EC) 16, 286, 453, 455, 460
- Regional Development Organisations (RDO) 448, 449, 462
- Regional Fisheries Board 383-4
- Religious segregation 40
- Reservoirs 381, 382
- Resort (tourist) 301, 320, 324
- Restrictive Practices Commission 262, 263
- Retailing 69, 158, 159, 246-66
- Ribbon development 154
- Rivers 313, 359
- Road haulage 290-1, 293
- Road Transport Act (1986) 292
- Roads, 272-4, 27, 285-6, 297, 298
- Rockall Dispute, The 397-8
- Roinn na Mára 413
- Rosslare 276, 293, 294, 319, 411
- Royal Irish Academy 115
- Royal Society for the Protection of Birds (RSPB) 414
- Royal Ulster Constabulary (RUC) 91, 92, 104, 105, 307
- Rugby Football 318
- Rural problems 145-65

- Salt 400
- School buses 288
- Sea-level 395
- Security 40, 396-7
- Sediment 369, 373, 374, 395, 400
- Sellafield 405-6
- Service industries 154, 162-4
- Sewage 154, 372, 376, 402, 403, 404
- Shannon Airport (Shannon Free Airport Development Corporation) (SAFDCO) 295-6, 444, 448
- Sheep 176, 180, 185, 188, 192, 195
- Shipbuilding 202, 203, 457
- Shipping 290, 294, 298, 327, 396
- Shopping centres 125, 159, 237, 238, 244, 250, 253, 255-9, 264
- Shopping 3, 97, 159-60, 242, 259
- Shrines 314, 315
- Silage 193, 377
- Siltation 411
- Single European Act 460
- Sinn Féin 28, 38, 44, 49, 50
- Sligo 7, 152, 319, 324, 447
- Slurry 377
- Smoke 423-9, 431-2, 437
- Smuggling 3, 176
- Social Democratic and Labour Party (SDLP) 38, 39
- Soils 188-9, 350, 363, 367, 433
- Solar panels 348, 349, 351
- Special Amenity Area Order 135
- Sport and Leisure 241, 302, 313, 407
- Stately homes 314-5
- Stop Line 133-4, 451
- Stormont (NI Parliament) 27, 33
- Strontium 405
- Subsidies, agriculture 176, 182, 192
- Subsidies, industry 219
- Subsidies, transport 275
- Sugar beet 12, 184
- Sulphur dioxide 345, 423, 425-32, 436
- Sunday trading 264-5

- Supermarkets 241, 243, 248, 250-1, 253, 255, 257
- Superstores 253, 254, 260
- Surfing 407

- Tallaght 68, 244, 259
- Taxation 44, 45, 212, 321, 322, 327,
- Taxis 163, 164, 277, 279
- Telephone system 159, 244
- Television 244
- Tenancy system 177
- Terrorism 99-106, 123-4, 256, 264, 265, 304, 307, 328, 333, 442
- Textiles 202, 224, 457
- Third level institutions 445, 448
- Tidal barrage 353, 401
- Tidal energy 351, 353, 401
- Tipperary 130
- Tourism 97, 182, 301-329, 408, 445
- Tractors 182
- Traditional music 315
- Traffic congestion 277-8, 284
- Tralee 115, 262
- Transport, 69, 158, 160, 163-4, 243, 271-98
- Troubles, The 40, 99, 105, 124, 216
- Turf Development Act 339
- Tyrone 7, 31, 104, 130, 316, 368

- Ulster Folk Museum 130
- Ulster Society for the Preservation of the Countryside (USPC) 131-2
- Ulster Transport Authority (UTA) 286-7
- Ulster Trust for Nature Conservation (UTNC) 134
- Ulsterbus 287, 288
- Underdeveloped Areas Act (1954) 147, 443
- UNESCO 320
- Unemployment 64, 95-6, 152, 161, 210, 226, 230, 240, 284, 442, 457
- Unionism 27
- Unionist 28, 30-7, 214
- University 35, 121, 125

- Uplands 172, 189, 193, 359, 382
- Uranium 346
- Urbanisation 44

- Vandalism 97

- Waste disposal 377-8, 401-4
- Water quality 19, 372-80, 401, 402, 414
- Water quality monitoring 373-8
- Water supply 154, 359, 360-2, 381
- Water Act (NI) (1972) 383
- Waterford 90, 93, 97, 222, 258, 276, 295, 319, 368, 447
- Wave energy 351, 352-3, 400-1
- Waves 395
- Welfare 158
- Wexford 7, 115, 126, 128, 130, 136, 157, 319, 410, 411
- Wheat 184
- Wholesalers 251
- Wicklow 135, 368, 408, 411
- Wildlife 193
- Wildlife Act (1976) 136
- Wildlife Order (NI) (1983) 387
- Wind power 333, 348, 351-2
- Wine shops 254

- Yachting 307, 411